ACTA PHYSICA AUSTRIACA / SUPPLEMENTUM III

ELEMENTARY PARTICLE THEORIES

PROCEEDINGS OF THE
V. INTERNATIONALE UNIVERSITÄTSWOCHEN
FÜR KERNPHYSIK 1966 DER KARL-FRANZENS-UNIVERSITÄT
GRAZ, AT SCHLADMING (STEIERMARK, AUSTRIA)
24th FEBRUARY—9th MARCH 1966

SPONSORED BY
BUNDESMINISTERIUM FÜR UNTERRICHT
STEIERMÄRKISCHE LANDESREGIERUNG
THE INTERNATIONAL ATOMIC ENERGY AGENCY AND
KAMMER DER GEWERBLICHEN WIRTSCHAFT FÜR STEIERMARK

EDITED BY

PAUL URBAN

GRAZ

WITH 57 FIGURES

1966

SPRINGER-VERLAG / WIEN · NEW YORK

Organizing Committee:

Chairman: Prof. Dr. PAUL URBAN
Vorstand des Institutes für Theoretische Physik,
Universität Graz

Committee Members: Dr. R. BAIER
Dr. P. BREITENLOHNER

Secretary: A. SCHMALDIENST

Acta Physica Austriaca / Supplementum I
Weak Interactions and Higher Symmetries
published in 1964

Acta Physica Austriaca / Supplementum II
Quantum Electrodynamics
published in 1965

ISBN 978-3-211-80755-2 ISBN 978-3-7091-5566-0 (eBook)
DOI 10.1007/978-3-7091-5566-0

Library of Congress Catalog Card Number: 66-26128

Titel-Nr. 9183

Preface

The great success of the experimental research on elementary particles and their qualities during the last years suggests giving a summary of the present situation also in the theoretical description of this important branch of physics. In spite of the precarious situation in this field of theoretical physics I believe I can fully account for this choice and must see that the number of participants and the general interest justify my opinion.

In organizing the proceedings it was our prime concern to reduce the delay in editing and also keep down the price. This was possible only through the assistance of the Springer-Verlag who chose photomechanical method working quicker and cheaper. Therefore we apologize for any mistakes and errors that may occur in the text and formulae.

I am very indebted to my secretary, Miss A. Schmaldienst and one of my assistants, Dr. H. Kühnelt, who did all the typing and correcting of the manuscripts with great patience and knowledge.

Graz, May, 1966 P. Urban

Contents

Dear Colleagues, Ladies and Gentlemen:

It gives me great pleasure to welcome you to our fifth International Meeting at Schladming, and it is very gratifying for me to see among our guests of honour so many distinguished members of our government and provincial authorities, giving proof of their serious interest in our endeavours. It is a great honour for me to welcome the representative of the Minister of Education, to whom we are indebted for sponsorship and grants which are the basis of our undertaking. We are further indebted for support to the provincial government whose representatives I herewith also welcome most heartily, and the Director of the Chamber of Commerce. It is a special pleasure for me to have among our guests of honour also the Vicechancellor of the University of Graz, the representative of the dean of our Faculty, and the representative of the Ministery of Commerce who have honoured our meeting with their presence.

I most heartily welcome Prof. Källén, who so often has contributed to the successes of our meetings. Furthermore I want to express my special thanks to all those, who have participated and also collaborated in our meetings in past and present. I trust they will enjoy staying together with us here in the beautiful mountains of Styria, as the surroundings are certainly contributing to the success of a meeting. When speaking of our host city Schladming, we must not forget its representative, the Mayor of the City, to whom we owe so much in the past. It will, no doubt, interest my listeners to hear that the number of participants of this year's meeting again is almost 200, representing as many as twenty nations. It is a great honour for me to have succeeded in gaining the cooperation of highly outstanding lecturers from abroad and I want to express my gratitude that they undertook the trouble of coming to us which will be rewarded by the benefit their knowledge will give us.

Now let me say a few words about this year's scientific program: We try to give a detailed survey of the principles of elementary particle theories which today attract such world-wide interest as theoretical basis of every fundamental research. In this field the concepts of symmetries play an outstanding part in the advancement of our knowledge disclosing the structure of the matter. As a consequence physicists have applied group theoretical methods in their research and our lectures here lay special stress upon such mathematical apparatus. Various aspects on this subject will be discussed in the course of this symposium and I trust that the result of our efforts will prove successful in bringing us a step further towards a solution of these very complicated problems.

In conclusion may I express my thanks and appreciations to all those who contributed by means of funds and grants towards the organization of this meeting. In this respect I have to mention especially the support we received from the Austrian Ministery of Education, the Provincial Government of Styria and other organizations such as the Chamber of Commerce and I am also indebted to the International Atomic Energy Agency in Vienna for grants received. I wish to thank you all once more for having taken the trouble to come here and make our fifth meeting on

"Elementary Particle Theories"

the success, we all wish it to have.

P. Urban, Graz

ON THE NONLINEAR SPINOR THEORY OF ELEMENTARY PARTICLES*

By

H. P. DÜRR

Max-Planck-Institut für Physik und Astrophysik
München

The nonlinear theory of elementary particles dates back to about 1953 when Heisenberg [1] made his first attempts to formulate a general theory of elementary particles in form of a fundamental spinor field coupled to itself. This first version of the theory was thought to be a model theory which still lacked features essential for a realistic theory. In 1958 another form of the spinor theory was suggested by W. Heisenberg and W. Pauli [2] where more realistic features were introduced. About this theory I wish to talk since most of the further investigations in Munich were done in connection with this theory [3]. I indicated that this theory contains more realistic features than the former model theory. It is still uncertain whether one can honestly hope that it may in the end produce and explain the richness of all elementary particle physics. Many of you, I guess, will have the impression after my lecture that the chances are rather small for such a hope, because there seem to be rather obvious shortcomings some of which I will point out to you during my lecture. I do not want to argue on this point, because at the present moment we do not have any solid mathematical arguments but only physical perceptions how these shortcomings can be

---

* Lecture given at the V. Internationalen Universitätswochen für Kernphysik, Schladming, 24 February - 9 March 1966.

overcome, since many things which you may consider obvious at a first glance, may look quite puzzling at a second, and even possibly wrong at a third glance. The more you work in elementary particle physics the more you become aware of the tremendous complexity of the problem, which forbids simple answers, and the more you feel the limitations of the tools which are available to you to tackle the dynamical problem. In any case: you may consider this theory as a model theory which we use to deal with some of the urgent problems of elementary particles and leave it to the future to determine how much it can explain and how much is beyond its scope. Of course, the danger is there, that we never will really find out.

I will subdivide my lecture in three parts:

In a first part I will state and explain the basic assumptions of the theory.

In a second part I will discuss an unconventional feature of the theory, the assumption of an indefinite metrio in Hilbert space, and two grouptheoretical pecularities of the theory connected with the assumption of an unsymmetrical vacuum state and the parity problem. Whereas the indefinite metric appears to be of crucial importance for a consistent dynamical theory of our kind, the grouptheoretical refinements are demanded to rise our theory to the level of a realistic theory.

In a third part I will indicate how we proceed to calculate approximately másses and coupling constants.

I. Basic postulates

Experiments at high energies have clearly demonstrated that elementary particles cannot be considered anymore to be "elementary" in the original sense, i.e. to possess something like an indestructible individuality. Elementary particles transform into each other in various ways, at least if certain conditions are fulfilled, which are described by a number of conservation laws.

We observe a vast number of elementary particles today. A large number of them are exceedingly unstable, and only a few of them are completely stable. These are the classical elementary particles, the electron and proton and also the mass zero particles photon, the neutrinos and perhaps the gravitons. Because of their stability it was common some time ago to consider these particles also to be more fundamental than the others. But with the arrival of the mesons, in particular the π-meson, which is unstable, there were strong indications that stability is no good criterion for the elementary character of a particle, although it may be of great practical importance in low energy physics. Actually this was already indicated much earlier by the existence of a great number of stable atomic nuclei, the deuterion, the α-particle up to lead and higher, which after the start of nuclear physics, about three to four decades ago, were never again considered to be elementary systems. We may even use, if you like, stable atomic configurations, e.g. the H-atom, as more absurd examples. In fact, one had learned from the investigation of atomic and nuclear systems that there is no fundamental difference between stable configurations and low lying excited states which are unstable. Small changes in the interaction of the particles may induce stable and unstable configurations to exchange their role. What seemed to be fundamental in this context was a field, which obeys a fundamental dynamical equation, e.g. the electron wave function which obeys a Schrödinger equation with Coulomb interaction. The stationary or quasistationary solutions of this equation then are connected with the stable or metastable configurations. Sometimes the fundamental field has some similarity in its character with the stable states. E.g. the groundstate of the H-atom, if we neglect the proton spin, is a doublet state and hence transforms similarly to the electron-spinor wave function. If we do take the proton spin into account it is a singlet state and the similarity disappears.

The main postulate of our theory now is to consider all elementary particles, resonance states and scattering states on the same level and regard them as different configurations of some general matter field. This matter field is supposed to obey a simple dynamical law. From the interaction of this fundamental matter field, as expressed by the dynamical law, all dimensionless constants, i.e. coupling constants and mass ratios, should be calculable.

In order to realize such a program we are confronted with a task similar to the one we would encounter in discovering quantum mechanics and the Schrödinger equation from the iron spectrom without having the knowledge of the spectrum of the H-atom, only that in our case the problem is uncomparitively more complicated by the fact that we deal with strong interactions and a relativistic dynamics where there is no natural limitation on the number of states we have to consider explicitly. The only reason which may not render this undertaking completely hopeless may lay in the fact, that the conditions necessary for such a fundamental theory may be so restrictive that we do not have much arbitrariness in its formulation. The ideal situation would be to have no arbitrariness left, at all.

I will now write down a dynamical equation for a matter field $\psi(x)$ which we have selected as a starting point for our theory. I will then discuss this Ansatz and indicate why we have given preference to it in our work.

We introduce a nonhermitean 4-component Weylspinor-isospinor field operator $\psi(x)$ in some linear space which obeys the following local nonlinear field equation

$$i\sigma^{\nu} \frac{\partial}{\partial x^{\nu}} \psi(x) = - \ell^2 :\sigma_{\mu}\psi(x)[\psi^*(x)\sigma^{\mu}\psi(x)]: \qquad (1)$$

and a similar equation for $\psi^*(x)$. The $\sigma^{\mu} = (I,\vec{\sigma})$ are the Pauli-matrices times a unit matrix in isospin space which we have not indicated explicitly. The ℓ^2 is a coupling constant of (conventionally) a length square. The $\psi(x)$ are quantized according to anticommutation rules. The anticommutator vanishes for space like distances, i.e.

$$\{\psi(x),\psi^*(x')\} = 0 \qquad \text{for } (x-x') \text{ space like} \qquad (2)$$

For timelike distances this anticommutator should be different from zero to exclude classical theories. In general, it will be a complicated operator. The anticommutator should be a c-number for equal times or - if ill-defined - in some appropriate limit $t-t' \to 0$. We do not necessarily require that it should behave for equal times as $\delta(\vec{r}-\vec{r}')$ as in case of canonical quantization.

The equation is invariant under the following transformations ($\varepsilon_{\mu\nu}, \vec{\beta}, \vec{\gamma}, \vec{\alpha}, \alpha_o$ are parameters):

Lorentz transformation $\quad \psi(x) \to \exp(\frac{1}{2}\varepsilon_{\mu\nu}\sigma^{\mu\nu})\psi(L^{-1}x) = $

$$= \exp(i\vec{\beta}.\vec{\sigma}+\vec{\gamma}.\vec{\sigma})\psi(L^{-1}x)$$

Isospin transformation $\quad \psi(x) \to e^{i\vec{\alpha}\vec{\tau}}\psi(x)$

Fermionnumber transformation $\quad \psi(x) \to e^{i\alpha_o}\psi(x) \qquad (3)$

GP-reflection $\quad \psi(\vec{r},t) \to C_\sigma C_\tau \psi^{*T}(-\vec{r},t)$

CP-reflection $\quad \psi(\vec{r},t) \to C_\sigma \psi^{*T}(-\vec{r},t)$

T inversion $\quad \psi(x) \to C_\sigma^{-1}\psi(\vec{r},-t)$ } c-numbers complex conj.

CPT $\quad \psi(x) \to \psi^{*T}(-x)$

These are also symmetries of the theory if the quantization is chosen in accordance with it, i.e. if the equal time anticommutator (or the corresponding limiting expression) transforms like the 4th component of a vector-isoscalar.

There is another invariance property of the equation, to which mainly Mitter [5] has called attention, and about which he will talk in some detail, namely the invariance of the equation under

scale transformation $\quad \psi(x) \to \eta^{1/2}\psi(\eta x) \qquad (4)$

and similar for $\psi^*(x)$. Again this transformation will only constitute a fundamental symmetry of the theory if the quantization is invariant under this symmetry. For a canonical theory this is certainly not the case, since in a canonical theory

$$\{\psi(x),\psi^*(x')\}_{t=t'} = \text{const}\ \delta(\vec{r}-\vec{r}')$$

is not invariant under the scale transformation (4). We may also state the result in this way: In a canonical theory $(\psi^*\psi)$ has the dimension of a $(\text{length})^{-3}$ (a density) and hence transforms according to

$$\psi(x) \rightarrow \eta^{3/2}\psi(\eta x) \tag{5}$$

under a change of scale of the coordinates. Under this "canonical" scale transformation, however, the equation (1) is now not invariant which is indicated by the appearance of the coupling constant of dimension $(\text{length})^2$. Only theories with dimensionless coupling constants would be invariant, e.g. quantum-electrodynamics with mass zero electrons. Scale transformation (4) on the other hand attributes a dimension $(\text{length})^{-1}$ to $(\psi^*\psi)$ which can only be fulfilled by a noncanonical quantization condition. The behaviour under scale transformation is closely related to the regularization problem. Mitter will discuss this point in more detail.

There are many questions which arise in connection with this particular formulation; in particular the following ones:

1. Why does one introduce local field operators, at all, and why should there exist anything like a local differential equation for these operators?

2. Why should it be formulated with spinor fields, and not in terms of other fields?

3. Why does one choose these particular spinor fields which allow only the symmetry transformations (3), and not more.

4. Do the conditions given define a dynamical theory, at all?

The first question why we consider the introduction of a local field operator as appropriate for the formulation of a general theory of elementary particles is important in the light of the asserted possibility to formulate such a theory in terms of a pure S-matrix theory.

The concept of a local field operator is closely related to the concept of causality. Causality incorporates "primitive causality" which is roughly expressed by the statement that the events in the future are "caused" by events at present, such that the future becomes predictable, at least to a certain extent, on the basis of observations at the present moment. This is a fundamental requirement of science. It is reflected in the formulation of dynamical laws in the form of a differential equation with respect to time. Furthermore, in a relativistic theory, to avoid contradictions, causality has to be refined to mean that action is only exchanged locally and is never propagated faster than the velocity of light. It requires, at least in classical physics, the formulation of the dynamical laws in terms of local fields which are locally coupled in a Lorentz-invariant time-dependent hyperbolic differential equation. In a quantum theory one has to be careful in addition, that through the uncertainty principle no acausal effects are produced. For two observations which cannot be connected by a light signal, i.e. have a space-like separation, the uncertainty principle should be irrelevant. Our formulation in terms of a Lorentz-invariant differential equation (1) with the microcausality condition (2) for the field operator, hence, is thought to be a direct reflection of the relativistic causality requirements.

In view of the many difficulties we have encountered in the past in relativistic quantum field theories many objections can and should be raised at this point. One principle objection refers to our application of the causality arguments - which can be easily formulated and understood for classical events, and with some care probably also for observables in a quantum theory - to our anticommuting field operator which is certainly not an observable. Our formulation, hence, seems to be a sufficient but not a necessary condition for causality; it appears to be too optimistical and hence too restrictive a frame for a realistic theory. In order to avoid contradictions later on we have to weaken our requirements. We will do this by allowing $\psi(x)$ to act possibly in a larger Hilbert space than in the Hilbert space of physical realizable states. The structure of this larger Hilbert space will be reflected in the quantization condition of our fields and the choice of a vacuum state. Without these additional conditions on the Hilbert space our theory is not fully defined. It is not clear in what sense the product of field operators at the same space-time point should be understood. In the equation we have already limited the meaning of the product in some way by using double dots, which shall mean a Wick product, i.e. a time-ordered product in which all vacuum expectation values are subtracted. Still our field equation does not uniquely define the dynamics as e. g. it is supposed in case of quantum electrodynamics or more conventional quantum field theories where canonical commutation rules are required for the field operators, and the physical realizable states are assumed to span the entire Hilbert space. Our equation seems at this stage only a rather formal vehicle by which causality and symmetry requirements are introduced into the theory without fixing it completely. This answers part of question 4. above. In a comparison with a pure S-matrix theory, a theory of our kind which introduces local field operators, has the advantage, that the analyticity properties of the S-matrix, which reflect the causality structure, can be generally deduced, and that it may also yield an understanding why quantum electrodynamics, a local quantum field theory, works so well.

The second question, why the fundamental field equation should be formulated in terms of a spinor field, finds an answer in the observation that elementary particles have half-integral as well as integral spin. Hence, in a fundamental formulation a spinor field with some half-integral spin has to be introduced. Of course, it is not necessary, that only a spinor field occurs. Another attractive possibility e.g. is to introduce at the beginning a spinor and a vector field both of mass zero. This would be a theory of the type of quantum electrodynamics [6]. From a simplicity point of view we prefer to introduce only a spinor field of spin 1/2. To express interaction the equation has to contain a nonlinear term in these field operators. The expression "nonlinear" is used here in the same sense as quantum electrodynamics is a nonlinear theory, i.e. simply in the sense of a theory with interaction. One may feel somewhat uneasy about the "simplicity" argument which has led here to a preference, because of its arbitrariness and because simplicity is a rather relative term. In many ways e.g. a spinorfield-vectorfield combination has simpler features. There are, however, some indications [7] that such a theory, if it represents the same symmetries with appropriate additional conditions which are necessary anyway to define the theory completely, eventually may lead to results identical to our theory due to the underdetermined nature of all field equations.

The third question refers to the symmetries we have selected for our equation. As you all know symmetries are connected with conservation laws. In the interaction of elementary particles we observe besides the classical conservation laws, the conservation of energy, momentum, angular momentum of center of mass (which are related to the relativistic invariance of the theory), a number of other conservation laws which are attributed to an internal symmetry group of the theory. Here we have, in particular, the conservation of charge-like quantum numbers as electrical charge, baryon number, lepton number or even electron and muon number, separately.

A great puzzle, however, is the existence of some approximate conservation laws, as e.g. the conservation of isotopic spin which is violated in about every 10^4 instant, and the conservation of strangeness and parity which are violated even to much less an extent. The stronger or weaker violation of certain conservation laws is the cause of the puzzling hierarchy of interactions which we phenomenologically term strong, electromagnetic, weak interactions etc.

I.e. if we succeed to really understand the reason for the violation of the conservation laws we will, at the same time, have an answer, why elementary particles participate in these very different kinds of interactions. The big question remains, how the approximate conservation laws should be treated in a fundamental theory.

The conventional way to treat approximate conservation laws would be by requiring a corresponding approximate symmetry of the theory. This is always a possible way for the description, but it appears to us quite unsatisfactory, since it would not render a real explanation, but only postpone the question: The observed phenomenological asymmetry would be interpreted in terms of a fundamental asymmetry; the degree of violation would be "godgiven" or - from the point of view of the theory - accidental. I.e. certain dimensionless constants would appear in the theory, e.g. the fine-structure constant or something related to it, which are there from the beginning and cannot be calculated from the theory, in principle. This is against the spirit of our theory and hence, we believe, it should be only used as a last retreat if there is really no other way out.

We hence wish to require some maximal symmetry for the fundamental theory to cut down arbitrariness. This may mean that all approximate conservation laws cannot have a symmetry counterpart in the fundamental theory, at all. The partial symmetrization must then have a dynamical origin. The appearance of approximately valid higher symmetries is well known in atomic physics, nuclear physics and in many-body physics. Bootstrap mechanisms in elementary particle physics may further intensify such tendencies.

However, many-body physics seems to indicate also the possibility that a fundamental symmetry may be masked if the ground-state, which in elementary particle physics is the vacuum state, is not invariant under this symmetry group. The breakdown of a fundamental symmetry, however, seems then to be always accompanied by long range correlations or particles of mass zero. One may, perhaps, also say roughly, that in case of long range correlations the isolation of a system under consideration and hence the verification of a conservation law for a limited system becomes a problem. I will come back to this later on. At this point we will only state, that violated conservation laws which are accompanied by long range interactions or particles of mass zero, may be connected with an unimpaired fundamental symmetry.

We now turn back to the consideration of our equation. The simplest spinor equation we can write down is an equation for a 2-component spinor field. It looks exactly like our equation (1) with the difference that the isospin space is absent. The nonlinear term is here the only possible local momentum-independent term. From a grouptheoretical point of view the internal symmetry is reduced in this case to a simple gauge group U_1, which may be connected with some charge like quantum number, e.g. baryon number. Hence, we certainly do not have enough internal symmetry structure to accomodate even the most important exact conservation laws, e.g. conservation of electrical charge. A recent investigation of this 2-component theory by Mitter and Rechenberg [8], on which Mitter will report later, seems also to indicate that this theory has some rather pathological features which may eventually render it unfit for an elementary particle theory. It is also well-known that the equation interpreted classically, i.e. with commuting spinor fields, is a free particle theory because the interaction term vanishes identically. In any case, the 2-component theory is barely a candidate for a realistic theory.

Hence we double the number of components to enrich the internal symmetry structure. We get now a group U_2 as the maximally possible internal symmetry group. This is our choice.

Hence we not only get an additional gauge group but a rotation group, i.e. actually more than what is reflected by exact conservation laws. However, here we might be just lucky, because the violation of the isospin group is, indeed, connected with particles of mass zero, the photon. Hence isotopic spin may be a candidate for a fundamental symmetry and then there will be no immediate objection to start with such a theory.

An objection to this theory will be that it still cannot represent lepton conservation, or electron and muon conservation. It is also disconcerting that strangeness and parity are not visible, at all, which are conserved to a much higher degree than isotopic spin. The absence of a strangeness degree of freedom becomes also apparent in the following way: Our fields are spinor-isospinors, i.e. are connected with spin 1/2 and isospin 1/2. Any products of these field operators hence will only transform like basis vectors of a representation with half integral spin - half integral isospin or integral spin - integral isospin. I.e. applied to a vacuum state in the conventional sense we cannot produce states with integral spin-half integral isospin or half integral spin-integral isospin, like e.g. K-mesons or Λ-particles.

To introduce any of these additional symmetries it seems we have to introduce more spinor components. E.g. we may not only double but triple the number of components of the original 2-component spinor field. Then we would obtain maximally U_3 as the internal symmetry group. Our spinor field would be a quark field, and hence simply allow the definition of strangeness or hypercharge. But SU_3 would be established as a fundamental group and we cannot see how its observed strong violation can be brought about, because no mass zero particle is known up to date which can be attributed to this violation. On the other hand if we do not require maximal symmetry of our equation but only $U_1 \times U_2$ in the quark formulation then the interaction term is not unique anymore, the violation is introduced explicitly and arbitrarily.

This is the main trouble we get into by enlarging the number of components further: The requirement of maximal symmetry, which excludes arbitrariness in the formulation of the theory, immediately leads to symmetries which are much too high for elementary particle physics, and hence a lot of handwaving has to be done to get rid of them. The inclusion of a lepton number e.g. would immediately introduce a symmetry group which transforms baryons and leptons into each other. This would say that in some approximation baryons and leptons should be considered identical, which is not at all what we want. The same difficulties arise if we would incorporate parity as a fundamental symmetry by doubling the number of components, i.e. by a transition from a Weyl to a Dirac theory. It would also have the additional disadvantage that another linear term with a mass, like in the Dirac theory, would be possible which would establish another length and together with the length of the coupling constant a noncalculable dimensionless number. On the other hand we find it attractive in our theory that a bare mass cannot be introduced on group theoretical grounds, and hence all physical masses have to arise from the selfinteraction of the fundamental field.

Hence we will stick to our formulation with its deficiencies. I will again summarize these grouptheoretical difficulties:

1. No obvious symmetry provided to distinguish baryons, electrons and muons.

2. No symmetry provided to introduce hypercharge separate from fermion number and to describe strange particles.

3. No discrete symmetry provided frr parity.

4. Exact isospin rotational symmetry.

If our theory should have a chance to become a realistic theory, and does not serve merely as a model, all these difficulties have to be met. In part II we will indicate how we hope to meet this challenge.

Before I continue to this part I wish to make a few remarks about the particular form of the equation. One may be, at first, surprised about the statement that the interaction term is unique. If we write the nonlinear term

$$:V_i\psi(x)[\psi^*(x)V_i\psi(x)]:$$

we see, that only

$$(V_i)(V_i) \equiv (\sigma_\mu I)(\sigma^\mu I) \tag{6a}$$

has all the required symmetries but also

$$(V_i)(V_i) \equiv (\sigma_\mu\vec{\tau})(\sigma^\mu\vec{\tau}) \tag{6b}$$

and hence an arbitrary linear combination. However, because of the antisymmetrical combination of the ψ in the interaction term one can immediately deduce by Fierz transformation (recoupling transformation) that the two terms are identical and hence can both be used alone or in any linear combination without changing the results. One combination

$$V = (V_i)(V_i) = \frac{1}{4}(3II + \vec{\tau}\vec{\tau})\sigma_\mu\sigma^\mu \tag{6c}$$

is particularly convenient, since here the direct and the exchange term are equal to each other. We will call this combination Fierz-symmetrical. The isospin factor has the simple explanation that there is no interaction if the spinor fields are antisymmetrical in isospin space ($\vec{\tau}\vec{\tau} = -3$). This also follows from the original form $\sigma_\mu\sigma^\mu = II - \vec{\sigma}\vec{\sigma}$ where the absence of the interaction in the symmetrical spin case ($\vec{\sigma}\vec{\sigma}=+1$) is expressed. The spatial wave function is always symmetrical since the operators are taken at the same point.

Besides the equation for the $\psi(x)$ there is always a similar equation for the $\psi^*(x)$ or $\psi^G(x)=C_\sigma C_\tau\psi^{*T}(x)$. We can arrive at a different way of writing if we combine the upper two components of $\psi(x)$ with the upper two components $\psi^G(x)$ to form another

4-component spinor $\chi(x)$. Then we find for the field equation for $\chi(x)$

$$\Gamma^{\nu}\frac{\partial}{\partial x^{\nu}}\,\chi(x) = \ell^{2} :\Gamma_{5}\Gamma_{\mu}\chi(x)[\chi^{*}(x)\Gamma_{5}\Gamma^{\mu}\chi(x)]: \tag{7}$$

and a similar one for $\chi^{*}(x)$. This, in fact, was the original form of the equation written down by Heisenberg and Pauli [2], who used Dirac γ^{μ} instead of our Γ^{μ}, which, however, are only algebraically but not physically equivalent.

There is some other way in which our equation (1) may be written which is convenient for general discussions [9] . We may introduce hermitean spinor field operators of double number of components in place of the nonhermitean field operators

$$\Psi(x) = \begin{pmatrix} \frac{1}{\sqrt{2}}(\psi(x) + \psi^{*T}(x)) \\ \\ \frac{-i}{\sqrt{2}}(\psi(x) - \psi^{*T}(x)) \end{pmatrix} \tag{8}$$

The equation then can be written in the form

$$D_{\alpha\beta}\Psi_{\beta}(x) = V_{\alpha\beta\gamma\delta} :\Psi_{\beta}(x)\Psi_{\gamma}(x)\Psi_{\delta}(x): \tag{9}$$

where $D_{\alpha\beta}$ is a hermitean, purely imaginary antisymmetrical operator

$$D_{\alpha\beta} = i\Gamma^{\nu}_{\alpha\beta}\,\frac{\partial}{\partial x^{\nu}} \qquad (\Gamma^{\nu}_{\alpha\beta} = \text{real and symmetrical}) \tag{10}$$

and $V_{\alpha\beta\gamma\delta}$ a vertex operator which is real and completely antisymmetrized in all index pairs. The $\Gamma^{\nu}_{\alpha\beta}$ and $V_{\alpha\beta\gamma\delta}$ can be explicitly given, which I will not do here.

I also will not mention a representation of the equation due to Gürsey [10], which introduces a 4-component field operator $\Phi(x)=(\begin{smallmatrix}\cdot&\cdot\\ \cdot&\cdot\end{smallmatrix})$, in place of $\psi(x)$, where spin matrices multiply from the left, isospin matrices multiply from the right. All operators (σ,τ,Φ) in this representation then look like quaternions.

The picture, however, is somewhat spoiled by the fact that the Φ is an anticommuting operator.

II. Special features

I will start to discuss one feature of the theory, the assumption of an indefinite metric in Hilbert space which is crucial for any nonlinear spinor theory of our type. Then I will shortly discuss hypotheses to meet our grouptheoretical difficulties.

A. Indefinite metric in Hilbert space

Despite of their simple structure, theories based on a field equation of our type have proved inaccessible with conventional methods because they belong to the group of nonrenormalizable theories. The interaction is here so singular that divergencies result in perturbation theory which cannot be eliminated by the normal subtraction and renormalization procedure. More precisely the divergencies seem to be connected with the local coupling of the operators, or the rather singular nature of the propagator on the light cone which in turn is connected with the singularity of the anticommutator on the light cone. For the vacuum expectation value of the anticommutator we assume a Källen-Lehmann [11] representation

$$S(z) = \langle 0|\{\psi(x),\psi^*(x')\}|0\rangle = -\int d\kappa^2 \rho(\kappa^2)\bar{\sigma}^\mu \frac{\partial}{\partial z^\mu}\Delta(z;\kappa) \qquad (11)$$

with $\rho(\kappa^2) \neq 0$ for $\kappa^2 \geq 0$ the mass density function.

The S(z) vanishes for space like distances due to the same behaviour of the Δ-function. Near the light cone we have

$$S(z) = \frac{\bar{\sigma}_\mu z^\mu}{2\pi}\varepsilon(z)\left[\rho_0\delta'(z^2) + \rho_1\delta(z^2) + \text{finite terms}\right] \qquad (12)$$

with

$$\rho_0 = \int d\kappa^2 \rho(\kappa^2)$$
$$\rho_1 = \int d\kappa^2 \kappa^2 \rho(\kappa^2) \qquad (13)$$

the zeroth and first moment of the mass density distribution. For equal times one gets

$$S(z)|_{z^0=0} = \rho_0 \delta(\vec{r}-\vec{r}\,') \qquad (14)$$

One can now demonstrate that the divergencies arise at small distances in connection with the δ and δ' singularities or corresponding singularities $P(1/z^2)$ and $P(1/z^2)^2$ in the propagators. To avoid these divergencies one has to require

$$\rho_0 = \rho_1 = 0 \qquad (15)$$

This, however, can only be satisfied if $\rho(\kappa^2)$ in (13) is not positive definite. But $\rho(\kappa^2)$ can only be negative if states of negative norm contribute in the sum over a complete set of intermediate states.

Since $\rho_0=0$ the quantization is not canonical anymore. As a consequence the generators of the symmetry transformations in the Hilbert space, e.g. the Hamiltonian, charge operator etc., cannot be expressed as space integrals over current operators which are constructed from field operators at the same space time point. On the other hand the Lagrangean from which our field equation may be formally deduced by variation has nothing to do with the physical Lagrangean related to the physical Hamiltonian, i.e. the generator of the time translations, and does not go over into the classical Lagrangean in the limit $\hbar \to 0$ according to the correspondence principle. Hence one has to be careful to count our theory among the Lagrangean theories. There is, however, a chance that the physical operators may be obtained from the formal ones by some kind of local smearing [12].

For the propagator function we can write

$$F(z) = \langle 0|T\psi(x)\psi^*(x')|0\rangle = \frac{i}{(2\pi)^4}\int d\kappa^2\rho(\kappa^2)\int d^4p\frac{\bar{\sigma}_\mu p^\mu}{p^2-\kappa^2+i\varepsilon}e^{-ipz} \quad (16)$$

where we may equally set because of (13)

$$\int d\kappa^2\rho(\kappa^2)\,\frac{1}{p^2-\kappa^2} = \int d\kappa^2\rho(\kappa^2)\left[\frac{1}{p^2-\kappa^2} - \frac{1}{p^2-\mu^2} - \frac{\kappa^2-\mu^2}{(p^2-\mu^2)^2}\right]$$

$$= \int d\kappa^2\rho(\kappa^2)\,\frac{(\kappa^2-\mu^2)^2}{(p^2-\mu^2)^2(p^2-\kappa^2)} \quad (17)$$

The propagator appears now to be regularized. Mitter will talk later in this seminar about this noncanonical quantization and its possible relation to scale invariance of the theory at very small distances or high momenta. Therefore I will limit myself to discuss some of the consequences of the indefinite metric.

Because of the interpretation of a norm of a state as its probability the introduction of an indefinite metric appears to lead immediately to a complete breakdown of the quantum mechanical probability interpretation. This, however, is not the case. A study of a number of simple models by Heisenberg [13], Sudarshan [14] and others [15] have shown that in a theory with indefinite metric by a suitable selection of states, which are termed physical states, a submatrix of the pseudo-unitary S-matrix may be constructed which is unitary and which connects only these physical states. For these physical states hence a quantumstatistical interpretation is possible. The important point is that the physical states are only an invariant subset of the complete set of states in the Hilbert space which we can construct by applying products of field operators on the vacuum. Observables are hermitean operators which do not connect physical and unphysical states. Since the field operators are not observables they can connect physical and unphysical states.

An example of a theory with indefinite metric is the Gupta-Bleuler theory of quantum-electrodynamics [16]. Here scalar photons with negative norm appear in the propagator function. Physical states are then defined as states which obey the Lorentz condition. If one imposes this subsidiary condition on the incoming states, the outgoing states fulfill the condition too, and one arrives at a consistent Lorentzinvariant theory with a normal probability interpretation. In this case, however, it may not be surprising, since we know that there exists the completely consistent quantum electrodynamics of Dirac and Schwinger [17] which does not involve any states of negative norm. There only the two transversal modes are quantized. However, in the Hamiltonian of this formulation there occurs a nonlocal interaction arising from the instantaneous, nonretarted Coulomb interaction between the sources, and the field operators are not localizable because the transversality is a nonlocal condition. Despite its apparent noncovariance we know that the theory is relativistically invariant.

This may not be a singular case. Sudarshan [14] suspects from his intensive studies with various simple models that in a Lagrangean type theory manifest covariance and positive definite metric may be complementary aspects. The introduction of an indefinite metric just allows us the luxury to formulate our theory in a manifest covariant form, i.e. in a form where only locally coupled field operators are admitted which is important for simplicity. By the final projection on the positive definite invariant subspace of the Hilbert space we arrive at a physically interpretable theory. This theory is Lorentz invariant and probably causal (upper half plane analyticity of the scattering amplitudes) if time reversal invariance holds. A formulation of the theory in the positive definite physical subspace should always be possible but may be exceedingly complicated. It will involve nonlocal interactions in the form of effective relativistic form factors, which nobody knows how to construct and which would be impossible to guess from the start, because they would contain already a lot of the dynamics. - We will hopefully assume that a similar situation also holds in our theory.

If we analyze, as an example, the expression $\frac{\overline{\sigma}_\mu p^\mu}{(p^2-\mu^2)^2(p^2-\kappa^2)}$ in the regularized propagator (14,13) we observe that three states contribute to this expression. One normal state $|N\rangle$ of norm +1 with mass κ, and two states $|D\rangle$ and $|G\rangle$ with the metrical tensor

$$\left|\begin{matrix} D \\ G \end{matrix}\right\rangle \langle D,G| = \begin{bmatrix} 0 & 1 \\ 1 & 0 \end{bmatrix} \tag{18}$$

The D is a norm zero dipole ghost state which is not an eigenstate of energy and momentum. The G is a norm zero ghost state which is an eigenstate of energy momentum with mass μ. Dipole and ghost states are not orthogonal to each other. If one relies on a certain analogy to the Leemodel [13] one may be inclined to define the physical states as eigenstates of energy and momentum, and hence only allow combinations of actual particles N and G as incoming states. The G admixture does not change the physical interpretation because its norm is zero, but it may now be chosen such that no D occurs as outgoing state, if that is not automatically the case. In this case a unitary S-matrix could be constructed which only involves N-particles. However, this has still to be studied in greater detail.

Perhaps the necessity of an indefinite metric in a local theory, or a nonlocal interaction if a positive definite metric is assumed, may be connected with the requirement that no particle or state should be singled out as more "elementary" than the others by the formulation. In a conventionally quantized local theory this requirements seem not to be fulfilled, because certain elementary particles are distiguished from the beginning. Hence the divergency problem may have a physical origin and arise from this unphysical distinction.

B. The unsymmetrical vacuum state

To build up the Hilbert space we have to define a groundstate or vacuum state. This vacuum state has to have certain properties to allow a particle description of the field theory. We do not know whether there are one or more possibilities for the choice of the groundstate within a given theory. The vacuum state is defined as a state of lowest energy. Conventionally one assumes, in addition, that it is invariant under all the symmetry groups considered. In this case the vacuum is really a vacuum, it has all properties zero. As a consequence a single particle state is an isolated system and a two particle system will only have interactions which strictly obey all conservation laws of the theory.

If, however, the vacuum is not symmetrical under all symmetry transformations, this is not true anymore. The one particle state, in this case, can only be considered an isolated system in an approximation in which we can neglect the residual interaction with the background. To be more specific: We will assume that the vacuum is not symmetrical under isospin transformations but symmetrical under all the other symmetry operations, in particular Lorentz transformations and CPT. As a consequence the particle isomultiplets will split up in mass, and in the interaction of these particles the approximate isospin attributed to these particles will not be conserved. As a consequence of the still valid divergence conditions for the isospin currents particles of mass zero will appear according to the Goldstone theorem [18]. The formal Goldstone particles are spin zero, charged particles in this case. There are, however, indications that the Goldstone theorem may be generalized, as to allow the photon or the Coulomb interaction to take over the role of the Goldstone particle. This has still to be shown.

There is another aspect of an isospin unsymmetrical vacuum which is of particular interest to us, since it may lead to the possibility of strange particles. If the groundstate carries isospin there might be new quasi particle excitation with an anormal isospin. This may be exemplified by a ferromagnet type model studied by Biritz and Heisenberg [19].

It assumes a linear lattice of fixed particles with spin, bent to a circle to have periodicity after N spins, and additional particles in some excited states. The Hamiltonian shall be of the form:

$$H = \alpha\left[N - \sum_{n=1}^{N} \vec{\sigma}_n \cdot \vec{\sigma}_{n+1}\right] + \beta \sum_{n=1}^{N} a_n^{*\lambda} a_n^{\lambda}$$

$$- \gamma \sum_{n=1}^{N} \left(a_n^{*\lambda} a_{n+1}^{\lambda} + a_{n+1}^{*\lambda} a_n^{\lambda}\right) - \varepsilon \sum_{n=1}^{N} \left(a_n^{*\lambda} \vec{\sigma}_{\lambda\mu} a_n^{\mu}\right) \cdot \vec{\sigma}_n \qquad (19)$$

The first term indicates an interaction between neighbouring lattice particles which tries to align their spins. The second and third term is a rest energy and a kinetic energy of the additional particles in the excited state. The fourth term finally describes a local spin interaction between the excited particle and the lattice particles.

If no excited particle is present and the lattice spins are all aligned we get the groundstate $|0\rangle$ with energy $E = 0$. If we flip one lattice spin at point n we get a state $|n\rangle$. The stationary states are given by

$$|k\rangle = \frac{1}{\sqrt{N}} \sum_{n=1}^{N} e^{ikn} |n\rangle \qquad \text{with } k = \frac{2\pi}{N} r \qquad r = 1,2....$$

with energy $E_k = 4\alpha(1-\cos k)$. They correspond to the Bloch spin waves and represent the mass zero particles of the Goldstone theorem. Their energy vanishes for momentum $k \to 0$.

If we have one particle in the excited state, there are two different situations:

1. $\varepsilon \ll \alpha,\gamma$ i.e. small spin-spin interaction.

At first we assume no Bloch waves present. The particle energy term will split up $\sim\varepsilon$ depending on its spin orientation relative to the lattice polarization. We get a doublet:

$$= \quad E = \beta \pm \varepsilon \quad \text{(one particle excitation)}$$

$$— \quad E = 0 \quad \text{(groundstate)}$$

If Bloch waves are present, there will be a complicated dressing of the particles and each state will be the starting point of a Bloch-wave continuum.

2. $\varepsilon >> \alpha, \gamma$ i.e. strong spin-spin interaction. In this case the excited particle will strongly couple to a lattice spin and produce a singlet or triplet multiplicity term

$$\begin{array}{ll} — \quad E \approx \beta + 3\varepsilon & \\ \equiv \quad E \approx \beta - \varepsilon & \text{(one particle excitation)} \end{array}$$

$$— \quad E = 0 \quad \text{(groundstate)}$$

Every term will again be a starting point of a Bloch-wave continuum. Hence in this case we get an anormal multiplicity: the excited particle rides on a spin wave. It, so to say, borrows a spin 1/2 from the groundstate.

A phenomenon similar to the one described above may lead to an understanding of the strange particle modes in our theory [20]. They are anormal multiplicity states. The analogy of the isospin situation to the ferromagnet case, however, is not so close, since there is no particle-antiparticle invariance in the ferromagnet. It seems that a study of an antiferromagnet, which is undertaken by Yamazaki [21] at the moment, may render a closer analogy.

Still much can be said on this point, in particular about the relation of this point of view on strange particles and the approximate validity of SU_3, but, despite of the great efforts made [20], this is all still in a very preliminary stage.

Also I do not want to go into a discussion of the possible existence of the photon in our theory [22] which, of course, is closely linked to the question of an isospin unsymmetrical vacuum. I may just add that our preliminary investigations on quantum electrodynamics seem to point at an interesting - if not crazy - way to establish leptons in our theory. It appears that baryons and leptons are not distinguished group theoretically but should be attributed to pole and dipole singularities in the propagator. Because of the norm zero of the G-states, connected with the dipole singularity, leptons do not interact directly with baryons. However, they seem to be both coupled to photons with equal strength due to Ward's identity. This would ensure that baryons and leptons are conserved separately. The leptons would appear as the regularizer of the baryons. However, there are still many traps for this hypothesis. In particular, weak interactions, which seem to connect baryons and leptons directly, may cause trouble. This has not been investigated.

C. The parity problem

In order to complete the discussion on propositions to eliminate the grouptheoretical difficulties in our theory I have shortly to comment on the parity problem [23].

Our theory contains only CP but not P as a space-reflection symmetry. One may be surprised, at first, that such a theory is capable of giving finite mass solutions for spinor particles, at all. There, however, is no trouble, in principle, which can already be seen from the general form of our propagation function [16]. Finite mass solutions appear as solutions of a Weyl equation (or Klein-Gordon equation)

$$(p^2-m^2)\phi(\vec{p}) = 0 \qquad (20)$$

and not of a Dirac equation. Of course, the Dirac equation is equivalent to this equation.

We define

$$\hat{\phi}(\vec{p}) \equiv \frac{\sigma_\mu p^\mu}{m} \phi(\vec{p}) \tag{21}$$

then it follows from (20)

$$\bar{\sigma}_\nu p^\nu \hat{\phi}(\vec{p}) = m\phi(\vec{p}) \tag{22}$$

We now define the double component Dirac spinor $\Phi(\vec{p}) = \begin{bmatrix} \phi(\vec{p}) \\ \hat{\phi}(\vec{p}) \end{bmatrix}$ which fulfills the Dirac equation

$$\beta(\gamma_\nu p^\nu - m)\Phi(\vec{p}) = 0 \tag{23}$$

with $\beta\gamma_\nu = \begin{bmatrix} \sigma_\nu & 0 \\ 0 & \bar{\sigma}_\nu \end{bmatrix}$ and $\beta = \begin{bmatrix} 0 & 1 \\ 1 & 0 \end{bmatrix} = \gamma_4$ in the new space spanned by ϕ and $\hat{\phi}$. Here we can now clearly identify the parity operation

$$\Phi(\vec{p}) \to \gamma_4 \ \Phi(-\vec{p})$$

as the transformation

$$\phi(\vec{p}) \to \hat{\phi}(-\vec{p}) = \frac{\bar{\sigma}_\mu p^\mu}{m} \phi(-\vec{p}) \tag{24}$$

We realize that the parity operation is connected with a symmetry between the spinor field and its invariant derivative. If we would apply it to the field operators themselves it would mean a certain symmetry between $\psi(x)$ (which is a right helicity spinor) and $\sigma_\mu \frac{\partial}{\partial x\mu} \psi(x)$ (which is a left helicity spinor.). But since the latter is proportional to the local operator $:\sigma_\mu \psi(x)[\psi^*(x)\sigma^\mu \psi(x)]:$ one may also state that parity is connected with a certain symmetry between ψ, ψ^5, on one side and ψ^3, ψ^7, on the other side. Up to now we were not able to explicitly establish such a symmetry, but investigations are still under way. If we assume that such a symmetry is contained in a certain approximation one can then write down the form of the approximately valid parity symmetrical theory which would follow from our original equation [23]. I will not go into this.

III. Calculation of masses and coupling constants

I will now turn to the more concrete part of the theory. This concrete part, however, is not a straightforward calculation procedure, by which numerical results have to be simply cranked out from the dynamical equation. Since our dynamical equation constitutes more something like a framework than a uniquely defined theory, the procedure by which we calculate physical data, i.e. the approximation method, becomes part of the definition of the theory. The situation seems to be similar as in an S-matrix theory where the analyticity requirements, dispersion relations and selfconsistency relations merely provide the bottle into which the physics is filled in afterwards drop by drop. Assuming certain physical data and reestablishing them in a selfconsistent way is at present a substitute for a more mathematical "boundary" condition on the system.

We will proceed in the following way [24]. The field equation for the operators

$$D_{\alpha\beta}\Psi_{\beta} = V_{\alpha\beta\gamma\delta} :\Psi_{\beta}\Psi_{\gamma}\Psi_{\delta}: \tag{9}$$

(for simplicity we use here the hermitean representation for the field operators, and include the spacial dependence formally in the index) we interprete as a short hand description for an infinite linear system of differential equations for matrix elements of time-ordered products of these field operators, the socalled τ-functions

$$\tau_{AB}^{(k)}(1\ldots k) \equiv \langle A|T\Psi_1\ldots\Psi_k|B\rangle \tag{25}$$

The following formal consideration is valid for all field equations of the form (9), e.g. also for the case where Ψ may be a quark field operator. The different group structure is only reflected in the choice of the differential operator $D_{\alpha\beta}$ and of the vertex operator $V_{\alpha\beta\gamma\delta}$ which is assumed not to contain derivatives.

By integrating the differential equation with the Green's function $G = D^{-1}$ (and dropping terms only important for mass zero solutions)

$$\Psi_i = G_{i\alpha} V_{\alpha\beta\gamma\delta} : \Psi_\beta \Psi_\gamma \Psi_\delta : \qquad (26)$$

(equal indices shall indicate summation over indices and integration over the corresponding space-time variables) we may also derive an infinite linear system of integral equations for the τ-functions. We are in particular interested in matrix elements $\langle 0 | \ldots | B \rangle$ which have a vacuum state on the left. We represent these τ-functions graphically by 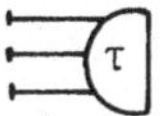 with as many lines as field operators are contained. Our integral equation can now be graphically represented by relations of the form:

+ terms $\sim \rho_0$ (27)

if G ≡  and V ≡  . Because D does not commute, in general, with the time ordering operator there are additional terms which are connected with the equal time anticommutator and hence are $\sim \rho_0$. These terms are absent or ill-defined in a noncanonical theory. The infinite set of integral equations for the τ-functions may be expressed by a single functional equation of the τ-function which is a functional of a classical anticommuting source function u:

$$\mathcal{T}_{AB}(u) = \langle A|Te^{iu_\alpha\Psi_\alpha}|B\rangle = \sum_k \frac{i^k}{k!}\, u_k\ldots u_1\,\tau^{(k)}_{AB}(1\ldots k) \qquad (28)$$

This functional equation has the form

$$\left\{\frac{\delta}{\delta u_i} + G_{i\alpha}V_{\alpha\beta\gamma\delta}\left[\frac{\delta^3}{\delta u_\beta\,\delta u_\gamma\,\delta u_\delta} + 3F_{\beta\gamma}\,\frac{\delta}{\delta u_\delta}\right] - i\rho_o u_\alpha\right\}\mathcal{T}_{AB}(u) = 0 \qquad (29)$$

Here $F_{\beta\gamma}$ is the propagation or contraction function

$$F_{\beta\gamma} = \langle 0|T\Psi_\beta(x)\Psi_\gamma(x')|0\rangle \qquad (30)$$

which, however, in (29) only occurs as F(0) which may be ill-defined, but is introduced to compensate a similar term in the $\delta^3/\delta u^3$. It arises because our interaction term was defined as a Wick product. Equation (29) is considered equivalent to our original wave equation (9). From it we can immediately deduce the general structure of the infinite system of integral equations for the τ-functions themselves. We have the following rules for the translation:

$$\frac{\delta}{\delta u_\alpha} \;:\; \tau^{(k)}(\ldots) \rightarrow i\tau^{(k+1)}(\alpha\ldots) \qquad (31)$$

$$u_\ell \;:\; \tau^{(k)}(\ldots) \rightarrow i\tau^{(k-1)}(\ell\ldots)$$

If we write the $\tau^{(1)}, \tau^{(3)}, \ldots \tau^{(2n+1)}$ as an infinite column vector, we have the structure

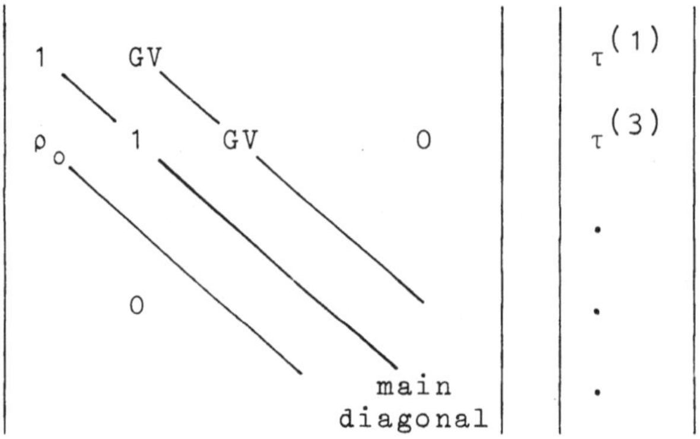

$= 0 \qquad (32)$

The problem of defining the theory is now connected with the problem how to solve this infinite set of coupled equations in order to give it the property of an eigenvalue equation. There are always an infinite number of solutions. We may e.g. obtain them by an iterative procedure, as was e.g. shown explicitly by Roos [25]. But they may not have a particle interpretation or may not fulfill important spectral conditions. An eigenvalue problem immediately arises if we are able to show that the higher τ-functions become irrelevant in a certain sense such that the system reduces effectively to a finite system.

We may think about to simply neglect a certain $\tau^{(k)}$-function with $k = 2n + 2$. Then the n lower functions would decouple from the rest and the system can be solved. If applied to the anharmonic oscillator, which has a very similar structure to our equation, it turns out, however, that this does not lead to a convergent result for $n \to \infty$ [26]. The reason is that e.g. in the single time formulation of the anharmonic oscillator the τ-functions correspond to the expectation values $\tau^{(2n)} = \langle 0 | q^{2n} | \omega \rangle$ which obviously have no tendencies to become small for $n \to \infty$. The situation is different if one considers matrix elements of Hermitean polynomials $\phi^{(2n)} = \langle 0 | H_{2n}(q) | \omega \rangle$. Because they oscillate more and more for higher n, their contribution to low lying states ω decrease. Hence the system of ϕ-functions may be broken off and converges for $n \to \infty$. This was rigorously proved for the single time anharmonic oscillator by Stumpf and Wagner [26]. A lot of investigations were done in this connection and are still continued [27]. The ϕ-functions are simply the expectation values of the Wick ordered product of the q-operators $\phi^{(2n)} = \langle 0 | :q^{2n}: | \omega \rangle$ and can be obtained from the τ-functions by the well-known Wick rule. The approximation method indicated is called the Tamm-Dancoff method.

We wish to apply this experience to our much more complicated case. It would suggest that we first transform the τ-function system to the ϕ-function system and then approximate it by neglecting higher ϕ-functions. Actually it seems that the more realistic correspondence to the anharmonic oscillator would be to consider the functions for which all vacuum expectation values vanish (ζ-functions or connected graphs). These, however, fulfill a very complicated set of equations, and we will not discuss it here [24].

The ϕ-functions we obtain from the τ-functions by applying the Wick rule, which we can simply express in terms of the generators. If we introduce the generator of the ϕ-functions

$$\Phi_{AB}(u) = \sum_k \frac{i^k}{k!} u_k \ldots u_1 \phi_{AB}(1 \ldots k) \tag{33}$$

the Wick rule is given by:

$$\Phi_{AB}(u) = \exp(-\frac{1}{2} u_1 F_{12} u_2)\, \mathcal{T}_{AB}(u) \tag{34}$$

where again F is the contraction or progagation function (30). The Φ_{AB} obeys the following functional equation

$$\{\frac{\delta}{\delta u_i} + G_{i\alpha} V_{\alpha\beta\gamma\delta} [\ ]_{\beta\gamma\delta} + [F_{o\alpha} - i\rho_o G_{o\alpha}] u_\alpha\} \Phi_{AB} = 0 \tag{35}$$

with

$$[\ ]_{\beta\gamma\delta} \equiv [\frac{\delta^3}{\delta u_\beta \delta u_\gamma \delta u_\delta} + 3 F_{\delta\ell} u_\ell \frac{\delta^2}{\delta u_\beta \delta u_\gamma} + 3 F_{\beta i} F_{\gamma j} u_i u_j \frac{\delta}{\delta u_\delta} + F_{\beta i} F_{\gamma j} F_{\delta\ell} u_i u_j u_\ell] \tag{36}$$

The ϕ-function system is still a linear system and has certain advantages over the τ-function system.

1. The singular expressions F(0) do not appear explicitly anymore.

2. The quantization does not enter through the equal time commutation rule which is proportional to ρ_o but through the form of the 2-point function F. The latter is of great impor-

tance for a noncanonical theory of our type where $\rho_0 = 0$. The τ-function system in this case could not exclude classical solutions. On the other hand the ϕ-function system has the same structure whether we quantize canonically or in any different way, since the ρ_0-term is cancelled by a corresponding term in F. The disadvantage of the system is, that the system can only be attacked, if the contraction function F is given. The contraction function plays the role of a potential in this set of equations. Of course, the F is not arbitrary but it has to fulfill a similar set of equations for the vacuum expectation values. This set of equations is nonlinear in F, and hence poses an exceedingly complicated problem. Mitter in his lectures will talk about this problem.

The system of ϕ-functions expressed in functional form by (35)can be perhaps considered meaningful in a mathematical sense, whereas the original equation for the field operators or the τ-function equation may have only formal significance. The symmetries of the field equation will appear also in the functional equation for Φ provided the contraction function F, which enters here, is invariant under these symmetries. Since the F-function is a vacuum expectation value it will only be invariant, if the vacuum state is invariant under the symmetry group. The violation of symmetries hence may be introduced by a nonsymmetrical F. There is another difference: The original field equation represents only a local condition. The Φ-equation also incorporates global conditions through the vacuum state. Roughly speaking the latter would be e.g. sensitive to a macroscopically curved space whereas the field equation would not.

From the functional equation (35) we again can deduce the general structure for the ϕ-function system. If we introduce a formal vector with the components $\phi^{(1)}, \phi^{(3)} \ldots \phi^{(2n+1)} \ldots$ we find a structure

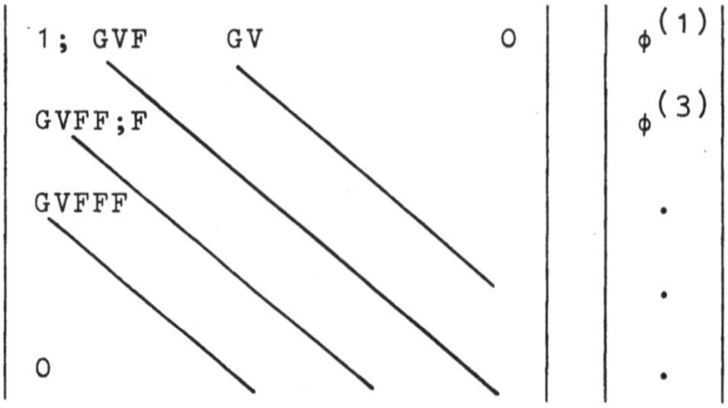

$$
\begin{vmatrix}
1; \text{GVF} & \text{GV} & & & 0 \\
\text{GVFF};\text{F} & & & & \\
\text{GVFFF} & & & & \\
& & & & \\
0 & & & &
\end{vmatrix}
\begin{vmatrix}
\phi^{(1)} \\
\phi^{(3)} \\
. \\
. \\
.
\end{vmatrix}
= 0 \qquad (37)
$$

More precisely if we represent ϕ-matrix elements $<0|\dots|B>$ by

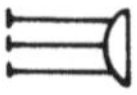

and the contraction function F by we can express the first lines in graphical form as

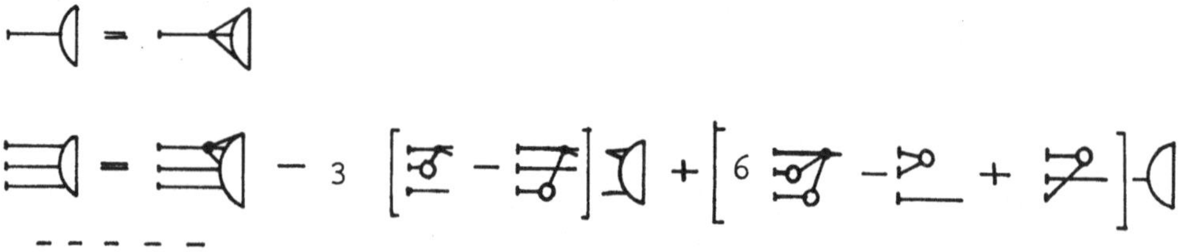

The New-Tamm-Dancoff approximation now consists in the prescription to break off this infinite system by neglecting the highest ϕ-function. We will apply this procedure for the calculation of the masses of the simplest elementary particles.

A. Fermion mass eigenvalue equation

We want to calculate the fermion masses in the lowest nontrivial NTD-approximation. For this we consider matrix elements $<0|\dots|F>$ with an odd number of operators, with F some fermion state which we characterize besides its internal quantum numbers by a 4-momentum p. These matrix elements represent, if p^2 is discrete, the 1-particle wave functions. To solve for these we neglect the 5-point function in (38) and obtain the finite system for the 1- and 3-point function

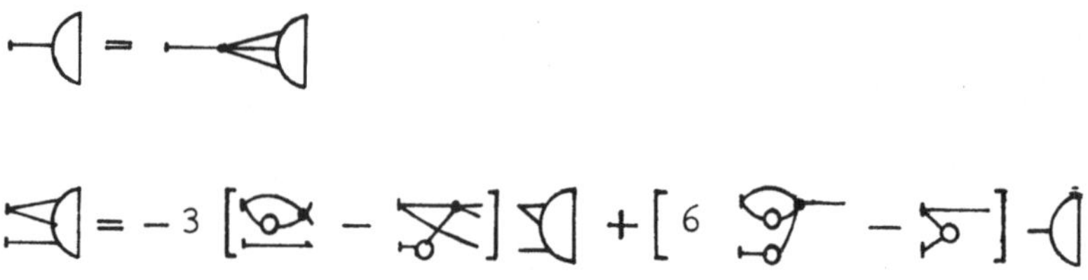

(39)

E.g. we may deduce from the second equation an expression for the 3-point function in terms of the 1-point function which then we insert into the first equation. This is already quite complicated. For simplicity reasons and to demonstrate our procedure we will further simplify the second equation by neglecting the 3-point function on the right hand side. Then we can write it

which we insert into the first equation to obtain the simplest fermion mass eigenvalue equation (for particles with nonzero mass)

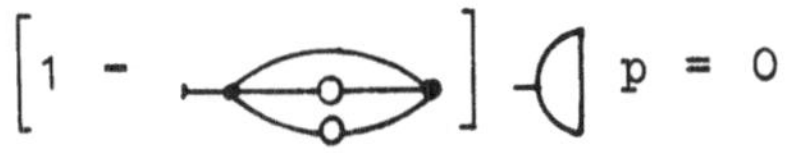

or

$$[1 - K_F]\ \phi(p) = 0 \tag{40}$$

Now the mass eigenvalues of the spin 1/2 fermions could be determined if the contraction function F() would be explicitly known, which is not the case. It is a solution of the nonlinear system for the vacuum expectation values. Hence we have to make an approximative assumption to get further. We will assume that F can be approximated by a single baryon pole at mass m. In addition its behaviour on the x-space lightcone should be adjusted in such a way that the conditions (15) $\rho_0 = \rho_1 = 0$, necessary for regularization of the theory, are fulfilled. We assume for F in momentum space:

$$F(p) = \frac{\bar{\sigma}_\mu p^\mu}{(p^2)^2(p^2 - m^2)} \tag{41}$$

I.e. the F-function appears as a baryon pole regularized by mass zero dipole ghosts. If we insert this F(p) in our fermion eigenvalue equation (40), the integral kernel $K_F(p^2/m^2,(m\ell)^2)$ will be an explicitly known function of p^2/m^2 and $(m\ell)^2$. In order now to remove the arbitrariness of our baryon mass m we require that the baryon states we put in, shall reappear as solutions of the fermion equation (40). This selfconsistency condition requires

$$K_F(\frac{p^2}{m^2} = 1, (m\ell)^2) = 1 \tag{42}$$

This, indeed, can be satisfied for a value of $m\ell \approx 5.8$. It fixes F(p) uniquely. The baryon mass m is now related to the coupling constant ℓ .

B. The boson mass eigenvalue equation

The first interesting result we obtain if we look at the boson eigenvalue equation in the same way. We consider

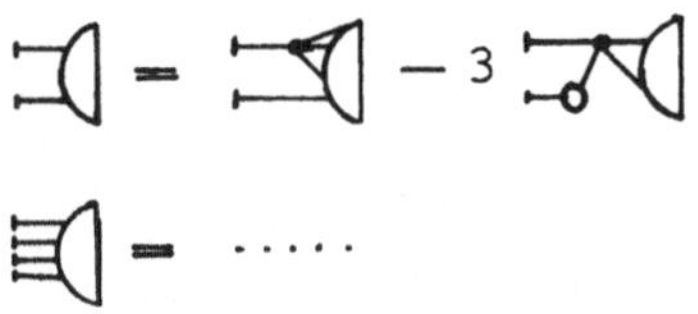

(43)

and neglect the 4-point function to get the simplest boson eigenvalue equation

or

$$[1 - K_B]\ \phi_B(k) = 0 \tag{44}$$

The kernel $K_B(k^2/m^2)$ is a well defined and explicitly known function of k^2/m^2 with k the boson 4-momentum, since $(m\ell)^2$ is fixed. The boson masses result from

$$K_B(\frac{\mu_B^2}{m^2}) = 1 \tag{45}$$

Because of the algebraical nature of K_B these are actually 6 eigenvalue equations, corresponding to mesons of spin = 0,1, isospin = 0,1 and deuteron ^1S and ^3S states. The mesons have quantum numbers like the η,π,ρ,ω; they are S-wave bound states of the fermion-antifermion system. (If parity is established they have necessarily all negative parity).

C. The baryon-meson coupling constants

The above system of ϕ-functions can also be applied to scattering states, and from the analytical structure of these equations the coupling constants can be deduced. We observe that in these equations kernels of the form

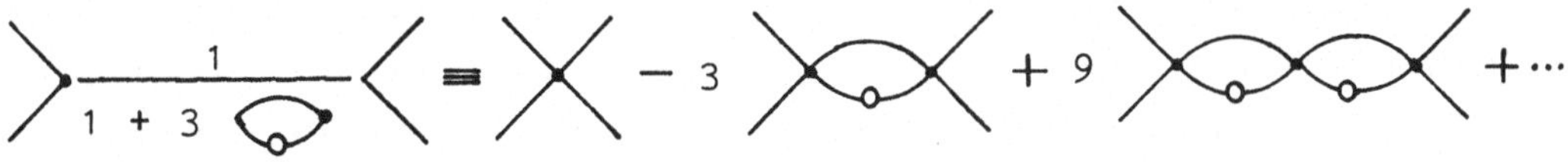

(46)

appear. In an expansion the first term corresponds to the original 4-fermion interaction. On the other hand these kernels have 1-particle singularities exactly at the approximate boson solutions (44). Hence they play the role of the boson propagators in this approximation. Their residue is connected with the fermion-boson coupling constants. For the π-nucleon pseudovector coupling e.g. we find [28]

$$f_\pi^2 = \frac{1}{4}\ell^2 \left[\frac{-m^2}{K'_\pi(\frac{\mu_\pi^2}{m^2})}\right] \approx 4\pi\left[\pi\,\frac{\mu_\pi^2}{m^2}\right] \qquad (47)$$

or for the pseudoscalar constant

$$g_\pi^2/4\pi \approx 4\pi$$

D. Higher approximations

Some higher approximations can still be pulled through. E. g. with the knowledge of the meson masses and coupling constants the baryon approximation may be corrected to include the meson contributions to their self mass. The value of $(m\ell)$ is somewhat reduced by this [20]. Some efforts are also made to solve really an integral equation by the use of Fredholm methods (Stumpf, Yamamoto) [29]. But this is already quite complicated. Research on this point is still fully in progress.

IV. Final remarks

I wish to finish my part of the lecture by reporting shortly on the numerical results we obtained. In connection with this I again want to point out that, in general, among the solutions of our eigenvalue equations there will be and are unphysical states which are connected with our indefinite metric in Hilbert space. The physical solutions are the solutions with positive norm, which have to be projected out in a selfconsistent way. To check the selfconsistency is still hard, but we believe it can be done if the projection is simply a projection on a subspace of Hilbert space which contains eigenstates of energy and momentum.

With this point in mind we find only acceptable solutions for bosons of spin 0, corresponding to η and π. Their masses are still somewhat too high compared with the average baryon mass m (e.g. π-meson about 0.3 m) but their mass ratio is excellent. In fact, the η was predicted on this basis in 1959. There is only one acceptable spin = 1 solution at mass zero for the photon, and this only if some assumptions in connection with an isospin degenerate vacuum are made. The ρ-and ω-mesons do not appear. We hope to get them if the neglected ^{3}D-contributions are taken into account. There are no deuteron solutions. Also the coupling constants of η and π were determined. The π coupling constant, as mentioned above, is pretty good. The η coupling constant is somewhat smaller but much larger than the SU_3 prediction. Experimentally the situation is here still quite uncertain. The Sommerfeld finestructure constant comes out pretty good, if the assumptions used with respect to the role of the leptons and the unsymmetrical vacuum proves acceptable. Also a number of calculations have been carried out on the strange particle masses and their coupling constants but they again involve features of the unsymmetrical vacuum which I have not dealt with in this lecture and which also require further understanding. Hence I will not mention them here.

References

1. W. Heisenberg, Nachr. Akad. Wiss. Göttingen IIa, 111 (1953) Zs. f. Naturforschg. 9a, 292 (1954)
2. W. Heisenberg, W. Pauli, preprint 1958 (unpublished)
3. H. P. Dürr, W. Heisenberg, H. Mitter, S. Schlieder, K. Yamazaki, Zs. f. Naturforschg. 14a, 441 (1959)
4. H. P. Dürr, Zs. f. Naturforschg. 16a, 327 (1961)
5. H. Mitter, Nuovo Cim. 32, 1789 (1964)
6. K. Johnson, private communication.
7. I. Bialynicki-Birula, Proc. of sem. on unified theories of elementary particles, Univ. Rochester, July 1963, Phys. Rev. 130, 465 (1963); D. Lurié, Proc. of seminar on unified theories of elementary particles, Max-Planck-Inst. f. Physik und Astrophysik München, July 1965.
8. H. Mitter, H. Rechenberg, in preparation.
9. H. P. Dürr, Proc. of seminar on unified theories of elementary particles, Univ. Rochester July 1963; H. P. Dürr and F. Wagner, preprint March 1966.
10. F. Gürsey, Nuovo Cim. 3, 988 (1956); 7, 411 (1958)
11. G. Källén, Helv. Phys. Acta 25, 417 (1952); H. Lehmann, Nuovo Cim. 11, 342 (1954); H. Umezawa and S. Kamefuchi, Prog. Theoret. Phys. 6, 543 (1951).

12. K. Sekine, Nucl. Physics 23, 245 (1961)

13. W. Heisenberg, Nucl. Physics 4, 532 (1957); 5, 195 (1958)

14. E. C. G. Sudarshan, Phys. Rev. 123, 2183 (1961);
H. J. Schnitzer, E. C. G. Sudarshan, 123, 2193 (1961).

15. K. L. Nagy, Nuovo Cim. 17, 925 (1960) (review)
L. A. Maksimov, Zhur. Eksp. Theoret. Fiz. 36, 140 (1959); 36, 465 (1959)

16. K. Bleuler, Helv. Phys. Acta 23, 567 (1950)
S. N. Gupta, Proc. Phys. Soc. (London) A 63, 681 (1950)

17. J. Schwinger, Quantum Electrodynamics, Dover Press, New York (1958);
J. Schwinger, The theory of coupled fields, Harvard lecture 1954.

18. J. Goldstone, Nuovo Cim. 19, 154 (1960);
J. Goldstone, A. Salam, S. Weinberg, Phys. Rev. 127, 965 (1962);
S. Bludman, A. Klein, Phys. Rev. 131, 2364 (1963).

19. H. Biritz, W. Heisenberg, in preparation;
see also W. Heisenberg, Proc. of seminar on unified theories of elementary particles, Munich, July 1965.

20. H. P. Dürr, W. Heisenberg, Nuovo Cim. 37, 1446 (1965); 37, 1487 (1965).

21. W. Heisenberg, K. Yamazaki, in preparation.

22. H. P. Dürr, W. Heisenberg, H. Yamamoto, K. Yamazaki, Nuovo Cim. 38, 1220 (1965)

23. H. P. Dürr, Zs. f. Naturforschg. 16a, 327 (1961)

24. loc. cit. 9).
see also W. Heisenberg, Introduction to the unified field theory of elementary particles, lecture S.S. 1965, to be published by Wiley & Sons Ltd.

25. H. Roos, dissertation Univ. Munich, in preparation.

26. H. Stumpf, F. Wagner, F. Wahl, Zs. f. Naturforschg. 19a, 1254 (1964);
F. Wagner, dissertation Univ. Munich 1966;
H. Rampacher, H. Stumpf, F. Wagner, Fortschr.d.Phys. 13,

385 (1965).

27. D. Maison, H. Stumpf, preprint March 1966;
Ch. Schwartz, Ann. of Phys. 32, 277 (1965).
28. J. Dhar, Y. Katayama, Nuovo Cim. 36, 533 (1965)
29. H. Stumpf, H. Yamamoto, Zs. f. Naturforschg. 10a, 1 (1965).

NONLINEAR SPINOR THEORY IN TERMS OF VACUUM EXPECTATION VALUES[†]

By

H. MITTER

Max-Planck-Institut für Physik und Astrophysik
München

. Equations derivable from the field equation

We shall consider in these lectures the system of equations for the vacuum expectation values of certain products of field operators (the n-point-functions) of a nonlinear, cubic spinor theory with the field equation

$$D_{\alpha\beta}\,\psi_\beta + V_{\alpha\beta,\gamma\delta}\,\psi^*_\beta\,\psi_\gamma\,\psi_\delta = 0 \qquad (1)$$

We shall use a rather abstract notation: the indices represent spin, isospin etc. indices as well as space coordinates. For indices appearing twice we use the summation convention including integration over space coordinates. $D_{\alpha\beta}$ is a local, hermitean differential operator of first order

$$D_{\alpha\beta} = i(\sigma_\mu)_{\alpha\beta}\,\delta(x^{(\alpha)} - x^{(\beta)})\,\frac{\partial}{\partial x_\mu^{(\alpha)}} \qquad (2)$$

σ^μ is a matrix transforming like a four vector under Lorentz-transformations. In the simple examples to be discussed, in which ψ describes one or two complex Weyl fields, σ^μ is essentially a Pauli spin matrix. Thus we shall consider

a) the two-component theory (only spin) $\sigma^\mu = (1,\vec{\sigma})$

[†] Lecture given at the V. Internationalen Universitätswochen für Kernphysik, Schladming, 24 February - 9 March 1966.

b) the four-component theory (spin + isospin) $\sigma^{\mu} = (1,\vec{\sigma}) \times I$ (I denotes the unity matrix in isospace). Sometimes we shall use also $\bar{\sigma}^{\mu} = (1,-\vec{\sigma})$ which is the inverse of σ^{μ}. V is shorthand for a local expression which we assume to be Fierz-symmetric

$$V_{\alpha\beta,\gamma\delta} = - V_{\beta\alpha,\gamma\delta} = - V_{\alpha\beta,\delta\gamma} \tag{3}$$

We shall consider the following forms (ℓ is a constant)

a) $V_{\alpha\beta,\gamma\delta} =$

$$= \ell^2 \delta(x^{(\alpha)} - x^{(\beta)}) \delta(x^{(\gamma)} - x^{(\alpha)}) \ \delta(x^{(\gamma)} - x^{(\delta)}) (\sigma^{\mu})_{\alpha\gamma} \ (\sigma_{\mu})_{\beta\delta} :: \tag{4a}$$

$$\text{b) } V_{\alpha\beta,\gamma\delta} = \frac{\ell^2}{4} \delta(x^{(\alpha)} - x^{(\beta)}) \delta(x^{(\gamma)} - x^{(\alpha)}) \delta(x^{(\gamma)} - x^{(\delta)}) \times$$

$$\times \ [(3(I)(I) + (\vec{\tau})(\vec{\tau}))\ (\sigma^{\mu})(\sigma_{\mu})]_{\alpha\gamma,\beta\delta} :: \tag{4b}$$

The dots denote the Wick-Product of all three field operators involved and are a simple-minded prescription to avoid some of the divergences connected with products of operators at the same point. One could of course consider different prescriptions here and it is clear, that eq. (1) has only a rather formal meaning, as long as we have not proved, that the theory given by the above mentioned prescriptions is meaningful. One should consider the whole procedure as a kind of consistency problem: One starts with some V and considers equations for n-point-functions. If there are no finite solutions one goes back and modifies V and repeats the whole procedure and so on. The group theoretical and interpretational aspects of (1) have been discussed in Dürr's lecture and is therefore not repeated here.

We consider a generating functional [1]

$$\underline{T} = T \exp i(u^*_{\alpha} \psi_{\alpha} + \psi^*_{\beta} u_{\beta}) \tag{5}$$

where u, u^* are "classical" spinor sources anticommuting with themselves, each other, with ψ and ψ^*. From $\underline{T}$ we can generate Green's functions by the procedure

$$\tau_{\alpha_1 \ldots \alpha_k | \beta_1 \ldots \beta_\ell} \equiv \langle 0 | T \psi_{\alpha_1} \ldots \psi_{\alpha_k} \psi^*_{\beta_1} \ldots \psi^*_{\beta_\ell} | 0 \rangle =$$

$$= \frac{1}{i^{k-\ell}} \frac{\delta^{k+\ell}}{\delta u^*_{\alpha_1} \ldots \delta u^*_{\alpha_k} \, \delta u_{\beta_1} \ldots \delta u_{\beta_\ell}} \langle 0 | \underline{T} | 0 \rangle |_{u=u^*=0} \tag{6}$$

As a consequence of translation invariance these functions depend only on $k + \ell - 1$ difference coordinates. They are antimetric under commutation of two variables of each group, because the T-product has this property by definition. It is simple to derive a system of coupled differential equations for the τ-functions by functional methods. We assume the equal time commutation relation

$$\{\psi_\alpha(x), \psi^*_\beta(y)\}|_{x^0=y^0} = \rho_0 \, \delta_{\alpha\beta} \, \delta_3(\vec{x} - \vec{y}) \tag{7}$$

Here ρ_0 is the lowest spectral moment of the propagator, as will be discussed later. This together with (1) implies

$$(D_{\alpha\beta} \frac{\delta}{\delta u^*_\beta} - V_{\alpha\beta,\gamma\delta} [\frac{\delta^3}{\delta u^*_\gamma \, \delta u^*_\delta \, \delta u_\beta} - 2\tau_{\delta|\beta} \frac{\delta}{\delta u^*_\gamma}] + i \rho_0 u_\alpha) \underline{T} = 0 \tag{8}$$

The second term in the square bracket is due to the Wick product. If another prescription is taken, we have to modify the equation correspondingly. The conjugate complex equation to (1) gives

$$(D_{\alpha\beta} \frac{\delta}{\delta u_\alpha} - V_{\alpha\beta,\gamma\delta} [\frac{\delta^3}{\delta u^*_\delta \, \delta u_\beta \, \delta u_\alpha} - 2\tau_{\delta|\beta} \frac{\delta}{\delta u_\alpha}] - i \rho_0 u^*_\gamma) \, \underline{T} = 0 \tag{9}$$

By functional differentiation we obtain an infinite, linear

system of differential equations for the τ's. As a consequence of (7) it contains many δ-functions, and in addition the propagator $\tau_{\alpha|\beta}$ at zero argument, which is expected to be singular. There are other functions which shall be more interesting for us. We define a functional $\underline{H}$ by the equation

$$<0|\underline{T}|0> = \exp \underline{H} , \quad \underline{H} = \ln <0|\underline{T}|0> \tag{10}$$

The $\underline{H}$ generates the correlated parts of the τ-functions, which are called η-functions [1] [2]

$$\eta_{\alpha_1 \ldots \alpha_k | \beta_1 \ldots \beta_\ell} = \frac{1}{i^{k-\ell}} \left. \frac{\delta^{k+\ell}}{\delta u^*_{\alpha_1} \ldots \delta u^*_{\alpha_k} \ldots \delta u_{\beta_1} \ldots \delta u_{\beta_\ell}} \underline{H} \right|_{u=u^*=0} \tag{11}$$

We have

$$\eta_{\alpha|\beta} = \tau_{\alpha|\beta}$$

$$\eta_{\alpha\beta|\gamma\delta} = \tau_{\alpha\beta|\gamma\delta} - \tau_{\alpha|\delta}\tau_{\beta|\gamma} + \tau_{\alpha|\gamma}\tau_{\beta|\delta} \quad \text{etc.} \tag{12}$$

In perturbation theory the η's are represented by the sum of all connected parts of all Feynman graphs with k incoming and ℓ outgoing lines. The system of equations for the η's is generated by

$$D_{\alpha\beta} \frac{\delta \underline{H}}{\delta u^*_\beta} - V_{\alpha\beta,\gamma\delta}\, e^{-\underline{H}} \left[\frac{\delta^3}{\delta u^*_\gamma \; \delta u^*_\delta \; \delta u_\beta} - 2\tau_{\delta|\beta} \frac{\delta}{\delta u_\gamma} \right] e^{\underline{H}} + i\rho_0 u_\alpha = 0 \tag{13}$$

Here the δ-function and therefore the equal-time commutation rule is present only in the lowest equation. In addition all terms involving the propagator at zero argument (tadpole terms) are absent. Thus in some sense the Wick-product "fits" to the η-system. The disadvantage is, that the system is an awfully nonlinear one. We have to face this and it will prevent us from believing in too "cheap" approximations.

It will be instructive to write down the lowest two equations explicitly. They read

$$D_{\alpha\alpha'}\ \eta_{\alpha'|\beta} - V_{\alpha\rho,\sigma\tau}\ \eta_{\sigma\tau|\rho\beta} = i\rho_o\ \delta_{\alpha\beta}\ ,$$

$$D_{\beta'\beta}\ \eta_{\alpha|\beta'} - V_{\sigma\tau,\rho\beta}\ \eta_{\alpha\rho|\sigma\tau} = i\rho_o\ \delta_{\alpha\beta}\ , \qquad (14)$$

$$D_{\alpha\rho}\ \eta_{\rho\beta|\gamma\delta} + V_{\alpha\rho,\sigma\tau}\big[\eta_{\sigma\tau\beta|\rho\gamma\delta} + 2\eta_{\sigma\beta|\rho\delta}\ \eta_{\tau|\gamma} - $$

$$- 2\eta_{\sigma\beta|\rho\gamma}\ \eta_{\tau|\delta} + \eta_{\sigma\tau|\gamma\delta}\ \eta_{\beta|\rho} + 2\eta_{\sigma|\delta}\ \eta_{\beta|\rho}\ \eta_{\tau|\gamma}\big] = 0 \qquad (15)$$

We have to keep in mind, that because of the local nature of our interaction the coordinates belonging to three of the indices of V (ρ,σ and τ) are equal. Thus the arguments of the highest function occuring in a given equation of the set are always on a hypersurface (three coordinates equal): only the boundary value of the function enters the equation. This shows apparently that the system does not determine the functions involved completely. We could for instance read each equation as a determining equation for the highest function involved: it expresses this function (at a special hypersurface) by functions with less coordinates and their derivatives. This leaves the propagator completely undetermined; there is also very much of the higher functions left open, since only their boundary values occur in the system, if we read it in this recursive way. The field equations are therefore rather a framework for a quantum field theory than the theory itself, if one looks upon them in this very formal way. One has however to keep in mind, that one has to be very careful in drawing such conclusions: the fact that the system in consideration is an infinite one might bring in essentially new features.

All further restrictions which are necessary to determine the functions in order that they describe a useful quantum field theory have to be deduced from physics. There are some of them implied by the groups which leave eq. (1) invariant; we have already stated some consequences of Lorentz invariance in this context. In case b) isospin transformations can be discussed correspondingly. Note however, that the behaviour of the vacuum state under these transformations enters always vitally. A priori we can never rule out the possibility, that the vacuum is not invariant against one of these transformations. We can only assume something about the vacuum and look, whether this leads to contradictions. Other important restrictions come from microcausality and from the physical demand, that the theory has a particle interpretation (that there are asymptotic conditions, that there is a unitary S-matrix connecting asymptotic states). If one starts from these assumptions alone (a procedure, which is to some extent complementary to the approach taken here) one obtains, as is well-known, another framework for a reasonable field theory and it remains to be shown, that these framework have a nonzero overlap.

Here we do not want to solve this complicated problem. We just want to discuss, how severe the restrictions imposed by the equations (13), implemented by some group theoretical considerations, really can be. Even in this approach we have to make simplifying assumptions in some places, since otherwise the domain to be covered by our investigation is too vast.

2. Possible Regularisations

In the equations as they have been written down there are terms which may contain divergences, for instance the product of three propagators in the eq. (15) for the four-point-function: If we try to form $\eta(xx|xy)$ which we need for the first equation, we encounter $\eta(x|y)\ \ \eta(y|x)\ \ \eta(x|y)$. This is a product of distributions at the same point and therefore a sin against the basic commandement of distribution theory ("Thou shall never multiply distributions"). This can be seen from the UKKL-representation*

$$\eta(x|y) \equiv S_c'(x,y) \equiv F(x,y) = \int \rho(\zeta)\, S_c(x-y,\zeta)\, d\zeta \qquad (16)$$

in which

$$S_c(x-y,\zeta) = \frac{1}{(2\pi)^4} \int e^{-i(p,x-y)} \frac{-i(\bar{\sigma}p)}{p^2-\zeta+i\varepsilon}\, dp$$

is the corresponding quantity for an interaction-free theory. If one inserts the representation naively and tries to compute $\eta^3(x|y)$, one encounters wild ultraviolet divergences, if one does the momentum integrals.

If now the theory in question makes sense at all this can happen in two ways: either there are corresponding divergent quantities in other terms, which cancel the ones just mentioned or the propagator and the other η-functions are in fact not as singular as one might guess from the representation above. We shall study only the second possibility here and I shall try to give a summary on some possible ways in which this can happen, without discussing for the moment the question, whether the system actually has solutions of the desired type (this point will however be touched later). Let me state, that this does not mean, that there are no other ways to give a meaning to the theory in question. Our approach is just to look for very smooth solutions. There might be more singular ones, and it could be that these are of physical interest as well.

If the propagator is less singular than the free one, definite metric, is out; this can be seen from the expansion of

---

+ The letters stand for Umezawa-Kamefuchi-Källén-Lehmann and give an outline of the historical sequence of papers, which deal with this representation (one could add also the names Gell-Mann, Low, Schwinger etc., [3].

the propagator at the light cone. Using the abbreviations*

$$\rho_n = \int \rho(\zeta)\zeta^n \, d\zeta \quad , \quad \rho_{L,n} = \int \rho(\zeta)\zeta^n \ln\zeta \, d\zeta \quad ,$$

$$x - y = z \quad , \quad s = -z_\nu z^\nu \tag{17}$$

we have

$$\eta(x|y) = \frac{-i}{4\pi^2} \left(\bar{\sigma} \frac{\partial}{\partial z}\right) \left[\frac{\rho_0}{s+i\varepsilon} + \frac{\rho_1}{4} \left(\ln (s+i\varepsilon) + \ln \frac{\gamma}{2} - 1\right) + \right.$$

$$\left. + \frac{\rho_{L,1}}{4} + \ldots . \right] =$$

$$= \frac{-i}{2\pi^2} (\bar{\sigma} z) \left[- \frac{\rho_0}{(s+i\varepsilon)^2} + \frac{\rho_1}{4(s+i\varepsilon)} + \right.$$

$$\left. + \frac{\rho_2}{32} \left(\ln(s+i\varepsilon) + 2\ln \frac{\gamma}{2} - \frac{3}{2}\right) + \frac{\rho_{L,2}}{32} + \ldots \right] \tag{18}$$

If ρ is positive-definite its moments are all different from zero and we get the same (or worse) singularities as for a free theory (which is obtained, if we put $\rho = \delta(\zeta - m^2)$). Note that the imaginary part of η, which is related to the vacuum expectation value of the anticommutator, contains δ'- and δ-functions as is easily seen from the relation

$$\frac{1}{(s+i\varepsilon)^n} = \frac{1}{s^n} + \frac{(-1)^n i\pi}{(n-1)!} \delta^{(n-1)}(s) \quad .$$

This shows again, why the product of propagators is ill-defined. At the same time we see, that this difficulty does not arise, if our spectral-function is such that

* We have to assume, that all the spectral representations, Fourier transforms, moments etc. used in this section are meaningful quantities. For the propagators discussed below this will be almost always the case, if the formulae are understood in the sense of the theory of distributions.

$$\rho_0 = 0 \ , \qquad \rho_1 = 0 \tag{19}$$

There are then only logarithmic singularities and the products can be defined. By power-counting in momentum space one may show, that also the propagator divergences in the higher equations are absent. If only the lowest momentum is zero, the theory is logarithmically divergent and resembles to some extend a renormalizable theory.

If the lowest momentum is zero, the quantisation is not the canonical one, since we have, as a consequence of the UKKL representation

$$\langle 0|\{\psi(x),\ \psi^*(y)\}|0\rangle|_{x_0=y_0} = \rho_0 \delta_3(\vec{x} - \vec{y})$$

Thus the canonical anticommutator is zero in our case. If there is no singularity at all at the light cone one may eventually prove, starting from equal times and using the nonlinear equation, that the anticommutator is zero everywhere, so that we are left with a kind of "classical" theory and there is no room for a quantum theory (this is Pauli's argument against noncanonical quantisation).

The argument does not go through, if there is a singularity at the cone, so we shall study only such cases. We should however, be aware of the fact, that there is always the danger to get a "classical" theory. In order to avoid it we have to show, that in the decomposition

$$\eta(x|y) = i(\bar{\sigma}\frac{\partial}{\partial z})\ (\mathrm{Ref} + i\ \mathrm{Imf})$$

the imaginary part is different from zero somewhere, since we have to identify

$$i(\bar{\sigma}\ \frac{\partial}{\partial z})\mathrm{Imf} = \frac{\varepsilon(x-y)}{2i}\ \langle 0|\{\psi(x),\ \psi^*(y)\}|0\rangle$$

There are some other physical requirements which are worth being mentioned here. One is microcausality. It states, that the anticommutator has to be zero for space-like arguments. This is always guaranteed automatically, if the UKKL- representation exists. Another important point is the spectrum condition. Wightman was able to show [4] that the condition, that the energy spectrum of the theory be positive, leads to the statement, that the propagator has to be analytic in the whole complex s-plane with exception of the time-like part of the real axis. Unfortunately some properties of the states have to be used for the proof, which are doubtful for theories with indefinite metric*, so that it is not clear, to what extent these analyticity properties are really necessary in that case. Nevertheless it might be useful to see, whether they are fulfilled in the examples to be discussed later.

There is another point, which has to be mentioned in context with noncanonical quantisation. In a canonical theory all the important physical operators as e.g. energy-momentum, angular momentum, total charge etc, are constructed from expressions involving products of fields and their derivatives at the same point in space-time. We have for example

$$P^{\mu} = \int d^3x \; T^{0\mu}(x) \quad ,$$

$$J^{\mu\nu} = \int d^3x (x^{\mu}T^{0\nu}(x) - x^{\nu}T^{0\mu}(x)) \quad ,$$

$$Q = \int d^3x \; j^0(x) \quad ,$$

where the local quantities can be obtained by the canonical formalism in the well-known way. In our case we would find e.g.

*The assumption that the eigenstates of the energy operator form a complete set, enters vitally. This is not true for some examples of theories with indefinite metric, e.g. for the Lee-model with dipole ghost states, [5].

$$j^{\mu}(x) \sim \psi^{*} \sigma^{\mu} \psi$$

(which had to be interpreted as the current belonging to the baryonic charge). Now the integral quantities must fulfill certain important commutation rules, which are the structure equations of the groups in question, e.g. for the Lorentz group

$$[P_{\mu}, P_{\nu}] = 0 , \quad \frac{1}{i}[J^{\mu\nu}, P^{\lambda}] = g^{\mu\lambda}P^{\nu} - g^{\lambda\nu}P^{\mu} \quad \text{etc.}$$

They, in addition, must have commutation relations with the fields

$$[P^{\mu}, \psi(x)] = \frac{1}{i} \partial^{\mu} \psi(x), \qquad [Q,\psi] = \lambda\psi \qquad \text{etc.}$$

These relations are usually proved using the canonical commutation relations of the fields. One might therefore wonder what happens in a noncanonical theory, where everything commutes at equal times. Obviously one can avoid accidents only, if one gives up the completely local constructions, which the canonical formalism suggests. Already if one uses operators from a thin time-slice rather than those at equal times there is a chance to obtain commutators which are different from zero*, since then the commutation relations at time-like intervals come in. Since the Hamiltonian is one of the operators in question it could turn out, that a local theory with noncanonical quantisation is to some extent equivalent to a noncano- one with canonical rules, if the former one really works in that way. We have at present, however, no indication, whether and how this happens in the relativistic case. There are only definite results in the Lee-model: there it has been shown [7]

* The idea, that one needs the fields within a finite time-slice, has been proposed by R. Haag, in a different context [6].

that one may introduce field operators depending explicitly on a time slice of extension Δt, which have canonical rules of the form

$$\langle 0|\{\psi_{V,r},\bar{\psi}^{(\Delta t)}_{V,r}\}|0\rangle \sim \frac{1}{\ln \Delta t}$$

As long as Δt is finite we have a canonical theory (there is a 1 rather than a δ-function because we have fixed sources). If we want local quantities ($\Delta t \to 0$) the theory goes over in the noncanonical one. Both theories (canonical one with Δt and noncanonical one) are equivalent.

But let us now look, what possibilities are open for noncanonical theories. This we shall do in the following way: we rewrite the propagator in the form

$$\eta(x|y) = \eta_{as}(x|y) + (\eta(x|y) - \eta_{as}(x|y)) \qquad (20)$$

where η_{as} denotes the asymptotic part of η for small distances (at the light cone) in coordinate space. Then we discuss possible forms of η_{as}, compute the corresponding spectral functions ρ^{as} in the UKKL-representation and look at their first two moments. Since the remainder $\eta - \eta_{as}$ is by definition less singular at the light cone we get in this way also an indirect information on the behaviour of the lowest moments of η. If for instance the first two moments of η_{as} are already zero this must be the case for η too (always supposed that the corresponding integrals exist).

In order to carry through our discussion we shall make use of the fourier transform of a distribution

$$f_c^{(\alpha)}(s) = \frac{1}{(2\pi)^4}\int e^{-i(p,x)}\,\tilde{f}_c^{(\alpha)}(p)\,d^4p \qquad (21)$$

As one can infer from the famous book by Gelfand and Schilow

[8] we have for

$$f_c^{(\alpha)}(s) = \frac{-2^{2\alpha}\ \Gamma(\alpha)}{16\pi^2(s+i\varepsilon)^{\alpha}}\ ,\ \tilde{f}_c^{(\alpha)}(p) = \frac{i\Gamma(2-\alpha)}{(-p^2-i\varepsilon)^{2-\alpha}}\ . \tag{22}$$

This function has a UKKL-representation with a spectral function

$$\rho^{(\alpha)}(\zeta) = \frac{1}{\Gamma(\alpha-1)\quad \zeta_+^{2-\alpha}} \tag{23}$$

These formulae hold for general, complex values of α with the exception of $\alpha = 0$ (which case will be discussed below) and α = negative integer, which is of no interest in this context. The distribution x_+^{λ} is defined by the following prescriptions:

$$x_+^{\lambda} = \begin{cases} x^{\lambda} & \text{for } x > 0 \\ 0 & \text{for } x < 0 \end{cases} \tag{24a}$$

At $x = 0$ the prescriptions are (with $\phi(x)$ a function regular at zero and sufficiently nice at infinity):

$$\text{Re}\ \lambda > -1\ :\quad \int_{-\infty}^{+\infty} x_+^{\lambda}\ \phi(x) = \int_0^{\infty} x^{\lambda}\ \phi(x)\ dx$$

$$\text{Re}\ \lambda > -n-1\ ,\quad n = 0,1,2,\ldots:\quad \int_{-\infty}^{+\infty} x_+^{\lambda}\ \phi(x)\ dx =$$

$$= \int_0^1 x^{\lambda}(\phi(x) - \phi(0) - x\phi'(0)\ \ldots - \frac{x^{n-1}}{(n-1)!}\ \phi^{(n-1)}(0))dx +$$

$$+ \int_1^{\infty} x^{\lambda}\ \phi(x)\ dx + \sum_{k=1}^{n} \frac{\phi^{(k-1)}(0)}{(k-1)!\ (\lambda+k)}$$

$$-n > \text{Re}\ \lambda > -n-1\ :\quad \int_{\infty}^{\infty} x_+^{\lambda}\ \phi dx =$$

$$= \int_0^\infty x^\lambda \left(\phi(x)-\phi(0)-x\phi'(0)\ldots - \frac{x^{n-1}\,\phi^{(n-1)}(0)}{(n-1)!}\right) \qquad (24b)$$

For negative integer λ the function has poles such that

$$\lim_{\lambda\to -n} \frac{x_+^\lambda}{\Gamma(\lambda+1)} = \delta^{(n-1)}(x) \qquad (24c)$$

Now we make the ansatz

$$\eta_{as} = i(\bar{\sigma}^\nu \frac{\partial}{\partial z^\nu})\, f_c^{(\alpha)}(s) = -i(\bar{\sigma}z)\, f_c^{(\alpha+1)}(s) \qquad (25)$$

and discuss, what happens for special values of α especially with respect to ρ_0^{as} and ρ_1^{as} .

a) $\alpha = 1$: In this case we obtain the behaviour of the free theory, as η^{as} is the propagator of a free particle with mass zero. The spectral function is the δ-function, as we see using (24c). We find $\rho_0^{as} = 1$, $\rho_1^{as} = 0$

b) $\alpha = 0$: This case is somewhat singular. Nevertheless we may use our formulae, if we take the second form (25) and use the fact, that z_μ in coordinate space means $i\frac{\partial}{\partial p^\mu}$ in momentum space. So we find, using again (24c)

$$\rho^{as}(\zeta) = \delta'(\zeta) \quad , \rho_0^{as} = 0, \quad \rho_1^{as} = -1$$

This is usually called a dipole state.

c) $\alpha = 1 - \beta$, $0 < \beta < 1$: Here the singularity is somewhat weaker than for the free theory. It is not weak enough, since we obtain

$$\rho_0^{as} = 0, \quad \text{but} \quad \rho_1^{as} = \infty$$

d) $\alpha = 1 + i\gamma$, γ real: This gives an oscillation around the free theory. One might argue, that the singularities can be oscillated away in this case. This is true for the worst singularity, but not for the next weaker one, since we have again $\rho_0^{as} = 0$ but $\rho_1^{as} = \infty$.

e) $\alpha = -\beta$, $0 < \beta < 1$. The first "fine" case, since we obtain $\rho_0^{as} = 0$, $\rho_1^{as} = 0$.

f) $\alpha = i\gamma$, γ real: This gives an oscillation around the dipole, which can take away the singularity, which was left in this case. We find $\rho_0^{as} = 0$, $\rho_1^{as} = 0$ in the sense of $\lim_{x\to\infty} x^{i\gamma}$. We shall come back to this later.

There remain some remarks to be added to this list. If we consider (23) it looks as if we obtain always typical mass zero situations in this way. This has to do only with our special selection of the asymptotic behaviour. If we take instead of $f_c^{(\alpha)}$ another function (with an arbitrary mass m)

$$g_c^{(\alpha)}(s) = \frac{-2^{\alpha}m^{2\alpha}}{8\pi^2} \frac{K_\alpha(m\sqrt{s+i\varepsilon})}{(m\sqrt{s+i\varepsilon})^{\alpha}} \tag{26}$$

then it turns out, that $g^{(\alpha)}$ has the same asymptotic behaviour as f for small distances. It has an analogous momentum representation and spectral function as f, with the only changes $\zeta \to \zeta - m^2$, $-p^2 \to -p^2 + m^2$. One can discuss then also the situation, where there are only logarithmic singularities at the origin ($\alpha = -1$) or even a still weaker case ($\alpha = -n$), where the first moments are, of course, always zero, as can be seen from the UKKL-representation directly.

Let us now discuss causality. Since all our asymptotic propagators can be UKKL-represented, they are all causal, since the representation has this property. By looking at the function f_c and splitting it in a real and imaginary part (i.e. looking for the corresponding f_p-function) this may also be verified directly.

The analyticity properties for complex values of s are clearly o.k. for real exponents. For complex exponents this is not so clear. One might also wonder, whether we do not run into difficulties with respect to imaginary masses, since ρ looks complex. You can, however, form simple superpositions with different imaginary parts of the exponent, for example

$$f^{(i\beta)} + f^{(-i\beta)} \sim 2 \cos (\beta \ln(s+i\varepsilon))$$

or more complicated expressions, for which ρ is real and you may even fulfil the analyticity demands, which Wightman wants. In the case f) you obtain a propagator, which has the form

$$\eta^{as}(x|y) = \frac{-i(\bar{\sigma}z)}{(s+i\varepsilon)} \phi(\ln s) \qquad (27)$$

where ϕ is a periodic function of its argument

$$\phi(\tau+\omega) = \phi(\tau) .$$

The Fourier decomposition leads back to the form with imaginary exponents (and Wightman analyticity can be fulfilled, if the Fourier series converges rapidly enough). The function (27) oscillates more and more rapidly with growing amplitude, if we approach the light cone. By varying the period we can shift the domain, where there is a violent oscillation closer and closer towards the light cone and can therefore approach better and better case b) without running into divergence difficulties. The same trick could be used with the exponent $1 + i\beta$ instead of $i\beta$ to make a renormalizable theory finite (but noncanonical).

Here is the appropriate place to mention another idea: it is that of asymptotic scale invariance[9]. Eq. (1) is invariand against the scale transformation

$$\psi(x) \to \psi'(x) = \lambda^{1/2}\psi(\lambda x) = U^{-1} \psi U \qquad (28)$$

It is clear, that the solutions we are looking for, e.g. the propagator, can certainly not have the corresponding covariance property, which would follow from the invariance of the vacuum state. If the latter were invariant, we had

$$\langle 0|T\psi\bar{\psi}|0\rangle = \langle 0'|T\psi\bar{\psi}|0'\rangle = \langle 0|U^{-1}T\psi\bar{\psi}U|0\rangle = \langle 0|T\psi'\bar{\psi}'|0\rangle$$

and therefore

$$\eta(x|y) = \lambda\eta(\lambda x|\lambda y)$$

If this were true we could have no discrete mass states (except mass zero or infinity), because δ-functions at finite masses contradict this equation (this can be seen easily in momentum space). By the way: this statement holds quite generally for any nonlinear equation, which does not contain a bare mass term or several coupling terms involving coupling constants of different dimension (such terms would violate scale invariance from the very beginning). In any such theory scale invariance has to be broken spontaneously if the theory yields finite rest masses.

In spite of this situation one can make some profit on scale invariance in a restricted sense. One should be able to forget about finite rest masses in the asymptotic region of very high energies (or, one should say better, small distances), so that the asymptotic behaviour of the n-point-functions near the light cone exhibits the invariance

$$\eta^{as}(x|v) = \lambda\eta^{as}(\lambda x|\lambda y)$$

If this is true for arbitrary, real values of λ^{*}, we obtain case b) as studied before, i.e. a theory more regular than perturbation theory only for renormalizable couplings (di-

---

* Imaginary values correspond to ordinary gauge transformations. Complex values correspond to a combination of gauge and scale transformations.

mensionless coupling constant). For superrenormalizable couplings asymptotic scale invariance would give stronger singularities. Thus the postulate of asymptotic scale invariance turns any theory of the type considered here into one of the renormalizable class, for nonrenormalizable theories at the price of indefinite metric.

Let us now look briefly on the oscillatory case (27), which gives a finite theory: here we have the invariance (29) still in a somewhat restricted sense; namely for a discrete, infinite set of values λ_n

$$\lambda_n = \exp\,(n\omega) \qquad n = 0,1,\ldots \tag{30}$$

where ω is the period of ϕ .

Before we conclude this very speculative chapter we have to add some remarks in order to relate our results to the special form of the propagator which has been used in Dürr's lecture. If the first two moments are zero we can rewrite the UKKL-representation in the following way:

$$\eta(x|y) = \frac{1}{(2\pi)^4} \int e^{-i(p,x-y)}(-i)(\bar{\sigma}p) \int \rho(\zeta)d\zeta \;\times$$

$$\times \left[\frac{1}{p^2-\zeta+i\varepsilon} - \frac{1}{p^2-M^2+i\varepsilon} - \frac{\zeta - M^2}{(p^2-M^2+i\varepsilon)^2}\right] =$$

$$= \frac{1}{(2\pi)^4} \int e^{-i(p,x-y)}(-i)(\bar{\sigma}p)\int \rho(\zeta)d\zeta \frac{(\zeta-M^2)^2}{(p^2-\zeta+i\varepsilon)(p^2-M^2+i\varepsilon)^2} \tag{31}$$

This form is more convenient, since it has the behaviour at large momenta already "built in", which is implied by the vanishing first two moments. If it is used in the calculation of graphs the divergences are also formally absent. The mass M^2 is completely arbitrary. This looks, as if we had subtracted a ghost and a dipole state at mass M^2. If such states are really present in the mass spectrum, the representation (31) simplifies even more: if we have

$$\rho(\zeta) = -\sigma_0 \, \delta(\zeta-M^2) - (\sigma_0 M^2-\sigma_1)\delta'(\zeta-M^2)+\sigma(\zeta) \qquad (32)$$

where σ_0 and σ_1 are the first moments of σ, which otherwise is arbitrary, the spectral conditions (19) are fulfilled. We can then write even

$$\eta(x|y) =$$

$$= \frac{1}{(2\pi)^4} \int e^{-i(p,x-y)}(-i)(\bar{\sigma}p) \int \sigma(\zeta)d\zeta \frac{(\zeta - M^2)^2}{(p^2-\zeta+i\varepsilon)(p^2-M^2+i\varepsilon)^2} \qquad (33)$$

since

$$x^2 \, \delta'(x) = x^2 \, \delta(x) = 0$$

In contrast to ρ in (31) σ is no longer bound to fulfil (19) and we could therefore, as a first and very rough step, approximate σ by a δ-function at just one mass without running into contradictions with (19). This has been done in all practical calculations so far, where M^2 has been taken be zero. If M^2 is taken to be very large we have the situation of a well-known invariant cutoff of the usual type. In coordinate space the propagator has a logarithmic singularity at the light cone in all these cases. Of course the behaviour of the propagator at large distances is quite sensitive to the mass of the subtractive states.

It is quite clear, that neither (31) nor (32) is the only possibility to regularize, since all we have done is playing around with spectral moments. We could arrive at the same behaviour also, if the negative parts of the spectrum are in the continuum. This might have advantages with respect to the physical interpretation of the theory (for instance with respect to questions of unitarity, which, apart from some rough and approximative investigations, are not settled at present). We shall, however, not go in further details here.

3. Restrictions from the Field Equations

In the following part of the lecture I wish to show, which restrictions the field equations together with some group theory impose on the η-functions. I shall do this by looking at the two-component theory. The results presented here have been obtained in collaboration with H. Rechenberg. We have begun to study also the four component theory along the same lines, but this more complicated case will not be discussed here. The idea of our approach is the following one: As was stated before the higher η-functions enter the system only at a certain boundary (for some equal arguments). We try therefore to obtain as much information as possible from the system on these boundary values themselves, i.e. as a first step we, for instance, do not study equations for $\eta(xy|zw)$ and $\eta(xyz|wuv)$, but rather for $\eta(xx|xy)$ (because this enters the propagator equation) and $\eta(xxy|xyy)$ (this enters the equation for $\eta(xx|xy)$) etc. With other words, we look on the subsystem of equations, which one obtains, if all coordinates, which are not differentiated upon, are put equal, so that there is only one essential difference z involved (as a next step one should then look for functions with two essential differences and so on). This means, that we study the most singular end of the theory in question, since possibly the η's have singularities on the various light cones. Clearly by the postulate that even the equations for these objects make sense we rule out a lot of the more singular solutions of the system. Apart from the fact, that it was our original goal to look for very smooth solutions we might however still obtain some information on our theory, if not else, then indirectly (e.g. by proving that there are no smooth solutions). Let me also remark here, that the situation with the η-functions is in general expected to be better than for the τ-functions, since some singularities have been pulled out by passing from τ's to η's (for instance in the Thirring-Johnson [10] model it turns out, that the η's in general are less singular than the τ's and they are in fact even zero, if many coordinates are equal)* .

The subsystem just mentioned contains some equations for objects which we really need, e.g. the functions mentioned above, and equations for less interesting objects as e.g. $\eta(xyy|yyy)$. Thus we have to treat the information from these latter equations, if we need it at all (this will not be the case for the two component theory) with even more suspicion. From the very beginning we expect the system to be underdetermined, since it contains objects (for instance $\eta(xx|yy)$) for which we cannot get equations in this simple way. Probably one could obtain information on them by consideration of more complicated limiting forms of equations. This will not be necessary for our purpose.

The subsystem has one very pleasant property (and this works only for spinor fields): it is a finite system, since the higher η-functions with many equal coordinates are zero. This happens because of the Pauli principle which states, that a product of too many factors ψ or ψ^* at the same point has to be zero because of overall antisymmetry. In the case of the two component theory the highest possible function involving only one coordinate difference is $\eta(xxy|xyy)$ and thus our subsystem consists only of two differential equations for the propagator and the four-point function $\eta(xy|yy)$. The latter equation involves the sixpoint function just mentioned, some other four point functions, namely $\eta(xx|yy)$ and $\eta(xy|xy)$,

---

* For a scalar field it can be shown, that there are solutions for this subsystem which correspond to scale invariance [11].

and the propagator. The equations have been written down before, see (14) and (15).

The question, how much freedom we actually have in this system is not trivial, since group theoretical considerations have to be taken into account and reduce the number of functions involved considerably. I shall indicate, in the special case of the four-point-function, how this happens. The object $\eta(xy|zw)_{\alpha\beta\,\alpha'\beta'}$ transforms under Lorentz-transformations like a product of two ψ's and two ψ^*'s. Therefore it can be expressed as a linear combination of basis elements of the algebra consisting of direct products of two Pauli matrices $\bar{\sigma}^\mu$. We can have two different forms depending on how we combine the indices. Thus we have

$$\eta(xy|z\,w)_{\alpha\beta\,\alpha'\beta'} = (\bar{\sigma}^\mu)_{\alpha\alpha'}(\bar{\sigma}^\nu)_{\beta\beta'}\,T_{\mu\nu} + (\bar{\sigma}^\mu)_{\alpha\beta'}\,(\bar{\sigma}^\nu)_{\beta\alpha'}\,\hat{T}_{\mu\nu}\ .$$

The coordinate dependence is in the tensors T respectively $\hat{T}$. It turns out, that it is possible to rewrite this ansatz in a much more convenient form. The second combination can be transformed into the first one by a Fierz transformation. In addition we split the tensors into their symmetric and antisymmetric parts. Having done all this we arrive at the form

$$\eta(xy|z\,w)_{\alpha\beta\,\alpha'\beta'} = (\bar{\sigma}^\mu)_{\alpha\alpha'}(\bar{\sigma}^\nu)_{\beta\beta'}\Big[g_{\mu\nu}\underset{\substack{a\\a'}}{S^\lambda_\lambda} + \underset{\substack{s\\a'}}{(A_{\mu\nu} + i\varepsilon_{\lambda\kappa\mu\nu}A^{\lambda\kappa})} +$$

$$+ \underset{\substack{s\\s'}}{(\hat{S}_{\mu\nu} - \tfrac{1}{4}\,g_{\mu\nu}\,\hat{S}^\lambda_\lambda)} + \underset{\substack{a\\s'}}{(\hat{A}_{\mu\nu} - i\varepsilon_{\lambda\kappa\mu\nu}\hat{A}^{\lambda\kappa})}\Big] \qquad (34)$$

Here we have*

---

* These tensors are the symmetric resp. antimetric parts of combinations of T and $\hat{T}$.

$$S_{\mu\nu} = S_{\nu\mu} \ , \ \hat{S}_{\mu\nu} = \hat{S}_{\nu\mu}$$

$$A_{\mu\nu} = - A_{\nu\mu} \ , \ \hat{A}_{\mu\nu} = - \hat{A}_{\nu\mu} \ .$$

Each term in (34) has a definite symmetry property upon commutation of the unprimed resp. primed indices, which is indicated by the letters under the formula. If we now consider a function involving only one relative coordinate z^μ the only possible tensors are $g_{\mu\nu}$ and $z_\mu z_\nu$. Thus there are no A-parts in this case. In addition we know that both

$$\eta(\underset{\alpha\beta}{xy}|\underset{\alpha'\beta'}{y\ y}) \qquad \text{and} \qquad \eta(\underset{\alpha\beta}{xx}|\underset{\alpha'\beta'}{y\ y})$$

have to be antimetric with respect to the primed indices. Thus we have

$$\eta(\underset{\alpha\beta}{xx}|\underset{\alpha'\beta'}{y\ y}) = (\bar{\sigma}^\mu)_{\alpha\alpha'}(\bar{\sigma}_\mu)_{\beta\beta'}a(z^2)$$

$$\eta(\underset{\alpha\alpha'}{xy}|\underset{\beta\beta'}{yy}) = (\bar{\sigma}^\mu)_{\alpha\alpha'}(\bar{\sigma}_\mu)_{\beta\beta'}b(z^2) = \eta(\underset{\alpha\beta}{xx}|\underset{\alpha'\beta'}{y\ x})$$

$$\eta(\underset{\alpha\beta}{xy}|\underset{\alpha'\beta'}{xy}) = (\bar{\sigma}^\mu)_{\alpha\alpha'}(\bar{\sigma}^\nu)_{\beta\beta'}[g_{\mu\nu}c(z^2) + (z_\mu z_\nu - \tfrac{1}{4} g_{\mu\nu})\ d\ (z^2)] \tag{35}$$

By similar arguments we find for the propagator

$$\eta(\underset{\alpha}{x}|\underset{\beta}{y}) = i(\bar{\sigma}z)_{\alpha\beta}\ f(z^2) \tag{36}$$

and for the six-point-function

$$\eta(\underset{\alpha\beta\gamma}{xxy}|\underset{\alpha'\beta'\gamma'}{x\ y\ y}) = (\bar{\sigma}^\mu)_{\alpha\alpha'}(\bar{\sigma}^\nu)_{\beta\beta'}(\bar{\sigma}^\rho)_{\gamma\gamma'} \times$$

$$\times\ i[i\varepsilon_{\mu\nu\rho\lambda}\ z^\lambda + g_{\mu\nu}\ z_\rho + g_{\nu\rho}\ z_\mu - g_{\mu\rho}\ z_\nu]g\ (z^2) \tag{37}$$

Thus we have six invariant functions a,b,c,d,f,g to determine If we insert these ansatzes in (14) and (15) and equate the coefficients of appropriate basis elements of the algebra we obtain the equations

$$s f' + 2f - 4\ell^2 b = i\rho_o \ \delta(z)$$

$$b' - 2\ell^2 \left[- 2g + f(- a+c + \frac{3s}{4} d) + \frac{s}{2} f^3\right] = 0 \qquad (38)$$

where a prime denotes the derivative with respect to the argument s. Any set of functions f,a,b,c,d which is consistent with these equations can be used for a further investigation of the theory in consideration, since it is consistent with the field equations. The next step would consist in choosing such a set and using it as boundary values for functions $\eta(xy|zw)$, $\eta(xyz|uvw)$ which are (again only partially) to be found from the equations (15) etc. for different coordinates.

From (38) we might guess that there could be even solutions with positive ρ. In this case the second equation contains divergences at the light cone because of the last term f^3. We could however invent functions a,c,d such that these divergences are compensated formally.

Once more it has to be noted however, that there is no guarantee that the functions obtained in this way establish a reasonable field theory, since we have not looked, how the Hilbert space looks like and what physical properties the functions describe. Thus our approach is nothing more than a very primitive and very formal starting point. We have still to fill more physics into field equations.

References

1. Functional techniques have been discussed by various authors. For renormalizable theories see e.g. K. Symanzik, Hercegovi lectures 1962, where more literature can be found.
2. The correlation functions η have been used in field theory by W. Zimmermann, Nuovo Cim. 13, 503 (1959).
3. For literature see footnote 10) in the paper by G.Källén, Acta Physica Austriaca, Suppl. II., 1965. The representation was used also by J. Schwinger in a lecture at

Harvard University 1954 (The Theory of Coupled Fields, unpublished lecture notes).

4. A. S. Wightman, Phys. Rev. 101, 860 (1956).
5. W. Heisenberg, Nucl. Phys. 4, 532 (1957).
6. R. Haag, in: Les Problèmes Mathématiques de la Théorie Quantique des Champs, Paris 1959.
7. K. Sekine, Nucl. Phys. 23, 245 (1961).
8. I. M. Gel'fand, G. E. Schilow, Verallgemeinerte Funktionen I, Berlin 1960.
9. H. Mitter, Nuovo Cim. 32, 1789 (1964).
10. K. Johnson, Nuovo Cim. 20, 773 (1961).
11. H. Mitter, Nuovo Cim. 38, 1040 (1965).

SPONTANEOUS BREAKDOWN OF SYMMETRIES[†]

By

P. SURÁNYI

Central Research Institute for Physics
Budapest

I. Introduction

Today's physics is characterized by the increasing importance of symmetry groups. Unfortunately most of these groups are connected only with an approximate symmetry. The generally accepted view about this problem is the following: the Hamiltonian of the system, H does not commute with the generators of the group /G_i, i = 1,...n/. On the other hand it is worth while to split off the Hamiltonian into two parts:

$$H = H_o + H' \tag{1.1}$$

where $[H_o, G_i] = 0$, denotes the invariant part of H. So the breakdown of the given symmetry is attributed to a perturbing term in the Hamiltonian. This means, that together with the symmetric one a new term has to be introduced into the Hamiltonian, which leads to the breakdown of the symmetry. On the other hand the structure of the general Hamiltonian is quite involved even without this term. For this reason there are people who look for another solution of the problem. As we know the solution of a field theory is equivalent to the solution of an infinite system of nonlinear integro-differential equations for the Green functions. In spite

[†]Lecture given at the V. Internationalen Universitätswochen für Kernphysik, Schladming, 24 February - 9 March, 1966.

of the invariant form of these equations in principle it is possible to obtain a non-symmetric solution of them. Such types of solutions are called spontaneous breakdown solutions. Theories of similar nature are well-known in solid state physics for a long time. Superconductivity, ferromagnetism etc. obtained a natural explanation by these theories. Several authors tried to explain different field theoretic problems using spontaneous breakdown solutions, e.g. the energy "gap" between nucleons and electrons [1,2] or muons and electrons [3,4,5] the breakdown of SU(3) symmetry [6,7] and CP symmetry [8]. The main problem connected with spontaneous breakdown is the following: any solution of a theory showing a lower symmetry than that of the equations must contain a zero mass boson state. This is the famous Goldstone theorem [9,10], and the corresponding bosons are often called Goldstone bosons. Some papers tried to make use of the theorem to generate the photons as Goldstone bosons [11,12]. Another problem of spontaneous breakdown theories: why does nature prefer these solutions when symmetric solutions also exist. Many authors [13] looked for stability conditions which would choose between symmetric and asymmetric solutions. These lectures will contain a general description of spontaneous breakdown theories (Section II), a discussion about Goldstone theorem (Section III), physical applications (Section IV) and at the end conclusions (Section V).

II. The Properties of Spontaneous Breakdown Theories

For sake of simplicity we choose a very simple field theory. We shall denote the operator corresponding to a two component field by ϕ_i. We do not specify the field equation satisfied by ϕ_i, only assume the invariance of it under the transformations of the two dimensional rotation group. The field operator transforms according to the two dimensional representation of the group. The only invariant constructed from the field is given by

$$I(x,y) = \sum_{i=1}^{2} \phi_i(x)\ \phi_i(y) \quad . \tag{2.1}$$

As usual we can construct a complex field from the two real fields $\phi = \phi_1 + i\phi_2$ and the group transformation acts on as

$$\phi \rightarrow \phi' = e^{i\alpha}\phi \tag{2.2}$$

where α is some real parameter. We know from elementary field theory, that we can construct a current from the Lagrangian of the theory

$$j_\mu(x) = \frac{\partial L}{\partial \partial_\mu \phi(x)}\ \phi(x) - \frac{\partial L}{\partial \partial_\mu \phi^*(x)}\ \phi^*(x) \ , \tag{2.3}$$

and the operator G

$$G = \int j_0(x)\ d^3x \tag{2.4}$$

generates the transformation of the group in the Hilbert-space:

$$e^{i\alpha}\phi = e^{i\alpha G}\phi\ e^{-i\alpha G} = U(\alpha)\phi\ U(-\alpha) \quad . \tag{2.5}$$

As is well known a very important consequence of the invariance of the Lagrangian is the time independence of G or with other words the equation

$$\partial_\mu j_\mu(x) = 0.$$

As we know all the informations about the theory are given in terms of vacuum expectation values of field operators

$$G_{ik,\dots\ell}(x,y,\dots z) = \langle 0|\phi_i(x)\ \phi_k(y)\dots\phi_\ell(z)|0\rangle \ . \tag{2.6}$$

We may work instead of (2.6) with the vacuum expectation values of products of ϕ and ϕ^*. If we perform a transformation on the vacuum expectation value of the type

$$G(x,y,\ldots z) = \langle 0|\phi(x)\ \phi^*(y)\ldots\phi(z)|0\rangle \tag{2.7}$$

then we obtain

$$G'(x,y,\ldots z) = \langle 0|U(\alpha)\phi(x)U(-\alpha)U(\alpha)\phi^*(y)U(-\alpha)\ldots U(\alpha)\phi(z) U(-\alpha)|0\rangle$$

$$= \langle 0|U(\alpha)\phi(x)\phi^*(y)\ldots\phi(z)U(-\alpha)|0\rangle\ . \tag{2.8}$$

On the other hand taking into account that $[G,H] = 0$ (where H is the Hamiltonian of the system) the eigenstates of H will be the eigenstates of G as well. So $G|0\rangle = C|0\rangle$, where C is a real constant, which can be choosen as C = 0. In such a way we obtain

$$U(\alpha)|0\rangle = |0\rangle$$

so

$$G'(x,y,\ldots z) = G(x,y,\ldots z)\quad . \tag{2.9}$$

Using eqs. (2.5) and (2.8) we can express G' in the following way:

$$G'(x,y,\ldots z) = \langle 0|e^{i\alpha}\phi(x)\ e^{-i\alpha}\ \phi^*(x)\ldots e^{i\alpha}\ \phi(z)|0\rangle =$$

$$= e^{i(n-m)\alpha}\ G(x,y,\ldots z) \tag{2.10}$$

where n is the number of ϕ fields, m is that of the ϕ^* fields. If we compare equations (2.9) and (2.10) we can conclude that

$$G = 0 \quad \text{if} \quad n \neq m$$

which means that the fields can appear only in $\phi\phi^*$ combinations in the vacuum expectation value. Looking at eq. (2.1) we can see that this is just the invariant combination of the fields. It is easy to generalize this statement for theories with different type of symmetries: only the invariant combinations of fields have a nonvanishing vacuum expectation value (more exactly combinations which are parts of invariants). We explain this simple statement on the example of twofold vacuum expectation values.

$$<0|\phi(x)\ \phi(y)|0> = <0|\phi^*(x)\ \phi^*(y)|0> = 0$$

$$<0|\phi(x)\ \phi^*(y)|0> = 2G(x,y) \neq 0 \qquad (2.11)$$

or expressing eq. (2.11) by ϕ_1 and ϕ_2

$$<0|\phi_1(x)\ \phi_2(y)|0> = 0$$

$$<0|\phi_1(x)\phi_1(y)|0> = <0|\phi_2(x)\phi_2(y)|0> = G(x,y) \qquad (2.12)$$

The immediate consequence of eq. (2.12) is the equality of masses of particles connected with fields ϕ_1 and ϕ_2. On the other hand our aim is to find a solution, which gives a different value for the masses of the two types of particles. Eq. (2.12) would mean the non-existence of a spontaneous breakdown solution. The essential point at the proof of this statement is the assumption of the existence of operators $U(\alpha)$, which "rotate" the fields and states in the two dimensional space. In such a way spontaneous breakdown is equivalent to the non-existence of the generator of the group G. Of course the current which is a local quantity must exist and the continuity equation must be fulfilled, only the space integral of the time component of the current does not exist.

Now it is clear we can achieve a spontaneous breakdown, if there is some non-invariant combination of field operators, which has a nonvanishing vacuum expectation value. The simplest choice is the following: the field operator itself has a non-vanishing vacuum expectation value. In usual field theory such expectation values are equal to zero. Of course this is necessary if ϕ_i carries some conserved discrete quantum number different from that of the vacuum. In any case following from translation invariance

$$<0|\phi_i(x)|0>=<0|\phi_i(0)|0>$$

is independent of x. Besides of that we can always introduce such rotated combinations of ϕ_1 and ϕ_2 that

$$<0|\phi_1|0> \neq 0$$

and

$$<0|\phi_2|0> = 0 \quad . \tag{2.13}$$

In the following step we prove the Goldstone theorem, the existence of a zero mass boson in the system, provided eq. (2.13) is fulfilled.

First we take the divergence of the current operator between two arbitrary physical states:

$$<p_1|\partial_\mu j_\mu(x)|p_2> = 0 \quad . \tag{2.14}$$

$<p_1|$ and $|p_2>$ are states with four momentum p_1 and p_2 respectively.

Using translation invariance we express the vacuum expectation value of j_0 from eq. (2.14)

$$<p_1|j_o(0)|p_2> = \frac{\underline{p}_1 - \underline{p}_2}{p_{10} - p_{20}} <p_1|\underline{j}(0)|p_2> . \tag{2.15}$$

We can see from eq. (2.15) that $<p_1|j_o|p_2>$ is always zero if $\underline{p}_1 = \underline{p}_2$ and $p_{10} \neq p_{20}$ that is to say the invariant masses of states 1 and 2, m_1 and m_2 are not equal to each other. Now we make use of the equal-time commutation relation

$$[j_o(\underline{x},t),\ \phi_2(\underline{y},t)] = \delta(\underline{x}-\underline{y})\ \phi_1(\underline{x},t)\ . \tag{2.16}$$

Taking the vacuum expectation value of eq. (2.16) and putting inside the commutator a closed system of physical states we obtain:

$$\sum_{\alpha,\underline{p}} \langle 0|j_o(\underline{x})|\alpha(\underline{p})\rangle\langle\alpha(\underline{p})|\phi_2(\underline{y})|0\rangle -$$

$$- \sum_{\beta,\underline{p}'} \langle 0|\phi_2(\underline{y})|\beta(\underline{p}')\rangle\langle\beta(\underline{p}')|j_o(\underline{x})|0\rangle = \delta(\underline{x}-\underline{y})\langle 0|\phi_1|0\rangle \neq 0\ . \tag{2.17}$$

If we take the three dimensional Fourier transform of eq. (2.17) we obtain the following relation:

$$(2\pi)^3 \sum_{\alpha}\langle 0|j_o(0)|\alpha(\underline{q})\rangle\ \langle\alpha(\underline{q})|\phi_2(\underline{y})|0\rangle -$$

$$- (2\pi)^3 \sum_{\beta}\langle 0|\phi_2(\underline{y})|\beta(-\underline{q})\rangle\langle\beta(-\underline{q})|j_o(0)|0\rangle =$$

$$= e^{i\underline{qy}}\ \langle 0|\phi_1|0\rangle\ . \tag{2.18}$$

If $\underline{q} \to 0$ the right hand side of eq. (2.18) tends to a finite value. On the other hand all the terms on the left hand side will contain factors like $\lim_{\underline{q}\to 0}\langle 0|j_o(0)|\alpha(\underline{q})\rangle$ which is equal to zero according to eq.(2.15) if the mass of the state α or β is not equal to zero. So we can see that the closed system of asymptotic states must contain a zero mass one particle state in the ϕ_2 channel. We can say that the renormalized mass of ϕ_2 particle, $m_2 = 0$.

Another important feature of spontaneous breakdown theories is the appearance of infinitely many inequivalent representations. This phenomenon can be explained as follows: As the field operators satisfy invariant equations but the matrix elements have noninvariant form, the only possibility is the noninvariance of the vacuum state under a rotation in our space. A vacuum state is defined here as the state having the lowest energy level and a direction in the two dimensional space. So there are infinitely many vacuum states, which are clearly inequivalent, because the unitary operator which would transform them among each other does not exist. We can build up a closed system of states on these vacuum states. As is possible to show the states belonging to two different directions of the vacuum state are orthogonal to each other, no measurement can connect the states of two "worlds".

At the end we can show a very simple-minded model. We write down a field equation for the ϕ_i field in the following form

$$(\Box - \mu^2)\ \phi_i(x) = g\ \phi_i(x)\ \phi_k(x)\ \phi_k(x) + j_i(x) \qquad (2.19)$$

where $j_i(x)$ is an external source.

If we take the vacuum expectation value of eq. (2.19) at $j_i = 0$ we obtain

$$-\ \mu^2\ \langle\phi_i\rangle_o = g\ \langle\phi_i\phi_k\phi_k\rangle_o \quad .$$

If we put a closed system of states between the operators and cut off the sum over intermediate states at the first contribution (the vacuum state) we obtain

$$-\ \mu^2\langle\phi_1\rangle_o = g(\langle\phi_1\rangle_o)^3$$

with solutions

$$\langle\phi_1\rangle = \begin{cases} 0 \\ \sqrt{(-\mu^2)/g} \end{cases} \qquad (2.20)$$

The second one being the symmetry breaking solution. The two-particle Green function can be obtained by taking the functional derivative of $\langle\phi_i\rangle_o = \Phi_i$

$$G_{ik}(x,y) = \frac{\delta\,\Phi_i(x)}{\delta j_k(y)}$$

which satisfies the equation

$$(\Box - \mu^2)\, G_{i\ell}(x,y) = g\, G_{i\ell}(x,y)\, \Phi_k\Phi_k + 2g\, \Phi_i G_{k\ell}(x,y)\Phi_k + $$

$$+ \delta_{i\ell}\delta(x-y)\ . \qquad (2.21)$$

We can project eq. (2.21) onto longitudinal and transversal parts with

$$G_{i\ell}(x,y) = \frac{\Phi_i\Phi_\ell}{\Phi^2}\, G_L(x,y) + (\delta_{i\ell} - \frac{\Phi_i\Phi_\ell}{\Phi^2})\, G_T(x,y) \qquad (2.22)$$

$$\Phi^2 \equiv \Phi_k\ \Phi_k$$

and obtain the equations

$$(\Box - \mu^2)\, G_L = 3g\Phi^2 G_L + \delta(x-y)$$

$$(\Box - \mu^2)\, G_T = g\Phi^2 G_T + \delta(x-y)$$

or using the symmetry breaking solution of eq. (2.20)

$$(\Box + 2\mu^2)\, G_L = \delta(x-y)$$

$$\Box\, G_T = \delta(x-y)\quad . \qquad (2.23)$$

So the zero mass particle appeared in the transversal mode, which is in our case the ϕ_2 particle. For the mass of ϕ_1 we obtained $m_1^2 = -2\mu^2$ (here μ^2 is only a parameter of the theory, $\mu^2 < 0$ is possible).

III. Problems connected with the Goldstone Theorem

The above proof of the Goldstone theorem lacks the mathematical rigour at some essential points:

1. We assumed the existence of the local current and
2. the existence of the Fourier transform of this current.

The above problems are much more general questions than that of ours and they are beyond our scope. There are other proofs of the theorem which are free of this troubles [14], but they encounter with different types of difficulties.

On the other hand, there are physical problems in connection with the statement of the theorem.

The proof starts from an unphysical assumption about the vacuum expectation value of the field. Instead of that we wish to start only from the fact, that the field equations are invariant, but the masses are not equal to each other. Nambu has pointed out, that the current conservation implies the existence of a pole in the matrix element of tne current, if the form factor does not vanish at zero momentum transfer; this can be shown very easily, if we consider the general structure of the current operator matrix element between states Ψ_1 and Ψ_2

$$(2\pi)^3\sqrt{2p_{10}}\ \sqrt{2p_{20}}\ \langle\Psi_1(\underline{p}_1)|j_\mu|\Psi_2(\underline{p}_2)\rangle =$$

$$= (p_{1\mu} + p_{2\mu})\ F_1(t) + (p_{1\mu} - p_{2\mu})\ F_2(t) \quad (3.1)$$

where we introduced the notation:

$$(p_{1\mu} - p_{2\mu})^2 = t\ .$$

Taking the divergence of this matrix element we obtain:

$$(m_1^2 - m_2^2)\ F_1(t) + t\ F_2(t) = 0\ . \quad (3.2)$$

Comparing equations (3.1) and (3.2) we obtain

$$\langle\Psi_1(\underline{p}_1)|j_\mu|\Psi_2(\underline{p}_2)\rangle = F_1(t)(p_{1\mu} + p_{2\mu} \qquad \frac{m_1^2-m_2^2}{t}(p_{1\mu} - p_{2\mu})) \tag{3.3}$$

which proves our statement provided we have a mass splitting. The only problem is that we do not know $F_1(0)$. We must not extrapolate from ordinary theories, where $F_1(0)$ is some charge and different from zero. Bassetto and Ciccariello [15] investigated the behaviour of $\langle 0|T(\phi_1(x)j_\mu(z)\phi_2(y))|0\rangle$ instead of the matrix element of the current. We can define $\Gamma_\mu(p_1,p_2)$ as

$$\Gamma_\mu(p_1,p_2) = \Delta_2^{-1}(p_2^2)\; F_\mu(p_1,p_2)\Delta_1^{-1}(p_1^2)$$

where $\Delta_i(p^2)$ is the propagator corresponding to field ϕ_i ,

$$F_\mu(p_1,p_2) = \int d^4x \int d^4y\; e^{ip_1x}\; e^{-ip_2y} \quad \langle 0|T(\phi_1(x)j_\mu(z)\phi_2(y)|0\rangle$$

It is easy to prove the generalized Ward identity for the vertex function Γ_μ

$$(p_{1\mu}-p_{2\mu})\;\Gamma_\mu(p_1p_2) = \Delta_1^{-1}(p_1^2) - \Delta_2^{-1}(p_2^2)\;. \tag{3.4}$$

$\Gamma_\mu(p_1,p_2)$ has the following structure:

$$\Gamma_\mu(p_1p_2) = (p_{1\mu}+p_{2\mu})F_1(p_1^2,p_2^2,t) + (p_{1\mu}-p_{2\mu})F_2(p_1^2,p_2^2,t)\,, \tag{3.5}$$

comparing eq. (3.4) and eq. (3.5) we obtain

$$(p_1^2-p_2^2)F_1(p_1^2,p_2^2,t) + t\;F_2(p_1^2,p_2^2,t) = \Delta_1^{-1}(p_1^2) - \Delta_2^{-1}(p_2^2)\;.$$

For $p_1^2 = p_2^2$

$$F_2(p_1^2,p_2^2,t) = \frac{\Delta_1^{-1}(p_1^2) - \Delta_2^{-1}(p_1^2)}{t} \tag{3.6}$$

If the masses m_1 and m_2 are different from each other then

$\Delta_1^{-1}(p^2) \neq \Delta_2^{-1}(p^2)$, so F_2 really contains a pole at $t = 0$. However we must emphasize, the existence of the pole is proved by eq. (3.6) for non-physical four-momenta of the other particles, taking part in the process.

IV. Physical Applications

As the first application of spontaneous breakdown theories I shall speak about the Nambu, Jona-Lasinio model of elementary particles [1,2]. The self interaction of a fermion field (essentially the nucleon field) is studied. The Lagrangian of the theory has the form

$$L = - i\bar{\psi}\gamma_\mu \partial_\mu \psi + g_0[(\bar{\psi}\psi)^2 - (\bar{\psi}\gamma_5\psi)^2] , \qquad (4.1)$$

where $g_0 > 0$.

As is easy to check,in addition to the vector current $V_\mu = \bar{\psi}\gamma_\mu\psi$ in such a theory the axial vector current $A_\mu = \bar{\psi}\gamma_\mu\gamma_5\psi$ is conserved as well. The conservation of the axial vector current is the result of the invariance of (4.1) under the chirality transformation

$$\psi \to e^{i\alpha\gamma_5}\psi \quad , \qquad \bar{\psi} \to \bar{\psi}\, e^{i\alpha\gamma_5} \quad .$$

A mass term (like $M\psi\psi$) would break the chirality transformation invariance, and so the axial vector conservation. Our aim is to obtain such self-consistent solutions of the theory, which break the continuous γ_5 transformation invariance and lead to finite mass particles. We shall show, that such solutions exist in a very primitive approximation (Hartree-Fock approximation) which reminds us to the variational method of quantum mechanics. At first we have to remark, that our theory defined by Lagrangian (4.1) is a highly divergent one, so the introduction of a cutoff of divergent integrals is necessary. One may hope, that in a more exact calculation the form factors play a similar role as the invariant cutoff.

The mass of the physical particle is determined from the equation

$$i\gamma p - \Sigma(p,m,g,\Lambda) = 0 \tag{4.2}$$

where Σ is the self-energy part, m is the physical mass, g is the physical coupling constant, Λ is the measure of the invariant cutoff. The self energy part Σ must satisfy the equation

$$\Sigma(p,m,g,\Lambda) = m, \quad \text{for} \quad \gamma p = m \quad . \tag{4.3}$$

Our Lagrangian (4.1) can be written as $L = L_o + L_{int}$, where L_o is the free part, L_{int} is the interaction part of (4.1). We shall write instead of that $L = L_o' + L_{int}'$, where $L_o' = L_o + L_s$, $L_{int}' = L_{int} - L_s$. Here L_s denotes the part of L_{int} which leads to self energy effects, L_{int}' does not lead to any self energy correction. We hope as in the variational method, that the special choice of L_s is not very essential, the eigenvalues are highly independent of the analytic form of L_s.

We choose $L_s = m\bar{\psi}\psi$ and calculate Σ in the first nonvanishing approximation. The corresponding Feynman diagram can be seen on fig. 1.

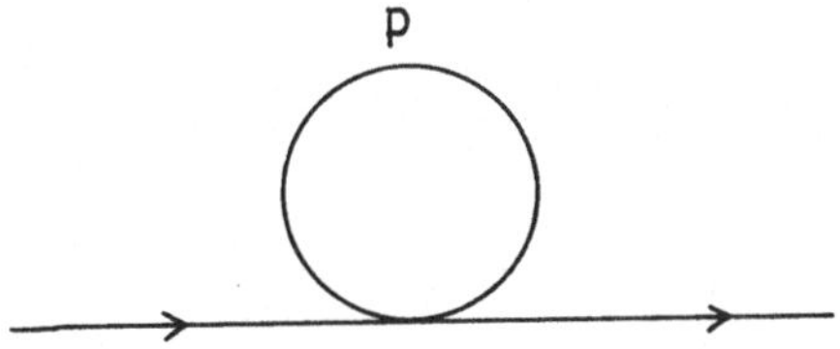

fig. 1

Using eq. (4.3) we may obtain

$$m = -\frac{g_o m}{2\pi^4} \int \frac{d^4p\ F(p,\Lambda)}{p^2-m^2+i\varepsilon} \tag{4.4}$$

where $F(p,\Lambda)$ is an invariant cutoff function. Eq. (4.4) has two solutions

1. $m = 0$, the trivial solution and
2. the solution of the implicit equation

$$1 = -\frac{g_o}{2\pi^4} \int \frac{d^4p\, F(p,\Lambda)}{p^2 - m^2 + i\varepsilon} \tag{4.5}$$

which is the symmetry breaking solution.
After a rotation of the integration contour in the p_o plane taking a simple $\Lambda^2 = p^2$ cutoff in the Euclidean space equ. (4.5) can be written as

$$\frac{2\pi^2}{g_o\Lambda^2} = 1 - \frac{m^2}{\Lambda^2} \ln\left(\frac{\Lambda^2}{m^2} + 1\right) .$$

From here we can see that the nontrivial solution exists only for $0 < 2\pi^2/g_o\Lambda^2 < 1$. The condition $g_o > 2\pi^2/\Lambda^2$, (the coupling is strong enough) makes possible the existence of a bound state with zero mass. This can be shown in the following way. We write down a Bethe-Salpeter equation for the vertey part in the approximation given by fig. 2 and take the pseudoscalar

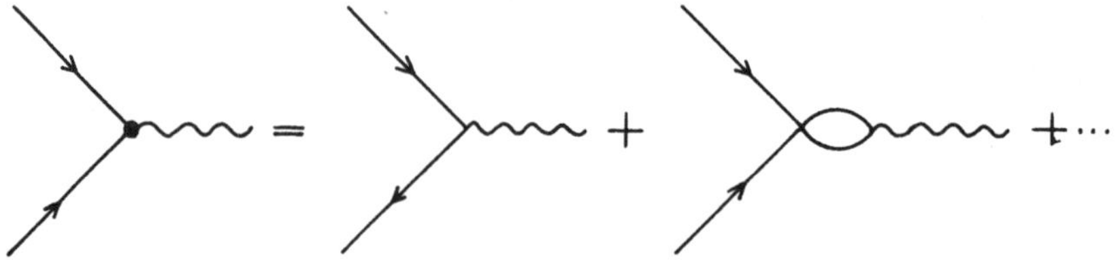

fig. 2

part of the equation

$$\Gamma(p+\tfrac{1}{2}q,\ p-\tfrac{1}{2}q) =$$

$$= g_o + g_o \int d^4q' \mathrm{Tr}(S_F(p+\tfrac{1}{2}q')\Gamma(p+\tfrac{1}{2}q', p-\tfrac{1}{2}q')\gamma_5 S_F(p-\tfrac{1}{2}q')\gamma_5) \tag{4.6}$$

where p + q/2 and p - q/2 are the momenta of fermions, Γ is the vertex part, $S_F(p)$ is the fermion propagator. Solutions of the homogeneous Bethe-Salpeter equation correspond to bound states (a pole of the solution of the inhomogeneous equation). On the other hand the homogeneous Bethe-Salpeter equation has a solution at $p^2 = 0$ if condition (4.5) is fulfilled. So again we obtained the Goldstone theorem in our approximation. The Goldstone particles are pseudoscalar bosons which can be identified as idealized pions. From the solution of the inhomogeneous vertex equation one can obtain the coupling constant of the pion to the nucleons as a dynamical quantity. $G^2/4\pi \simeq 15$ may be obtained if $\Lambda^2 \simeq m^2$ is chosen, which is a quite admissible value. A very interesting feature of the model, that G is independent of g_o in our approximation.

A more realistic model can be obtained [2] , if we introduce the isotopic spin. The Lagrangian will have the following form:

$$L = - i\bar{\psi}\gamma_\mu \partial_\mu \psi + g_o [(\bar{\psi}\psi)^2 - \sum_{i=1}^{3} (\bar{\psi}\gamma_5 \tau_i \psi)^2] \qquad (4.7)$$

where now ψ denotes the two component nucleon field. Then similarly to the preceeding calculation we can find the triplet zero mass pions as Goldstone bosons. If we wish to obtain a finite mass value for the pions, we must spoil the original γ_5 invariance of the Lagrangian. The simplest way to do it is to put into (4.7) a small mass term $L_m = m_o\bar{\psi}\psi$. The self consistent equation can be obtained easily

$$m = m_o + g_o \,\mathrm{Tr}\, S_F^{(M)}(0) = m_o + I(m) \,.\, m \qquad (4.8a)$$

where

$$I(m) = \frac{-20g_o}{(2\pi)^4} \int \frac{d^4p}{p^2-m^2+i\varepsilon} F(p,\Lambda) \quad .$$

If we use the same special cutoff as in the isoscalar model I(m) can be expressed as

$$I(m) = \frac{5g_o}{8\pi^2} \int_{4m^2}^{f(\Lambda^2,m^2)} d\kappa^2 (1 - \frac{4m^2}{\kappa^2})^{1/2} , \qquad (4.8b)$$

where $f(\Lambda^2,m^2)$ is a simple rational function of Λ^2 and m^2. It is interesting to find out the magnitude of m_o which is necessary to produce the observed pion mass. If we write down the homogeneous equation for the vertex function we obtain a condition for the existence of the solution

$$1 = \frac{5g_o}{8\pi^2} \int_{4m^2}^{f(\Lambda^2,m^2)} \kappa^2 d\kappa^2 \, (1 - \frac{4m}{\kappa^2})^{1/2} \frac{1}{\kappa^2-\mu^2} \, . \qquad (4.9)$$

We have to solve eq. (4.9) for μ^2, the mass of the bound state. It is clear $\mu^2 = 0$, if $m = 0$ (this can be seen using the self consistency condition (4.8b)). Using the observed values $\mu^2/4m^2 \sim 1/200$ we obtain $m_o \sim$ a few MeV. One may think about the possibility, that nucleons have the same order of bare mass as electrons and the big observed mass of nucleons is a result of a spontaneous breakdown of the γ_5 symmetry.

An extension of the Nambu-Jona-Lasinio model to SU(3) symmetry was given by Baker and Glashow [3] and Glashow [4]. The proposed Lagrangian of the model is the following

$$L = i\bar{\psi}\gamma_\mu\partial_\mu\psi + g(\bar{\psi}\psi)^2 + f(\bar{\psi}\gamma_5\psi)^2 + g'(\bar{\psi}\lambda_i\gamma_5\psi)^2 + f'(\bar{\psi}\lambda_i\psi)^2. \qquad (4.10)$$

Here ψ is a triplet field (e.g. the field of quarks), the λ_i matrices are the generators of SU(3) in the triplet representation. L is invariant under transformations

$$\psi \to \gamma_5\psi \quad \text{and} \quad \psi \to e^{i\lambda_i\alpha_i}\psi \quad .$$

The diagrams contributing to Σ_{ij} the self energy operator are given in fig. 3 .

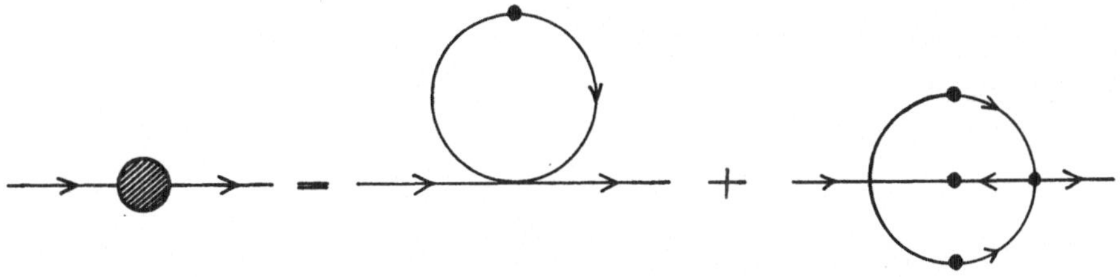

fig. 3

The dots denote all possible corrections to propagators and all possible proper corrections to the vertex part. If we use again the Hartree-Fock approximation and hold only the first diagram on fig. 3 then we obtain the following self consistent equations.

$$m_1 = C_1 m_1 F(m_1^2) + C_2\{m_2 F(m_2^2) + m_3 F(m_3^2)\}$$

$$m_2 = C_1 m_2 F(m_2^2) + C_2\{m_3 F(m_3^2) + m_1 F(m_1^2)\}$$

$$m_3 = C_1 m_3 F(m_3^2) + C_2\{m_1 F(m_1^2) + m_2 F(m_2^2)\} \quad . \qquad (4.11a)$$

In eq. (4.11a) m_1, m_2 and m_3 denote the physical masses of the three quark fields,

$$C_1 = 3g + g' + 3f' + f \ , \quad C_2 = 4g + 2g' - 6f'$$

$$F(m^2) = \frac{\Lambda^2}{16\pi^2}\{1 - \frac{m^2}{\Lambda^2}\ln(1 + \frac{\Lambda^2}{m^2})\} \ ,$$

if we use the simple cutoff $p^2 = \Lambda^2$. Introducing the variables $x = m_1/\Lambda$, $y = m_2/\Lambda$, $z = m_3/\Lambda$ and the function $f(x) = 1-x^2 \times \ln(1 + 1/x^2)$ we obtain equations (4.11b)

$$x = \alpha x\ f(x) + \gamma\{y\ f(y) + z\ f(z)\}$$

$$y = \alpha y\ f(y) + \gamma\{z\ f(z) + x\ f(x)\}$$

$$z = \alpha z\ f(z) + \gamma\{x\ f(x) + y\ f(y)\} \quad . \tag{4.11b}$$

The SU(3) symmetry of Lagrange function (4.10) is reflected in the permutation symmetry of equations (4.11).

It is easy to transform eqs. (4.11b) into the following form

$$x + y + z = G\ h(x) = G\ h(y) = G\ h(z) \tag{4.11c}$$

where

$$h(x) = x\ f(x) - \beta x\ , \beta = (\alpha-\gamma)^{-1}\ , \ G = (\alpha^2-\gamma^2)\gamma^{-1}$$

The function h(x) can be seen on fig. 4. We can obtain different values for x,y,z if h(x) attains the same value at two

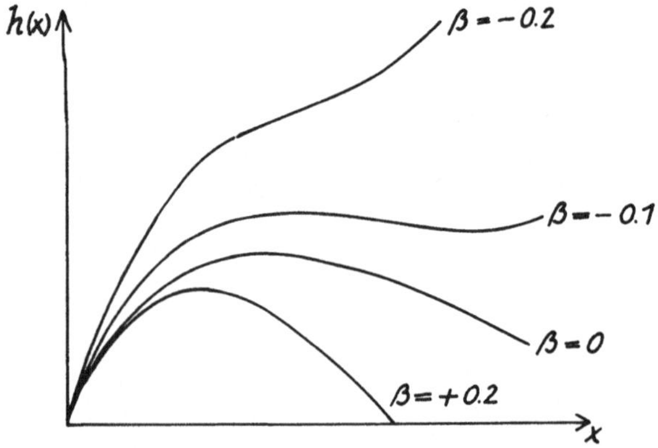

fig. 4

different x. This is possible at β values $-0.2 \lesssim \beta \lesssim +1$.

It is interesting to examine the regions on the G,β plane where different types of solutions of eqs. (4.11c) are allowed. We can give a simple table of possible solutions:

1) perturbative solution x = y = z = 0, U(3) and γ_5
2) γ_5 invariance is broken x = y = z > 0 , U(3)
3) γ_5 and SU(3) are broken x = y ≠ z , U(2) × U(1)

4) SU(3) and SU(2) are broken $x \neq y \neq z$, $U(1) \times U(1) \times U(1)$

The last column contains everywhere the remaining symmetry. Solution 1) exists for every value of G and β. As to the other solutions the stronger is the symmetry breaking the smaller is the allowed region in the G,β plane for the given type of solution.

A great mystery of elementary particle physics is the mass difference of muons and electrons in spite of their identical interactions. Many papers treated the μ-e problem by the help of spontaneous breakdown theories [6,7,8]. The basic idea of these papers is the following:

We start from a Lagrangian

$$L = -\bar{\psi}(x)(-i\gamma_\mu \partial_\mu + m_o)\psi(x) + e_o\bar{\psi}(x)\gamma_\mu\psi(x)A_\mu(x) + L_{Maxwell} , \tag{4.12}$$

where ψ denotes a doublet constructed from the μ and e fields. Lagrangian (4.12) has an U(2) invariance group. If we break only SU(2) the charge conservation remains valid. Using the self consistent method for diagrams given on fig. 5 we can

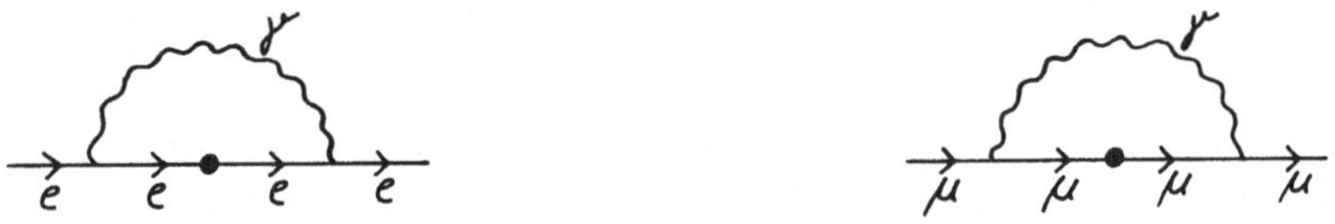

fig. 5

obtain different types of solutions:

1) the perturbative solution $m_\mu = m_e = 0$

2) $m_\mu \neq 0$, $m_e \neq 0$, $\mu \nrightarrow \gamma + e$

3) $m_\mu \neq 0$, $m_e \neq 0$, $\mu \rightarrow \gamma + e$ decay allowed .

The main problem of the theory is that it contains two free parameters, so the electron and muon masses cannot be fixed.

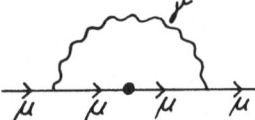

The models we have listed up to now could not give mass formulae without free parameters. We could not predict in SU(3) the observed type of symmetry breaking:

i) SU(3) breaks down to SU(2) × U(1), where the privileged direction in the triplet space is the hypercharge axis.

ii) the approximate validity of the Gell-Mann-Okubo mass formulae.

Cutkosky and Tarjanne [16] tried to solve this problem using a self interaction model of an octet of vector mesons. Working in ladder approximation with a cutoff they looked for solutions with small deviations from the symmetric one, for some values of the coupling constant. SU(3) is broken into SU(2) × U(1). Due to the symmetry of the problem the privileged axis is not determined uniquely, three equivalent solutions exist, in which I spin, U spin and V spin remain to be conserved quantities, respectively. Similar results can be obtained in entirely different models of symmetry breaking [14]. In addition it was possible to show that the dominant symmetry breaking term in the mass matrix has an octet transformation character. This result leads to an approximate Gell-Mann-Okubo mass formula.

More general results of similar nature are listed in the excellent review of self consistent unitary symmetries by Cutkosky [17].

A bootstrap theory is very convenient to treat the spontaneous breakdown problem. In what follows we present a very simple calculation due to Wong [18] as an illustration of these ideas. The members of a vector octet are assumed to be bound states of pairs of the same vector mesons (without octet-singulet mixing). Only an F type coupling is allowed $j^i_\mu \sim \partial_\mu V^j_\rho V^k_\rho f_{ijk}$. The p-wave scattering amplitude in the octet channel is represented in the N/D form. Here N and D are matrices in the octet space. The matrices are assumed to be reduced to I = 0, I = 1/2, I = 1 subspaces. A simple pole approximation is adopted for N. The place of the pole is the same in all channels, the residues are determined by the exact

SU(3) Clebsch-Gordan coefficients. The following channels are open

$$I = 0 \qquad (K^*, \bar{K}^*) \to (K^*, \bar{K}^*)$$

$$I = 1/2 \qquad (\rho K^*, \omega K^*) \to (\rho K^*, \omega K^*)$$

$$I = 1 \qquad (\rho\rho, K^*\bar{K}^*) \to (\rho\rho, K^*\bar{K}^*)$$

The D-functions are determined as usual

$$D = I - \frac{(s+s_o)}{\pi} \int_{(m_1+m_2)^2}^{\infty} ds' \left(\frac{q'^3_{12}}{\sqrt{s'}}\right) \frac{N}{(s'+s_o)(s'-s_o)}$$

where I is the unit matrix, m_1 and m_2 are the masses of vector mesons in the corresponding channel, q_{12} is their momentum in the CMS system, s is the invariant four-momentum squared, s_o is the place of the pole of N.

We require the existence of poles of the D matrix in all the channels at the appropriate mass values of ω, ρ, and K^*:

$$D^{(0)}(m^2_\omega) = 0$$

$$\det D^{(1/2)}(m^2_{K^*}) = 0$$

$$\det D^{(1)}(m^2_\rho) = 0 \,. \qquad (4.13)$$

The degenerate solution exists for all s_o values. Besides that a nondegenerate solution exists for $1.1\ m^2 \leqslant s_o \leqslant 3.0\ m^2$. The solutions satisfy the Gell-Mann-Okubo mass formula within a few percent for any s_o value in this interval. If we expand the solutions around the mean mass value and hold only the first order terms we obtain a homogeneous linear system of equations for the deviations. These equations have a solution which satisfy the exact Gell-Mann-Okubo mass formula for

$s_o \sim 1.8\ m^2$. The deviations of the dynamical coupling constants from the ones obtained in exact symmetry are less than 1 %.

The above listed physical applications of spontaneous breakdown theories do not exhaust the literature of this field. It is worth while to mention still some directions.

i) The Baker-Johnson quantum electrodynamics without divergences [19] .The theory starts from the zero bare mass electron-photon Lagrangian. The electrons obtain a finite mass through the spontaneous breakdown of the continuous transformation invariance.

The Dyson equations for the mass operator and for the vertex part have finite solutions in every step of the iteration procedure. We shall not discuss the Baker-Johnson electrodynamics here, because it would need a separate lecture.

ii) Another interesting application is the Björken-Guralnik electrodynamics [12, 11] which starts from a self interacting electron Lagrangian. The photon is generated as a Goldstone boson as a result of the spontaneous breakdown of Lorentz-invariance. It is possible to show, that this theory is equivalent to usual quantum electrodynamics. As we heard in the lecture of Prof. Dürr the nonlinear spinor field theory of Heisenberg generates the photons in a similar way.

V. Conclusions

We saw many applications of spontaneous breakdown theories. Now I should like to list here some common problems arising at the physical application of these theories. If we assume the existence of solutions with broken symmetry (which has not been proven in an exact theory) there remains the question of choice between the symmetric and unsymmetric solutions. The solution is called stable if a small external perturbation does not give rise to essential deviations in the parameters of the theory. For practical purposes we have not any such condition yet. Some conditions were given by different authors but they usually lead to different results depending on the approximation used.

One of the most serious problems of the practical applications of spontaneous breakdown theories is the lack of zero mass bosons, which are associated with broken symmetries. A possible way out of this problem would be if it happened that Goldstone bosons do not interact with physical (mass shell) particles.

The following difficulty is connected with the introduction of weak interactions into spontaneous breakdown theories [20]. Take e.g. the spontaneous breakdown of SU(3) symmetry. The divergence of the eight currents $j^i_\mu(x)$ is equal to zero. As we mentioned in Section II this leads to the following structure of the two particle matrix elements of the current

$$\langle p_1|j^i_\mu|p\rangle = [(p_{1\mu} + p_{2\mu}) - (p_{1\mu} - p_{2\mu}) \frac{m_1^2 - m_2^2}{t}]F(t) \quad (5.1)$$

where

$$t = (p_1 - p_2)^2 , \quad m_1^2 = p_1^2 , \quad m_2^2 = p_2^2 \quad .$$

It is easy to see that in a current-current interaction picture a similar matrix element is relevant at the kaon decays $K \to \pi\mu\nu$ and $K \to \pi e\nu$. An experimental check of eq. (5.1) is possible because the ratio of form factors multiplying the sum and difference of momenta can be measured. The experimental errors are very big, still it seems the results are not compatible with the form of the current matrix element given by eq. (5.1).

A possible solution of this problem may be the following: The current of physical particles is different from the conserved current; the weak interaction Hamiltonian contains the current of physical particles. Of course, the spontaneous breakdown theory is useful only in the case if we are able to give a unique recipe for deriving this non-conserved current from the theory. Otherwise our theory would be equivalent to a theory with an explicite symmetry breaking term in the Hamiltonian.

The author is very much indebted to Prof. P. Urban for making him possible to give this lecture at the winter school, Schladming.

References

1. Y. Nambu, G. Jona-Lasinio, Phys. Rev. 122, 345 (1961).
2. Y. Nambu, G. Jona-Lasinio, Phys. Rev. 124, 246 (1961).
3. M. Baker, S. Glashow, Phys. Rev. 128, 2462 (1962).
4. Th. Maris, V. Herscovitz, G. Jacob, Nuovo Cim. 34, 946 (1964); Nuovo Cim. 40, 214 (1965).
5. R. Arnowitt, S. Deser, Phys. Rev. 138, B712 (1965).
6. see ref.[3]
7. S. Glashow, Phys. Rev. 130, 2132 (1963).
8. G. Marx, Phys. Rev. Lett. 14, 334 (1965); Phys. Rev. 140, B.1068 (1965).
9. J. Goldstone, Nuovo Cim. 19, 154 (1961).
10. J. Goldstone, A. Salam, S. Weinberg, Phys. Rev. 127, 965 (1962).
11. G. S. Guralnik, Phys. Rev. 136, B1404 (1964).
12. G. S. Guralnik, Phys. Rev. 136, B1417 (1964).
13. K. Nakamura, Progr. Theor. Phys. 33, 251 (1965); K. Maki, Progr. Theor. Phys. 31, 698 (1964); D. Ito, Y. Kurata, Progr. Theor. Phys. 31, 1174 (1964).
14. G. Domokos, P. Surányi, Journal of Nuclear Physics (Moscow) 2, 503 (1965).
15. A. Bassetto, S. Ciccariello, Nuovo Cim. 40, 874 (1965).
16. R. E. Curkosky, P. Tarjanne, Phys. Rev. 132, 1354 (1963).
17. R. E. Cutkosky, Lecture presented at the Seminar on "High Energy Physics and Elementary Particles" Trieste, 1965.
18. C. Y. Wong, Phys. Rev. 138, B246 (1965).

19. K. Johnson, M. Baker, R. Willey, Phys. Rev. 136, B1111 (1964).
20. M. Suzuki, Phys. Rev. 136, B769 (1964).

THEORY OF SPONTANEOUS SYMMETRY BREAKDOWN AND THE APPEARANCE OF MASSLESS PARTICLES†

By

Y. FRISHMAN and A. KATZ

Department of Physics, Weizmann Institute, Rehovoth
Israel

1. Introduction

Broken symmetry solutions appeared in field theory when Nambu and Jona-Lasinio [1] tried to fashion a theory of strongly interacting particles analogous to the Bardeen, Cooper and Schrieffer [2] theory of superconductivity. The mass of the baryons was supposed to be completely dynamical - a result of spontaneous symmetry breakdown. The mass of the pions (analogs of the collective excitations of superconductivity) turned out to be zero. Soon afterwards Goldstone [3] conjectured that massless particles appear whenever spontaneous symmetry breakdown occurs. Arguments leading to this conclusion were later presented by Goldstone, Salam and Weinberg [4].

The above mentioned authors, as others after them [5] - [9] announce their result in terms of a "symmetry transformation depending on a continuous parameter" which leaves the Hamiltonian or Lagrangian invariant but changes the vacuum. We adopt the same definition of a broken symmetry in section 2 and show by means of a counter example that massless particles need not arise. Another counter example was given by Lopuszanski and Reeh [10].

---

† This is an abbreviation of a longer paper [12] by the same authors, as presented by Y. Frishman at the V. Internationalen Universitätswochen für Kernphysik, Schladming, 24 February - 9 March 1966.

A closer inspection of the arguments used by Goldstone, Salam and Weinberg [4] and by others [5] - [9], [11] shows that they involve the infinitesimal transformations rather than finite transformations depending on a continuous parameter. The condition needed is not the invariance of the Hamiltonian under finite transformations but rather that it commutes with the generator of infinitesimal transformations. The former does not imply the latter in a system with an infinite number of degrees of freedom. This is illustrated in our example in section 2, where the Hamiltonian is invariant under the finite transformation but fails to commute with the infinitesimal generator. Following these remarks we define in section 3 the infinitesimal generator and discuss its properties. For a more detailed discussion the reader is refered to a forthcoming paper [12].

In section 4 we redefine broken symmetry in terms of the infinitesimal transformations. We adopt the usual "symmetry breaking condition", namely the existence of some quantity A which is changed by δA under the infinitesimal transformation and for which $\langle 0|\delta A|0\rangle \neq 0$. This condition is stronger than the original requirement that finite transformations change the vacuum, and is equivalent to that adopted in axiomatic field theory, namely that the automorphism induced on the algebra of observables by the finite transformations cannot be represented by a unitary transformation in the separable Hilbert space of physical states [13] - [15]. The broken symmetry can, however, be represented as a unitary transformation over a non-separable Hilbert space, and so we regard it here. In section 4 we also give sufficient conditions under which the energy spectrum of our system features no "gap", namely that every neighbourhood of the origin in the momentum-energy space contains points of the spectrum other than the one corresponding to the ground state. The arguments of Goldstone, Salam and Weinberg [4] seem convincing enough to this effect. We generalize to generators which do not commute with the Hamiltonian but are constants of the motion which depend on time explicitly as polynomials, and also to cases where the generators do not commute with the total momentum, but the sequence of successive commutators terminate (For proofs the reader is again refered to ref.[12]) Our generalizations allow us to deal with generators of Galilei and Lorentz transformations as well.

It now remains to decide whether the fact that our system features no "gap" means that there are branches of excitation with energy tending to zero together with the momentum or whether there are only states other than the vacuum with zero momentum and energy. The former case corresponds to massless particles and the latter to "spurious states". It is possible that both massless particles and spurious states exist. Kastler, Robinson and Swieca [15] make enough postulates (see section 4) to make the vacuum unique and spurious states impossible. But in model field theories which are not exactly solvable it is usually unclear whether all the axioms are satisfied and the question remains open. An example where the symmetry is broken but only spurious states results can be found in ref. (12). Sufficient conditions for the existence of massless particles, regardless of the existence of spurious states, are presented in section 5. These conditions concern the expansion of the generator as an integral over a density, the commutator of the generator with the Hamiltonian and the form of the symmetry breaking conditions [29]. It turns out that in the general case the generator breaks into two parts, one of which can be expanded while the other projects onto discrete eigenstates of linear momentum. It turns out that if the whole symmetry breaking is due to that part which cannot be expanded, massless particles need not exist. The general results are presented in section 5. Proofs may be found in ref. [12]. As for the commutator of the generator with the Hamiltonian, the actual calculation in a given field theory is very tricky. This is so because in such a commutator bilinear products of local relativistic fields at the same space-time point are usually involved, and an evaluation using canonical commutation relations is ambiguous and leads to contradictions [16] - [18]. It seems to us that this point is the main source of uncertainty about whether various field theoretical models do or do not lead to the appearance of massless particles.

Let us remark that a generator which is the charge of a conserved current need not always be a constant of the motion, due to surface terms which it is not always permissible to discard [19], [20].

We prove in section 5 the existence of massless particles for one set of sufficient conditions. Proofs for other sets of sufficient conditions, as well as cases where massless particles do not appear, can be found in ref.[12].

In section 6 we present examples in which the symmetry breaking condition is realized [30]. Due to our formulation of this condition, we can include the result announced by Streater [21] as a generalization of the Goldstone theorem, to be a corollary of it. This result states that spontaneous mass splitting entails the spectrum has no "gap".

Another important corollary which we would like to mention here (for details see ref.[12]) states that no gap is possible in the type of theories of electrons and muons interacting symmetrically with the electromagnetic field, which consider differences in their two point functions or their three-point functions with the electromagnetic field, as spontaneous [22].

2. Broken Symmetry without massless Particles

By "broken symmetry" we mean a unitary transformation $U(\alpha)$ depending on a continuous parameter α which leaves the Hamiltonian H of our system invariant but changes the ground state. We take our system to possess unbroken translational symmetry, which means that the ground state, as well as the Hamlitonian is translationally invariant. In the following we treat relativistic and non-relativistic theories at the same time.

We would like to show now that a broken symmetry in this sense does not imply the existence of massless particles. To show this we construct an example in which a symmetry is broken but no massless particles exist. By "massless particles" we mean here modes of excitation with energy tending to zero together with momentum (For a more precise definition see section 5). Our example is essentially a model for a ferromagnet. We consider an infinite sequence of spin operators $\vec{s}_i$, $i = 1,2,\ldots \infty$, $\vec{s}_j$ being associated with lattice point j. Consider the states in which all but a finite number of spins point in a direction in the y - z plane which makes an angle θ with the z axis, the remaining ones (a finite number) pointing in the opposite direction. This set of states form the basis of a separable Hilbert space, which we denote $\mathcal{H}_\theta$. Ω_θ denotes the state with all spins pointing in the θ-direction. We denote by $\mathcal{H}$ the direct sum of all $\mathcal{H}_\theta$ [23].

$$\mathcal{H} = \oplus \mathcal{H}_\theta, -\pi < \theta \leq \pi \tag{2.1}$$

A state ψ in $\mathcal{H}$ has components ψ_θ in $\mathcal{H}_\theta$. All but a denumerable set of ψ_θ are the zero vectors in their respective $\mathcal{H}_\theta$. The squares of the norms $||\psi_\theta||^2$ of the others (a denumerable set) must add to a finite value $||\psi||^2$, the norm of ψ. $\mathcal{H}$ is a nonseparable Hilbert space.

As our symmetry operation we choose spin rotations in y - z plane (i.e. around the x axis). A rotation by an angle α replaces each basis vector of $\mathcal{H}_\theta$ by the corresponding basis vector of $\mathcal{H}_\theta'$, where $\theta' = \theta + \alpha (\mathrm{mod}\ 2\pi)$, the correspondence obtained by rotating each spin of the basis vector of $\mathcal{H}_\theta$ by an angle α. This transformation, which we denote by U(α), is unitary and satisfies

$$U(\alpha)\, U(\beta) = U(\alpha+\beta) \tag{2.2}$$

$$U^{-1}(\alpha) = U^{+}(\alpha) = U(-\alpha) \tag{2.3}$$

$$U(\alpha)\ \mathcal{H}_\theta = \mathcal{H}_{\theta'} \qquad \theta' = \theta+\alpha \pmod{2\pi} \tag{2.4}$$

$$U(\alpha)\ \Omega_\theta = \Omega_{\theta'} \qquad \theta' = \theta+\alpha \pmod{2\pi} \tag{2.5}$$

$$U^+(\alpha)\ S_{ir} U(\alpha) = \mathcal{U}_{rp}(\alpha)\ S_{ip} \tag{2.6}$$

where r in S_{ir} denotes a component index which takes the values 1,2,3 (for x,y,z) and the summation convention is implied $\mathcal{U}_{rp}$ is the 3 × 3 matrix describing the rotation of a space vector by an angle α about the x-axis.

Define P_o to be the projection operators on all states Ω_θ in $\mathcal{H}$. Thus $P_o\ \Omega_\theta = \Omega_\theta$ for all θ , while $P_o\psi = 0$ for ψ orthogonal to all Ω_θ . Because of eq. (2.5) P_o is rotationally invariant:

$$U^+(\alpha)\ P_o\ U(\alpha) = P_o \tag{2.7}$$

We are now in a position to define the Hamiltonian of our example, which we choose as

$$H = C\ (1-P_o) \tag{2.8}$$

C is a positive constant. H has two eigenvalues: 0 and C. All Ω_θ are eigenstates of zero energy, and any state orthogonal to all the Ω_θ has energy C. Since P_o is rotationally invariant, so is also H. In each separable Hilbert space $\mathcal{H}_\theta$ there is a translationally invariant ground state Ω_θ , all other states in $\mathcal{H}_\theta$ may be expanded in terms of (unnormalizable) excitations with definite wave number out of the first Brillouin zone of the reciprocal lattice. All these excitations correspond to energy C and there are no modes whose energy tends to zero with the wave number.

In the above example rotational symmetry is spontaneously broken, since the Hamiltonian (eq. (2.8)) is invariant under rotations while the ground state Ω_θ is not (eq. 2.5)). However, no massless particles arise. The Heisenberg ferromagnet [24] provides an example of a broken rotational symmetry with massless particles (the spin waves). Thus, broken symmetry neither implies nor excludes massless particles.

A connection between broken symmetries and massless particles will be found below, subject to further conditions, involving the generators of infinitesimal transformations (to be defined in the following). One of these conditions will turn out to be that the generator is a constant of the motion. This condition, distinct from the invariance of the Hamiltonian under finite transformations, is violated in our example. The generator of rotations about the x-axis turns out to be $S_x = \sum_i S_{ix}$, which fails to commute with H of eq. (2.8).

3. Generators of infinitesimal Transformations

Let us further consider the example of the previous section. The symmetry operation is spin rotation about the x-axis; a rotation by an angle α transforms a state $\psi_o \in \mathcal{H}_o$ into a corresponding $\psi_\alpha \in \mathcal{H}_\alpha$, $\psi_\alpha = U(\alpha)\psi_o$. Since the norm $\psi_\alpha - \psi_o$ for any $\alpha \neq 0$ is $\sqrt{2}$ (ψ_α is orthogonal to ψ_o for $\alpha \neq 0$), the dependence of $\psi_\alpha = U(\alpha)\psi_o$ on α turns out not to be continuous in the topology of Hilbert space. Hence expressions involving $d\,U(\alpha)/d\alpha$ turn out to be meaningless, and the usual way of defining the generator of infinitesimal transformations by differentiating the finite transformations can not be applied.

When the derivative $dU(\alpha)/d\alpha$ ceases to exist, there is still a way open for discussion of an infinitesimal transformation. This way is to describe the transformation in the Heisenberg picture rather than in the Schrödinger picture [23]. In this way we transform the operators according to

$$S_{ir} \to S_{ir}(\alpha) \equiv U^+(\alpha)\, S_{ir} U(\alpha) = \mathfrak{U}_{rp}(\alpha)\, S_{ip} \qquad (3.1)$$

and leave the state vectors unchanged. This accomplishes the same transformation of all matrix elements and all physical results.

We find that $S_{ir}(\alpha)$ is indeed differentiable with respect to α , and the same goes for any operator in Hilbert space which is constructed of the spin operators. Hence the left handside of eq. (3.1) is differentiable. Had $U(\alpha)$ been differentiable we would have defined the generator Q by

$$dU(\alpha)/d\alpha \equiv iQU(\alpha) \quad ; \tag{3.2}$$

in that case, differentiation of both sides of eq. (3.1) would yield

$$dS_{ir}(\alpha)/d\alpha = i[Q,S_{ir}(\alpha)] \quad . \tag{3.3}$$

In the general case $dU(\alpha)/d\alpha$ has no meaning. We now adopt eq. (3.3) as the definition of Q! This is consistent with the former definition, eq. (3.2) whenever it applies, but can render Q meaningful also where eq. (3.2) has no meaning.

In our example

$$\frac{dS_{ir}(\alpha)}{d\alpha} = \mathcal{U}'_{rp}(\alpha)\ S_{ip} = \mathcal{U}'_{rp}(0)\ S_{ip}(\alpha) \quad . \tag{3.4}$$

$\mathcal{U}'_{rp}(\alpha)$ is the derivative of $\mathcal{U}_{rp}(\alpha)$ with respect to α and we have used the property $\mathcal{U}_{rp}(\alpha+\beta) = \mathcal{U}_{rs}(\alpha)\ \mathcal{U}_{sp}(\beta)$. Now, the Q which satisfies eq. (3.3) is obviously

$$S_x = \sum_i S_{ix} \quad . \tag{3.5}$$

This quantity is not an operator in $\mathcal{H}$. When acting on Ω_o it yields

$$S_x\ \Omega_o = \sum_i \tfrac{1}{2}(S_{ix} - S_{iy})\Omega_o \quad . \tag{3.6}$$

Each term in the last sum is a vector in $\mathcal{H}_o$ of length 1/2 and they are all orthogonal. Thus $S_x \Omega_o$ is not normalizable and therefore outside $\mathcal{H}$. We get a non normalizable vector whenever S_x is applied to any state in $\mathcal{H}$. On the other hand, the quantity which results from the application of S_x on any vector in $\mathcal{H}$ is well defined in terms of its components relative to the basis vectors in $\mathcal{H}$. It belongs to a linear space $\mathcal{F}$ of vectors with finite components (which we do not attempt to render topological). The commutator in equ. (3.3) is the difference of two terms each of which, when acting in $\mathcal{H}$ yields a result in $\mathcal{F}$; their difference is again an operator in $\mathcal{H}$.

Up to now we have considered a specific model, namely the spin model of the previous section. However, all these arguments can be generalized to the more general case. In general we consider a unitary transformation $U(\alpha)$ depending on the continuous parameter α and satisfying

$$U(0) = 1 \tag{3.7}$$

$$U(\alpha)\, U(\beta) = U(\alpha + \beta) \tag{3.8}$$

Vectors in Hilbert space transform according to

$$\psi(\alpha) = U(\alpha)\psi \tag{3.9}$$

again, we may identify the generator of the infinitesimal transformations through the Heisenberg picture, where operators are transformed according to

$$A \to A(\alpha) \equiv U^{+}(\alpha)\, A\, U(\alpha) \tag{3.10}$$

and the states are unchanged. The equation

$$\frac{d}{d\alpha} A(\alpha) = i\,[Q, A(\alpha)] \quad , \tag{3.11}$$

where $A(\alpha)$ is differentiable by assumption, is used as the definition of Q. For a further treatment of this general case, as well as discussion of the freedom left in determining Q and its properties under the finite transformations $U(\alpha)$ the reader is refered to a forthcoming paper [12].

4. The Symmetry breaking Condition and the Spectrum

We would like now to refine the definition of a broken symmetry and reexpress it in terms of the infinitesimal generator.

Consider the quantity

$$F_A(\alpha) = F_{A(\alpha)}(0) \equiv \langle 0|U^+(\alpha)\ A\ U(\alpha)|0\rangle \tag{4.1}$$

where $|0\rangle$ is a translationally invariant ground state and A is a Hermitian operator with the property that $A(\alpha) = U^+(\alpha) \times AU(\alpha)$. If the symmetry is not broken, that is if $U(\alpha)|0\rangle = |0\rangle$ then $F_A(\alpha)$ is independent of α for any A. Thus the condition

$$\frac{d}{d\alpha} F_A(\alpha) \neq 0 \tag{4.2}$$

for some A and α is a sufficient condition for symmetry breaking. We refer to the last equation as "the symmetry breaking condition". It is this condition rather than the original definition which is involved in the discussion of the existence of massless particles.

Assume now that $dF_A(\alpha)/d\alpha = 0$ for all A and α but the symmetry is broken

$$U(\alpha)|0\rangle = |\alpha\rangle \neq |0\rangle \quad . \tag{4.3}$$

Then it can be shown, that one may construct a new unitary

operator $W(\alpha)$ with the following properties: $W(\alpha)$ induces on the operators A the same transformation as $U(\alpha)$

$$W^{+}(\alpha)\; A\; W(\alpha) = A(\alpha) \quad , \tag{4.4}$$

but it leaves the vacuum invariant:

$$W(\alpha)|0\rangle = |0\rangle \quad . \tag{4.5}$$

$W(\alpha)$ is thus a unitary operator in $\mathcal{H}_o$ and also in each $\mathcal{H}_\alpha$! The transformation $A \to A(\alpha)$ is thus achieved by a unitary operator which does not change the vacuum, namely by a transformation which does not break the symmetry. The cases of a real symmetry breaking are those in which the symmetry breaking condition eq. (4.2) is realized. For an explicit construction of the operator $W(\alpha)$ and further discussion about the mapping $A \to A(\alpha)$ the reader is refered to ref. [12].

The symmetry breaking condition eq. (4.2) can be rewritten in terms of the generator Q as

$$\frac{d}{d\alpha} F_A(\alpha) = i\langle 0|[Q,A]|0\rangle \neq 0 \tag{4.6}$$

Let us add the following requirements on Q:

1) Q is a constant of the motion.
2) Q depends explicitly on time and most as a polynomial.
3) The sequence of Q^ν, defined by

$$\begin{aligned} Q^o &\equiv Q \\ Q^1_{r_1} &\equiv i[P_{r_1}\,,\,Q^o] \\ Q^2_{r_1 r_2} &\equiv i[P_{r_2}\,,\,Q^1_{r_1}] \\ &\ldots\ldots\ldots\ldots \\ Q^{\nu+1}_{r_1\ldots r_{\nu+1}} &\equiv i[P_{r_{\nu+1}}\,,\,Q^\nu_{r_1\ldots r_\nu}] \quad , \end{aligned} \tag{4.7}$$

terminates at a finite $\nu = n$, namely $Q^\nu \equiv 0$ for $\nu > n$.

Then it is possible to show (for details see [12]), that

under these conditions, together with the symmetry breaking condition, the energy spectrum of our system features no "gap", namely that every neighbourhood of the origin in the momentum energy space contains points of the spectrum other than the one corresponding to the ground state $|0\rangle$.

It now remains to be decided whether these points are limit points of branches of excitation for which the energy tends to zero with momentum, and hence correspond to massless particles, or whether they are states other than the ground state corresponding to the origin in the momentum energy space, in which case they are refer to as "spurious states". It is also possible that both massless particles and spurious states exist.

If one postulates, that $\mathcal{H}_0$ is the representation space of a faithful irreducible representation of an algebra of local operators, spurious states are guaranteed not to exist, and our conditions on the spectrum lead to the appearance of massless particles. This was postulated by Kastler, Robinson and Swieca [15]. However, when dealing with specific dynamical models, it is not obvious whether one ends with an irreducible representation or a reducible one. It may thus be useful to obtain sufficient conditions for the appearance of massless particles irrespectively of the existence of spurious states. We formulate such conditions in the following section. Let us remark, that conditions 2 and 3, which allow the generator to depend explicitly on time and not commute with $\vec{P}$ make it possible for our considerations to apply to the generators of Galilei and Lorentz transformations. This will be used in the following.

5. Sufficient Conditions for the Appearance of massless Particles

Let us start with the following simple set of sufficient conditions [29]:

1) The generator Q may be represented as an integral over

a density

$$Q = \int q(\vec{x})\, d^3\vec{x} \tag{5.1}$$

with $q(\vec{x})$ satisfying

$$i[\vec{P}, q(\vec{x})] = \frac{\partial}{\partial \vec{x}} q(\vec{x}) \quad ; \tag{5.2}$$

no explicit dependence of $q(\vec{x})$ on $\vec{x}$ is allowed.

2) The generator Q is a constant of the motion which does not depend explicitly on time. This means $[H, Q] = 0$.

3) The symmetry breaking condition eq. (4.6) is satisfied for $A = A(\vec{x}, t)$ which satisfies

$$i[\vec{P}, A(\vec{x}, t)] = \frac{\partial}{\partial \vec{x}} A(\vec{x}, t) \quad , \tag{5.3}$$

$$-i[H, A(\vec{x}, t)] = \frac{\partial}{\partial t} A(\vec{x}, t) \ . \tag{5.4}$$

Under conditions 1) and 2) we have $[H, Q] = [\vec{P}, Q] = 0$ and the symmetry breaking condition becomes

$$\langle 0|[Q, A(\vec{x}, t)]|0\rangle = C \neq 0 \quad . \tag{5.5}$$

The proof for the appearance of massless particles under these conditions is as follows: Let $dE(\vec{k},\omega)$ be a spectral measure in terms of which the total momentum P and the Hamiltonian H can be expanded simultaneously,

$$H = \int \omega \, dE(\vec{k}, \omega) \tag{5.6}$$

$$\vec{P} = \int \vec{k} \, dE(\vec{k}, \omega) \tag{5.7}$$

$dE(\vec{k},\omega)$ associates with each measurable set of points in the

$\vec{k},\omega$ space a projection operator in the Hilbert space of states of the system, the projection operator being the integral of $dE(\vec{k},\omega)$ over the set of points in question, $dE(\vec{k},\omega)$ possesses the property

$$\int dE(\vec{k}, \omega) = 1 \tag{5.8}$$

Consider the expression

$$<0|[q(\vec{x}), A(\vec{y}, t)]|0> = \int \rho(\vec{k}, \omega) e^{i\{\vec{k}(\vec{x}-\vec{y})+\omega t\}} d^3 \vec{k} d\omega . \tag{5.9}$$

An expression for $\rho(\vec{k},\omega)$ may be obtained by the insertion, in the two terms of the commutator, of 1 in the form of $\int dE(\vec{k},\omega)$ This gives

$$\rho(\vec{k}, \omega) d^3 \vec{k}\, d\omega =$$

$$= <0|q(\vec{0})\, dE(\vec{k},\omega)\, A(\vec{0},0)|0> - <0|A(\vec{0},0)\, dE(-\vec{k},-\omega)\, q(\vec{0})|0>. \tag{5.10}$$

This equation shows that the support of $\rho(\vec{k},\omega)$ is contained in the support of $dE(\vec{k},\omega)$ or of $dE(-\vec{k}, -\omega)$.

If we now integrate both sides of equation (5.9) over $d^3\vec{x}$ and use eq. (5.5) we get

$$<0||Q, A(\vec{y},t)||0> = C =$$

$$= (2\pi)^3 \int \rho(\vec{k},\omega)\, \delta^3(\vec{k})\, e^{-i(\vec{k}\vec{y}-\omega t)} d^3\vec{k} d\omega \tag{5.11}$$

and thus

$$(2\pi)^3 \rho(\vec{k},\omega)\, \delta^3(\vec{k}) = C\, \delta(\omega)\, \delta^3(\vec{k}) . \tag{5.12}$$

The support of the right hand side is the origin $\vec{k} = \vec{0}$, $\omega = 0$, that of the left hand side is the intersection of the support of $\rho(\vec{k},\omega)$ with the line $\vec{k} = \vec{0}$. Comparing both sides of eq. (5.12) shows that $\vec{k} = 0$, $\omega = 0$ is contained in the support of $\rho(\vec{k},\omega)$, but it cannot appear as an isolated point, since if it did, $\rho(\vec{k},\omega)$ would be as singular as a 4-dimensional δ-function and contradict the equality in eq. (5.12). In fact, an inspection at eq. (5.12) shows that the support of $\rho(\vec{k},\omega)$ cannot be more concentrated than a three-dimensional surface. Thus, there exists a three-dimensional surface which passes through $\vec{k} = 0$, $\omega = 0$ and is contained in the support of $\rho(\vec{k},\omega)$. This surface either belongs to the spectrum or to the reflection of the spectrum through $\vec{k} = 0$, $\omega = 0$, or is made up of two parts one of which belongs to the spectrum and the other to its reflection. In each of the above three cases, the spectrum contains a three dimensional surface which contains the origin. We take the existence of such a surface as our definition of the existence of massless particles. Hence their existence is established by our proof.

Let us now discuss the scope of our assumptions. We start from the first, namely the possibility of expanding the generator as an integral over a density satisfying eq. (5.2). This assumption certainly excludes cases when Q does not commute with $\vec{P}$ (with the series of successive commutators terminating after a finite number. See section 4. We would like to include such cases in our treatment, since, for example, Galilei and Lorentz generators are of that kind. Now it can be shown that under the assumptions made in section 4, it is always possible to split the generator Q into two parts

$$Q = Q' + Q_o \tag{5.13}$$

Q_o can not be expanded as an integral over a density, while Q' can be such expanded. Q_o is of the form $Q_o = E_o\ Q\ E_o$, where E_o is the projection operator onto normalizable states of definite momentum. Q can be expanded as

$$Q' = \int q(\vec{x},\vec{x})\ d^3\vec{x} \tag{5.14}$$

with $q(\vec{x},\vec{y})$ being a polynomial in $\vec{y}$ and satisfying

$$i[\vec{P},\ q(\vec{x},\vec{y})] = \frac{\partial}{\partial\vec{x}}\ q(\vec{x},\vec{y})\ . \tag{5.15}$$

Q' has no matrix elements between normalizable states of definite momentum and energy. For a detailed discussion and proofs the reader is again refered to ref. [12].

As for the second assumption, we are going to allow an explicit polynomial dependance of the generator Q on time. As for the third we shall allow A to depend also explicitly on space and time, but again only as a polynomial.

The three following weaker conditions are also sufficient for the existence of massless particles:

1') The generator Q may be represented as

$$Q = \int q(\vec{x},\vec{x},t,t)d^3\ \vec{x} \tag{5.16}$$

with $q(\vec{x},\vec{y},t,s)$ being an explicit polynomial in $\vec{y}$ and s, a Heisenberg operator in t - with

$$-i[H,q(\vec{x},\vec{y},t,s)] = \frac{\partial}{\partial t}\ q(\vec{x},\vec{y},t,s)$$

and satisfying

$$i[\vec{P},\ q(\vec{x},\vec{y},t,s)] = \frac{\partial}{\partial\vec{x}}\ q(\vec{x},\vec{y},t,s) \tag{5.17}$$

2') The generator Q is a constant of the motion.

3') We have

$$\langle 0|[Q,\ A(\vec{x},\vec{x},t,t)]|0\rangle = P(\vec{x},t) \tag{5.18}$$

where $P(\vec{x},t)$ is a polynomial in $\vec{x}$ and t and $A(\vec{x},\vec{y},t,s)$ depends on $\vec{x},\vec{y},t,s$ the same way $q(\vec{x},\vec{y},t,s)$ does.

The proof of the existence of massless particles under these conditions is a slight generalization of the former proof, and can be found in ref. [12]. Other sets of sufficient conditions, especially involving Q_o can also be found in ref. [12], as well as a discussion of the conditions under which massless particles need not appear.

6. Corollaries

In this section we present examples in which the symmetry breaking condition is realized. This may serve as an illustration for the use that can be made of the freedom to choose A or $A(\vec{x},t)$ in the symmetry breaking condition and the relation which this quantity bears to the fundamental fields in a field theory. For further discussion see ref. [12],[30].

a) Breaking of Galilean invariance:

The fundamental fields in this case are creation and annihilation operators of Fermions or Bosons $\psi^+(\vec{x},t)$ and $\psi(\vec{x},t)$. The mass density and mass current may be expressed as

$$d(\vec{x},t) = m\,\psi^+(\vec{x},t)\,\psi(\vec{x},t) \quad , \tag{6.1}$$

$$\vec{j}(\vec{x},t) = \frac{\hbar}{2i}\{\psi^+(\vec{x},t)\,\frac{\partial\psi(\vec{x},t)}{\partial\vec{x}} - \frac{\partial\psi^+(\vec{x},t)}{\partial\vec{x}}\,\psi(\vec{x},t)\} \quad . \tag{6.2}$$

m is the mass of a particle. The generator Q of the Galilei transformation may be expanded as in eq. (5.16) with

$$\vec{q}(\vec{x},\vec{y},t,s) = s\vec{j}(\vec{x},t) - \vec{y}\,d(\vec{x},t) \tag{6.3}$$

One has

$$i[Q_r, j_{r'}(\vec{x},t)] = t \frac{\partial}{\partial x_r} j_{r'}(\vec{x},t) + \delta_{r,r'} d(\vec{x},t) \tag{6.4}$$

and hence

$$i\langle 0|[Q_r, j_{r'}(\vec{x},t)]|0\rangle = \delta_{rr'}\langle 0|d(\vec{x},t)|0\rangle = \delta_{rr'}d' \neq 0 \tag{6.5}$$

for a translationally invariant ground state $|0\rangle$ which corresponds to a finite mass density d'. If the system is Galilean invariant that is if $\vec{Q}$ is a constant of the motion, conditions 1'), 2'), 3') are satisfied. The spectrum contains modes of zero mass excitation (e.g. sound, zero sound, etc.)

b) Mass splitting:

Consider a symmetry transformation between two relativistic fields $\psi_1(x)$ and $\psi_2(x)$.

$$i[Q, \psi_1(x)] = \psi_2(x) \tag{6.6}$$

$$i[Q, \psi_2(x)] = -\psi_1(x) \quad , \tag{6.7}$$

where x stands for $\vec{x}$,t. One has

$$i\langle 0|[Q, \psi_1(x)\ \psi_2(x+y)]|0\rangle =$$

$$= \langle 0|\psi_2(0)\ \psi_2(y)|0\rangle - \langle 0|\psi_1(0)\ \psi_1(y)|0\rangle \quad . \tag{6.8}$$

We may consider $\psi_1(x)\ \psi_2(x+y)$ as the $A(\vec{x},t)$ of condition 3), with y being a parameter. Thus if the two point functions of $\psi_1(x)$ and $\psi_2(x)$ are not identical, the symmetry breaking condition 3) is satisfied for some fixed value of y. If Q satisfies also conditions 1) and 2) or 1') and 2'), massless particles are implied. In particular this example applies to the case the two fields corresponding to different masses. This is important in relation with theories which consider the electron-muon mass splitting [25] - [27] or the splitting between the various members of an SU(3) multiplet [25]as spontaneous.

It is usually very difficult to check whether conditions 1), 2) or 1'), 2') are really satisfied. This is because products of relativistic fields at the same space-time point, which are in most cases involved in the density of Q are poorly defined and lead to contradictions [16] - [18].

Streater [21] and Fuchs [22] consider the results of this example as a generalization of the Goldstone theorem. We see that they are rather corollaries of it. This was also pointed out by Guralnik [28].

Acknowledgement

One of us (Y. F.) would like to acknowledge the John F. Kennedy Memorial Foundation of the Weizmann Institute, Rehovoth, Israel, for a Kenedy Memorial Prize which enabled him to participate in the Elementary Particle Seminar at Schladming, Austria.

References and Footnotes

1. Y. Nambu and G. Jona-Lasinio, Phys. Rev. 122, 345 (1961)
2. J. Bardeen, L. N. Cooper and S. R. Schrieffer, Phys. Rev. 108, 1175 (1957)
3. J. Goldstone, Nuovo Cim. 19, 154 (1961)
4. J. Goldstone, A. Salam and S. Weinberg, Phys. Rev. 127, 965 (1962)
5. S. A. Bludman and A. Klein, Phys. Rev. 131, 2364 (1963)
6. G. Jona-Lasinio, Nuovo Cim. 34, 1790 (1964)
7. A. Klein and B. W. Lee, Phys. Rev. Lett. 12, 266 (1964)

8. P. W. Higgs, Phys. Lett. 12, 132 (1964)
9. G. S. Guralnik, Phys. Rev. Lett. 13, 295 (1964)
10. J. Lopuszanski and H. Reeh, Phys. Rev. 140, B926 (1965)
11. W. Gilbert, Phys. Rev. Lett., 12, 713 (1964)
12. Y. Frishman and A. Katz, to be published
13. R. Haag, Nuovo Cim. 25, 287 (1962)
14. R. F. Streater, Proc. Roy. Soc. 287, 510 (1965)
15. D. Kastler, D. W. Robinson and A. Swieca, preprint (Univ. of Illinois)
16. J. Schwinger, Phys. Rev. Lett. 3, 296 (1959)
17. K. Johnson, Nuclear Physics, 25, 431 (1961)
18. K. Johnson, Phys. Lett. 5, 253 (1963)
19. G. S. Guralnik, C. R. Hagen and T. W. B. Kibble, Phys. Rev. Lett. 13, 585 (1964)
20. R. V. Lange, Phys. Rev. Lett. 14, 3 (1965)
21. R. F. Streater, Phys. Rev. Lett. 15, 475 (1965)
22. N. Fuchs (Phys. Rev. Lett. 15, 911 (1965)) gets, as a result of assuming a difference in the two point functions of two fields as spontaneous, that there are "states $|0'\rangle$ with momentum and energy zero which however are not Lorentz invariant. His treatment of these $|0'\rangle$ appears incorrect because of the points raised in ref.[4] sec. III
23. A. Katz: "Is the Heisenberg picture better than the Schrödinger picture?"preprint.
24. The Hamiltonian of the Heisenberg ferromagnet is a spin-spin interaction

$$H = - \sum_{ij} J_{ij} \vec{S}_i \vec{S}_j$$

S_i being the spin operator associated with the i-th point of the lattice, and J_{ij} are positive, usually assumed to be a constant $J > 0$ for nearest neighbours and zero otherwise. (See W. Heisenberg, Z. Physik, 49, 619 (1928)). The ground state of this Hamiltonian is obviously a state with all spins pointing in the same direction and as such not rotationally invariant. The fact that the spectrum contains massless excitations is demonstrated in almost every book on solid state physics.

25. M. Baker and S. L. Glashow, Phys. Rev. 128, 2462 (1962)
26. Th. A. J. Maris, V. E. Herscovitz and G. Jacob, Nuovo Cim. 34, 946 (1964); a second paper by the same authors is to be published in Nuovo Cim.
27. R. Arnowitt and S. Deser, Phys. Lett. 13, 256 (1964); Phys. Rev. 138, B712 (1965)
28. G. S. Guralnik, Preprint (University of Rochester).
29. A. Katz and Y. Frishman: "Massless Particles rather than spurious States following broken Symmetry", to be published in Nuovo Cimento.
30. Y. Frishman and A. Katz, Phys. Rev. Lett. 16, 370 (1966)

CURRENT ALGEBRAS AND CALCULATIONS IN BROKEN SYMMETRIES

By

J. W. MOFFAT

Department of Physics
University of Toronto, Canada

Introduction

In our efforts to formulate a satisfactory theory of particle physics, we can search for a fundamental system of equations the solutions of which will tell us all about the complex phenomena of nuclear forces. Failing this, we can seek to understand the algebraic structure of nature, and try to abstract as many relations as possible from our present knowledge, leaving the discovery of the "key" dynamical equations till later. Time will tell whether this latter type of approach succeeds only in "skating on the surface" of the problem. The business of strong interaction physics should consist of calculating cross sections, phase shifts etc. Only through our ability to do such calculations can we hope to understand the mysteries of high-energy phenomena. However, it may be that at present the problem of strong interactions is comparable to the days of atomic spectroscopy when exact sum rules played an important role in unraveling the complexities of the energy levels of the atom.

In the following, we shall study the connection of higher symmetries to the Lie algebra of the space integrals of the time components of currents. Gell-Mann has advocated the use of these exact algebraic properties as a way of clarifying the origins of the problems in particle physics. The hadron

Lecture given at the V. Internationalen Universitätswochen für Kernphysik, Schladming, 24 February - 9 March 1966.

current operator relationships are abstracted from a field theory model based on the unitary symmetric Yukawa interactions. The general methods of S-matrix theory, or field theory, can then be applied, according to one's tastes, and the structure of particle physics can be studied without too much emphasis on specific Lagrangians or scattering theory.

In Sections I and II, we begin by considering simple field theory models which give rise to algebraic structures. We develop the symmetry properties of baryons and mesons on the basis of the unitary symmetric Lagrangian describing Yukawa interactions. The group generators of SU(3) and properties of the hadron current conservation are studied in Section III. In Section IV, the role of current algebras and convergent dispersion relations is studied. The Goldberger-Treiman relation and the Adler-Weisberger relation are derived and considered in detail. The problem of vector meson interactions with the hadron current and the derivation of mass relations from current algebras and field theory, are discussed in Section V. In Section VI, a field theory model is used to calculate symmetry breaking corrections to mass sum rules in SU(3). Finally, in Section VII, the problem of weak interactions and its relation to the current algebra corresponding to the group SU(3) $\times$ SU(3) is studied within the context of the universality of weak decay processes.

I. Field Theory Models and Currents

All significant attempts so far to describe elementary particles in terms of a mathematical theory have been based on three main approaches:

1. Field theory,
2. Dispersion relations,
3. Exact and approximate symmetries.

Non of these approaches has been entirely successful up till the present time, since we do not possess a satisfactory dynamical theory of strong interactions. However, all three methods have something to teach us, and they can be combined so that a pleasing feature of one can be applied to the others.

Although the details of field theory can be criticized and it has doubtful validity in applications to strong interaction phenomena, we shall begin our description of the symmetry group of baryons and mesons in the framework of field theory, where the Lagrangian density of the strong interactions is expressed in terms of certain local fields which describe the "elementary" baryons and mesons.

The simplest models in which the field theory scheme leads to an algebraic structure are the Fermi-Yang model [1] and the Sakata [2] or quark [3] model. The Fermi-Yang model is based on the Lagrangian

$$\mathcal{L} = - \overline{N}\gamma_\alpha \partial_\alpha N - (\partial_\alpha V_\beta - \partial_\beta V_\alpha)^2/4 - \mu_o^2 V_\alpha V_\alpha/2 - \\ - ig_o V_\alpha \overline{N}\gamma_\alpha N - M_o \overline{N}N \quad , \qquad (1.1)$$

where N means (p,n) and V_α denotes a massive neutral vector meson coupled to the nucleon current [4]. The total isotopic spin current is just given by

$$\mathcal{J}_{i\alpha} = i \overline{N}\gamma_\alpha (\tau_i/2) N \qquad (i = 1,2,3) \quad , \qquad (1.2)$$

and (1.2) is conserved when we ignore the nucleon mass in (1.1). The components I_i of the isotopic spin are

$$I_i = - i \int \mathcal{J}_{i4} \, d^3x \qquad (1.3)$$

which obey the familiar commutation relations

$$[I_i, I_j] = i \, \varepsilon_{ijk} I_k \qquad (1,4)$$

where ε_{ijk} are the structure constants of the Lie algebra corresponding to the special unitary group SU(2).

The Sakata model is based on the assumption that all strongly interacting particles are made out of N, Λ, $\overline{N}$ and $\overline{\Lambda}$. [2] The quark model [3] assumes the existence of a triplet of fermion fields corresponding to three spin 1/2 quarks i.e. a doublet u and d with charges 2/3 and -1/3 respectively, and a singlet s with charge -1/3. The Lagrangian for this model is

$$\mathcal{L} = - \overline{\psi}\gamma_\alpha \partial_\alpha \psi - \mathcal{L}_V - ih\, V_\alpha \overline{\psi}\gamma_\alpha \psi \tag{1.5}$$

where $\mathcal{L}_V$ is the free Lagrangian for a neutral vector meson field V_α. To this Lagrangian, we may add a quark mass term $M_{oQ}\overline{\psi}\psi$. In this model the vector current density of the quarks is given by

$$\mathcal{J}_{i\alpha} = i\overline{\psi}\, (\lambda_i/2)\, \gamma_\alpha \psi \qquad (i = 1...8) \tag{1.6}$$

Here the λ_i satisfy the commutation relation

$$[\lambda_i, \lambda_j] = 2if_{ijk}\lambda_k \tag{1.7}$$

In this case, the f_{ijk} are the structure constants of the Lie algebra corresponding to the special unitary group SU(3) [5] - [7] .

After we have extracted as many interesting relations as we can from the field theory model these relations can be transferred to the methods of dispersion relations and S-matrix theory (unless you happen to be a die-hard field theorist!). In the following, we shall base our development of unitary symmetries on yet another model, which is a familiar one in particle physics, namely, the Lagrangian describing the simple Yukawa interactions between baryons and pseudoscalar mesons.

II. Unitary Symmetry and Yukawa Interactions

The algebraic system of the eight operators $\hat{F}_i$ (i=1...8) of SU(3) defines a linear eight-dimensional vector space A [5][7]. The octets of baryons, pseudoscalar mesons and vector mesons are described in this space by B_i, Π_i and $V_{i\alpha}$, respectively. The baryons form the components of the vector B_i:

$$B_1 = \frac{1}{\sqrt{2}}(\Sigma^+ + \Sigma^-),\ B_2 = \frac{i}{\sqrt{2}}(\Sigma^+ - \Sigma^-),\ B_3 = \Sigma^0\ ,$$

$$B_4 = \frac{1}{\sqrt{2}}(p + \Xi^-),\ B_5 = \frac{i}{\sqrt{2}}(p - \Xi^-),\ B_6 = \frac{1}{\sqrt{2}}(n + \Xi^0)\ ,$$

$$B_7 = \frac{i}{\sqrt{2}}(n - \Xi^0),\ B_8 = \Lambda \qquad (2.1)$$

The pseudoscalar mesons form the components of Π_i:

$$\Pi_1 = \frac{1}{\sqrt{2}}(\pi^+ + \pi^-),\ \Pi_2 = \frac{i}{\sqrt{2}}(\pi^+ - \pi^-),\ \Pi_3 = \pi^0\ ,$$

$$\Pi_4 = \frac{1}{\sqrt{2}}(K^+ + K^-),\ \Pi_5 = \frac{i}{\sqrt{2}}(K^+ - K^-)\ ,$$

$$\Pi_6 = \frac{1}{\sqrt{2}}(K^0 + \overline{K}^0),\ \Pi_7 = \frac{i}{\sqrt{2}}(K^0 - \overline{K}^0),\ \Pi_8 = \eta\ . \qquad (2.2)$$

We shall specify a rotation in the eight-dimensional vector space A by the eight components of a rotation vector ρ_i. The transformation of the eight component baryon vector B_i is given by

$$B \to R_\rho B = \exp(i\rho_i F_i)\ B \qquad (2.3)$$

where the F_i denote 8 × 8 matrices. The rotation around the "2" axis by an angle π is the isotopic spin charge-symmetry operation, while rotations around the "3" axis are intimately related to phase (gauge)transformations of the first kind (constant phase). The latter is also true of rotations around the "8" axis.

The field theory model that we are perhaps most familiar with is the simplest Yukawa interaction describing the coupling between meson and baryon fields

$$\mathcal{L} = - \bar{B} (\gamma_\alpha \partial_\alpha + M_{oB}) B - \frac{1}{2} (\partial_\alpha \Pi_i \partial_\alpha \Pi_i + M_{ob}^2 \Pi_i \Pi_i) +$$

$$+ i G_o \bar{B} \gamma_5 [\alpha D_i + (1 - \alpha) F_i] B \Pi_i \qquad (2.4)$$

Here we have accounted for the double occurrence of the representation $\underline{8}$ in the direct product

$$\underline{8} \times \underline{8} = \underline{1} + \underline{8} + \underline{8} + \underline{10} + \overline{\underline{10}} + \underline{27} \qquad (2.5)$$

The antisymmetric coupling is formed with the aid of the 8 × 8 matrices $F_i^{jk} = - if_{ijk}$. The F matrices are imaginary and antisymmetric with respect to the chosen basis [5]. At any time t, we have

$$[\hat{F}_i, \hat{F}_j] = if_{ijk} \hat{F}_k \qquad (2.6)$$

The symmetrical coupling is formed with the help of the 8 × 8 matrices D_i^{jk} defined by $D_i^{jk} = d_{ijk}$. The D matrices are real and symmetric and the operators $\hat{F}$ and $\hat{D}$ satisfy the relation

$$[\hat{F}_i, \hat{D}_j] = if_{ijk} \hat{D}_k \qquad (2.7)$$

The Lagrangian (2.4) is a scalar in the eight-dimensional space of rotations of SU(3). Let us study the term $\bar{B}\gamma_5 F_i B \Pi_i$.

Under the rotation R_ρ, we get

$$\overline{B}\, e^{-i\rho_j F_j}\, F_i\, e^{i\rho_j F_j}\, B \tag{2.8}$$

For infinitesimal ρ_i, we find

$$B \to e^{i\rho_i F_i}\, B \sim (1 + i\rho_i F_i)\, B \tag{2.9}$$

and, in view of (2.6) and (2.9), (2.8) becomes

$$\overline{B}\, (F_i - \rho_j f_{ijk} F_k)\, B \tag{2.10}$$

Therefore, the Yukawa coupling is invariant if the octet of meson fields Π_i transforms like

$$\Pi_i \to \Pi_i - f_{ijk}\rho_j\, \Pi_k \tag{2.11}$$

under infinitesimal rotations.

For arbitrary ρ there exists a unitary operator $U(\rho)$ which induces the rotation

$$U(\rho) = \exp\, (i\rho_i \hat{F}_i) \tag{2.12}$$

Thus for infinitesimal ρ it follows from (2.9) and (2.12) that

$$B(x) \to U^{-1}(\rho)\, B(x)\, U\, (\rho) = B(x) + i\rho_i F_i\, B(x) \tag{2.13}$$

$$\Pi_i(x) \to U^{-1}(\rho)\, \Pi_i(x)\, U(\rho) = \Pi_i(x) - f_{ijk}\rho_j \Pi_k(x) \quad . \tag{2.14}$$

Eqs. (2.13) and (2.14) lead to the relations

$$[B(x), \hat{F}_i] = F_i \; B(x) \tag{2.15}$$

$$[\Pi_i(x), \hat{F}_j] = if_{ijk} \; \Pi_k \tag{2.16}$$

The operator $\hat{F}$ which satisfies (2.15) and (2.16) is given by

$$\hat{F}_i = i \int_\Sigma \bar{B} \, \gamma_\mu F_i B d\Sigma_\mu - f_{ijk} \int_\Sigma \Pi_j \partial_\mu \Pi_k \; d\Sigma_\mu \tag{2.17}$$

where Σ is an arbitrary space-like hypersurface and $d\Sigma_\mu$ is determined by

$$d\Sigma_\mu = \frac{d^4x}{dx_\mu} \quad , \; d\Sigma_4 = - id^3x \tag{2.18}$$

Let us now choose Σ to be a constant t plane. This gives

$$\hat{F}_i = \int (B^+ F_i B + f_{ijk} \; \Pi_j \; \partial_t \; \Pi_k) d^3x \tag{2.19}$$

The total unitary F-spin for the baryon-meson system is defined by the relation

$$\hat{F}_i = - i \int \mathcal{F}_{i4} \; d^3x \tag{2.20}$$

where the total F-spin current density $\mathcal{F}_{i\mu}$ of the baryons and mesons is of the form

$$\mathcal{F}_{i\mu} = i\bar{B} \; \gamma_\mu F_i B - f_{ijk} \Pi_j \; \partial_\mu \; \Pi_k \tag{2.21}$$

The components of B_i and Π_i are unrenormalized Heisenberg field operators, and B_i and Π_i satisfy the equal time canonical commutation rules

$$\{B_i(\underline{x}, t), B_j^+(\underline{x}', t)\} = \delta_{ij}\, \delta(\underline{x} - \underline{x}') \tag{2.22}$$

$$[\Pi_i(\underline{x}, t), \dot{\Pi}_j(\underline{x}', t)] = i\delta_{ij}\, \delta(\underline{x} - \underline{x}') \quad , \tag{2.23}$$

while all other anticommutators and commutators vanish.

With the help of these commutation rules and (2.19) it is easily proved that (2.15) and (2.16) are indeed satisfied. A calculation also verifies that the definition of the unitary spin given by (2.19) satisfies the commutation relation (2.6).

The commutator of $\mathcal{F}_{i4}(\underline{x}, t)$ and $\mathcal{F}_{j4}(\underline{x}', t)$ must vanish for $\underline{x} \neq \underline{x}'$, as required by micro-causality. Gell-Mann[6] postulates the relation

$$[\mathcal{F}_{i4}(\underline{x}, t), \mathcal{F}_{j4}(\underline{x}', t)] = - f_{ijk}\, \mathcal{F}_{k4}(\underline{x}', t)\, \delta(\underline{x} - \underline{x}'). \tag{2.24}$$

This relation can be verified in our field theory model by using (2.21) and the canonical commutation relations (2.22) and (2.23). One should be careful that explicit commutation does not give misleading results. Schwinger [8] has considered the commutator $[j_i(\underline{x}, t), j_4(\underline{x}', t)]$ which, on the face of it, may be expected to vanish for $i = 1,2,3$. But the vacuum expectation value of this commutator is a non-zero quantity times $\partial_i \delta(\underline{x} - \underline{x}')$. However, the integrated commutator does vanish, since the gradient of the delta function does not contribute because of symmetry.

III. Current Conservation and Group Generators

Let us suppose that the Lagrangian $\mathcal{L}$ is defined by

$$\mathcal{L} = \mathcal{L}_s + \mathcal{L}_b \tag{3.1}$$

where $\mathcal{L}_s$ is invariant under the infinitesimal transformations of SU(3), while $\mathcal{L}_b$ is not. With the aid of (2.19), (2.22) and (2.23), it can be shown that

$$[\hat{F}_i, \mathcal{L}_s] = 0 \quad , \tag{3.2}$$

where $\mathcal{L}_s$ is described by the symmetric Lagrangian (2.4). We postulate that the breaking part of the Lagrangian $\mathcal{L}_b$ contains no gradients of field operators and it is then the negative of the Hamiltonian density $\mathcal{H}_b$. It follows that

$$\partial_\mu \mathcal{F}_{i\mu}(\underline{x}, t) = - i \, [\hat{F}_i(t), \mathcal{H}_b] \quad . \tag{3.3}$$

The integrated formula is

$$\partial_t \hat{F}_i = \int \partial_\mu \mathcal{F}_{i\mu} \, d^3x = - i \, [\hat{F}_i, \int \mathcal{H}_b \, d^3x] \quad . \tag{3.4}$$

We generate the divergences of the currents by studying the transformation properties of the breaking Hamiltonian $\mathcal{H}_b$.

Let us introduce the de Swart [9] notation for the generators of the group

$$\hat{F}_1 = I_1, \quad \hat{F}_4 = K_1, \quad \hat{F}_6 = L_1, \quad \hat{F}_8 = M,$$
$$\hat{F}_2 = I_2, \quad \hat{F}_5 = K_2, \quad \hat{F}_7 = L_2,$$
$$\hat{F}_3 = I_3. \tag{3.5}$$

We then form the "ladder" operators

$$I^\pm = I_1 \pm iI_2,$$
$$K^\pm = K_1 \pm iK_2,$$
$$L^\pm = L_1 \pm iL_2. \tag{3.6}$$

We note that I_3 and M are two commuting operators and they are denoted by the vector $\underline{E} = (I_3, M)$. The commutation relations are

$$[\underline{E}, I^{\pm}] = \pm \underline{a}\, I^{\pm} ,$$
$$[\underline{E}, K^{\pm}] = \pm \underline{b}\, K^{\pm} ,$$
$$[\underline{E}, L^{\pm}] = \pm \underline{c}\, L^{\pm} , \tag{3.7}$$

$$[I^+, I^-] = 2\underline{a}.\underline{E} ,$$
$$[K^+, K^-] = 2\underline{b}.\underline{E} ,$$
$$[L^+, L^-] = 2\underline{c}.\underline{E} , \tag{3.8}$$

where we define the unit vectors

$$\underline{a} = (1, 0), \qquad \underline{b} = (\tfrac{1}{2}, \tfrac{1}{2}\sqrt{3}), \qquad \underline{c} = (-\tfrac{1}{2}, \tfrac{1}{2}\sqrt{3}) . \tag{3.9}$$

We also have the commutation relations

$$[I^-, K^+] = L^+, \qquad [K^-, I^+] = L^-,$$
$$[I^+, L^+] = K^+, \qquad [L^-, I^-] = K^-,$$
$$[K^+, L^-] = I^+, \qquad [L^+, K^-] = I^-. \tag{3.10}$$

The transformation properties of $\mathcal{H}_b$ are assumed to have the property

$$\partial_t \hat{F}^{A^{\pm}} = - i\ [\hat{F}^{A^{\pm}}, \int \mathcal{H}_b\ d^3x] \sim A^{\pm} , \tag{3.11}$$

where A = I, K, L. The assumption of the broken eightfold way is that $\mathcal{H}_b \sim F_8$ and with the aid of the commutation rules (3.7) it is observed that (3.11) is satisfied for A = K,L.

IV. Broken Symmetries, Current Algebras and Dispersion Relation Sum Rules

It is a vexing feature of nature that it is not perfectly symmetric under the unimodular transformations of SU(3). In some cases the symmetry is badly violated as exhibited by the kaon-pion mass difference (∿360 MeV) and the apparent smallness of the renormalized coupling constant $G^2/4\pi$ for NKΛ and NKΣ compared to NπN. One of the unpleasant features of approximate symmetry physics is that it is not always clear when to expect quantitative agreement and when not to expect it.

In several cases, SU(3) symmetry is remarkably good. In the decay ratio

$$\frac{K^* \to K\pi}{\rho \to 2\pi} = \frac{1}{2} \quad , \tag{4.1}$$

where the phase space corrections have been included, the agreement with experiment is good. On the other hand, the decay ratio [10]

$$\frac{Y_1^*(1385) \to \Sigma\pi}{Y_1^*(1385) \to \Lambda\pi} = \frac{1}{7} \tag{4.2}$$

does not agree so well with the experimental value $\sim\frac{1}{12}$.

Gell-Mann [5] has emphasized that the non-linear relation (2.24) is valid however badly the symmetry is violated. It is postulated that the algebra generated by the hadron currents is imprinted on nature and leaves an indelible impression despite the bulges and kinks that occur in the space of rotations of elementary particles. The electromagnetic and weak fields act as perturbations and can be used to probe the hadron currents. In this way information can be obtained about the nature of strong interactions without predisposing a knowledge of the Lagrangian.

This type of reasoning becomes particularly relevant when we consider the consequences of SU(6) [11] or U(6) × U(6) symmetry [12]. These schemes are badly broken and they do not appear to possess a Hamiltonian that is invariant under the rotations of the group. It fares even worse with the intrinsically broken groups like $\tilde{U}(12)$ which exemplify attempts to combine the internal symmetries with the Poincaré group [13]. In the current algebra approach there is no need to worry about whether or not strong interactions are SU(6) symmetric; only unitary symmetry is assumed from the outset.

The vector and axial vector current operators obey equal-time commutation relations corresponding to the algebra of U(6) × U(6) [12]. A quark model

$$\psi^+ A\lambda_i \psi \quad ,$$

where A is any Hermitian Dirac matrix and the λ's are the nine matrices corresponding to U(3), would close commutation when 144 S, P, T, A and V densities are included. The space integrals of the 144 densities generate the algebra of the compact group U(12).

The commutation relations of the generators K_i:

$$[K_i, K_j] = i A_{ijk} K_k \quad ,$$

where the A_{ijk} are real and antisymmetric, can be made to correspond to the compact group G. In a non-compact-group $\tilde{G}$, the generators have some of their signs changed and the total antisymmetry is lost. A non-compact algebra always possesses infinite-dimensional representations, because for finite matrices the structure constants are totally antisymmetric.

In the simple noncompact group SO(3,1) corresponding to the homogeneous Lorentz group the infinitesimal generators Ω_i and N_i (i = 1,2,3) satisfy

$$[\Omega_i, \Omega_j] = i \, \varepsilon_{ijk} \, \Omega_k \quad ,$$

$$[\Omega_i, N_j] = i \, \varepsilon_{ijk} \, N_k \quad ,$$

$$[N_i, N_j] = -i \varepsilon_{ijk} \, \Omega_k \quad .$$

We can define the Ω_i and N_i in terms of the symmetrical energy-momentum tensor $T_{\mu\nu}$:

$$\Omega_i = i \, \varepsilon_{ijk} \int x_j \, T_{k4} \, d^3x$$

$$N_i = - \int x_i \, T_{44} \, d^3x - t \, P_i$$

where $P_i = i \int T_{i4} d^3x$ is the momentum operator and $\mathcal{H} = - T_{44}$. Schwinger [14] has derived the result:

$$[\mathcal{H}(\underline{x}), \mathcal{H}(\underline{x}')] = \partial_i \, \delta(\underline{x} - \underline{x}') \, [T_{i4}(\underline{x}) + T_{i4}(\underline{x}')]$$

Dashen and Gell-Mann [12] have considered approximate symmetry under the operators of an algebra for stationary and quasi-stationary states of hadrons at rest in the Lorentz frame in which we perform the space integrals. They required that

a. The operators of the algebra act on states in such a way that they concentrate on a few low lying states with energies close together.

b. The matrices of the operators split approximately into irreducible representations.

Taking commutation relations between two states a sum rule is obtained when we perform an expansion in a complete set of intermediate states. In terms of a. , the sum rule is "saturated" by low-lying intermediate states and the result is a relation between matrix elements.

An algebra satisfying a. and b. is called a "good symmetry" [12]. The positive parity, nonchiral U(6) × U(6) of operators

$K(\lambda_i \sigma_j \frac{1+\beta}{2})$ and $K(\lambda_i \sigma_j \frac{1-\beta}{2})$ $(j = 0,1,2,3)$ was considered a "good symmetry" of baryon and meson states at rest.
However, Coleman [15] has pointed out that the condition

$$\langle n | \mathcal{J}_z^5 | 1 \rangle = 0 \quad ,$$

where $\mathcal{J}_z^5$ denotes the z component of the axial vector current, for all many particle states n leads to a contradiction, since the Hamiltonian then commutes with the symmetry group. However, Gerstein and Lee [16] have suggested that many particle states may cancel and thus lead back to something satisfactory.

Gerstein [17] and Gerstein and Lee [16] and Bergia and Lannoy [18] have studied the $U(3) \times U(3)$ algebra in the limit of infinite momentum ($\beta = 1$, $\gamma = 0$) and obtained SU(6) predictions. The infinite momentum limit can probably be circumvented in a more careful analysis of the frame-dependence-problem.

It is very interesting that the algebra of equal time commutation relations can lead to the results of a particular symmetry group without explicitly assuming it beforehand.

The current algebra per se does not provide us with dynamical information. We must supplement the algebra with a dynamical formalism either in the language of dispersion relations or in the language of field theory. At this point, we recognize once again that the basic problem of strong interaction physics is a perennial one. We do not have a unique, convergent method for performing calculations in strong interactions. This is the fundamental problem of modern physics. Approximate symmetries help us to organize and classify certain data, like the particle spectrum, and they also provide us with certain approximate sum rules. But we are still faced with the challenge of conquering the Himalayan peak - the solution of the dynamical problem of strong interactions.

The use of current algebras and highly convergent dispersion relations for form factors has proved a useful tool [6] [19][20][21]. Let us review these methods. We begin by considering the derivation of the Goldberger-Treiman relation [22]. This relation is based on the hypothesis that the divergence of the axial vector current is dominated by the pion pole near $s = -k^2 = 0$. This is a stronger statement than the notion that the divergence of the axial vector current is dominated by the one-pion pole near $s = -M_\pi^2$. For nucleon β-decay the matrix element of the axial vector current $\mathcal{J}_\alpha^5$ is

$$<N|\, \mathcal{J}_\alpha^5 |N> = \overline{W}_N \, [i\gamma_\alpha \, F(s) + k_\alpha \beta(x)] \, \gamma_5 \, (\tau/2) \, W_N \quad , \tag{4.3}$$

where W_N is the nucleon spinor, k_α is the four-momentum transfer and $s = -k^2$. The divergence of (4.3) is obtained by multiplying the right-hand side by $-ik_\alpha$. By using the Dirac equation, we get

$$<N|\, \partial_\alpha \mathcal{J}_\alpha^5 |N> = \overline{W}_N i\gamma_5 (\tau/2) \, W_N \cdot [2M_N F(s) + s\beta(s)] \quad . \tag{4.4}$$

At $s = 0$, we have

$$F(0) = -G_A/G \quad . \tag{4.5}$$

The matrix element of $\partial_\alpha \mathcal{J}_\alpha^5$ between the vacuum and a one-pion state is expressed by Gell-Mann and Lévy [23] as

$$<0|\partial_\alpha \mathcal{J}_\alpha^5 |\pi> = M_\pi^2 \, (2f_\pi)^{-1} \, \phi \quad , \tag{4.6}$$

where ϕ is the renormalized pion wave function and the decay constant f_π is obtained from the rate of $\pi^+ \to \mu^+ + \nu$:

$$\Gamma_\pi = G^2 \, M_\pi M_\mu^2 \, (1 - M_\mu^2/M_\pi^2)^2 \, (f_\pi^2/4\pi)^{-1} \, (64\pi^2)^{-1} \quad . \tag{4.7}$$

If we study the diagram

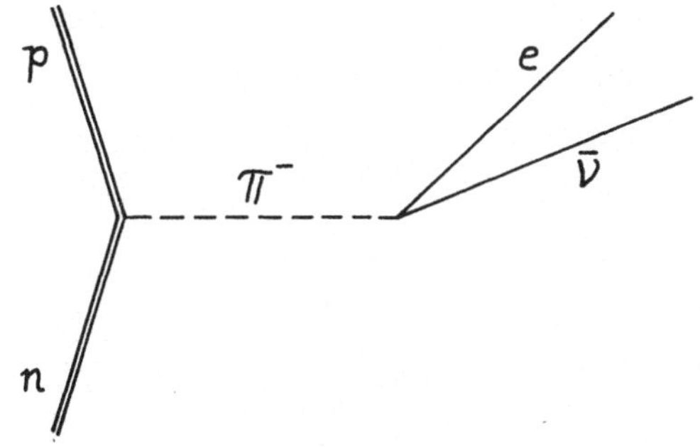

Fig. 1

we see that the matrix element (4.4) is dominated by the pion pole at $s = -M_\pi^2$ resulting from the virtual emission of a pion that undergoes leptonic decay. The pole strength is just the product of M_π^2/f_π and the coupling constant for pion-nucleon scattering $G_{\pi N \pi}$. Let us assume that $2M_N\, F(s) + s\beta(s)$ satisfies an unsubtracted dispersion relation

$$2M_N\, F(s) + s\beta(s) = \frac{(G_{\pi N\pi}/f_\pi)\, M_\pi^2}{M_\pi^2 - s} + \int \frac{B(\kappa^2)\kappa^2\, d\kappa^2}{\kappa^2 - s - i\varepsilon} \quad . \qquad (4.8)$$

For s = 0 the sum rule results

$$2\, M_N(-G_A/G) = G_{\pi N\pi}/f_\pi + \int B(\kappa^2)\, d\kappa^2 \quad . \qquad (4.9)$$

Thus if the dispersion relation (4.8) is highly convergent and dominated by the lowest mass state at small s, we arrive at the Goldberger-Treiman relation

$$2M_N(-G_A/G) \approx G_{\pi N\pi}/f_\pi \quad , \qquad (4.10)$$

which is in good agreement with experiment.

The linear, homogeneous dispersion relations for the form

factors can be used to find the dependence of the current matrix elements on s, but they do not permit a determination of the scale of these matrix elements. Gell-Mann [6] suggested that constants such as $-G_A/G$ can be calculated by using the equal time commutators (2.24). These non-linear relations fix the scale of the matrix elements. The universality of weak interactions then becomes realized through the equality of these relations for the hadron system and the lepton system.

This programme was carried out in practice by Adler and Weisberger [24] to calculate the coupling constant renormalization r_A. The simplest derivation is based on a method proposed by Fubini and Furlan [20][25].

The $\Delta S = 0$ part of the hadron current responsible for leptonic decays is

$$j_\mu = G \cos\theta \, (\mathcal{J}_{1\mu} + i \, \mathcal{J}_{2\mu} + \mathcal{J}^5_{1\mu} + i \, \mathcal{J}^5_{2\mu}) \, , \qquad (4.11)$$

where θ is the Cabibbo angle [26] and $G \approx 1.02 \times 10^{-5}/M_N^2$. Due to the conservation of the vector current the coupling constant G is unrenormalized. The renormalized axial vector coupling constant is defined by writing the matrix element of j_μ as

$$\langle N | j_\mu | N \rangle = (\frac{M_N}{k_o}) \, G \cos\theta \, \overline{W}_N \, (\gamma_\mu + r_A \, \gamma_\mu \gamma_5) \, \tau^+ \, W_N \, . \qquad (4.12)$$

The axial vector currents satisfy the equal time commutation relations in the algebra of SU(3) × SU(3) [27]:

$$\left[\mathcal{J}^5_{i4}(\underline{x}, t), \mathcal{J}^5_{j4}(\underline{x}', t) \right] = - f_{ijk} \, \mathcal{J}_{k4}(\underline{x}, t) \, \delta(\underline{x} - \underline{x}') \, . \qquad (4.13)$$

The chiralities of the currents are defined by the isotopic spin components

$$\mathcal{I}^{\pm}(t) = - i \int d^3x \, (\mathcal{J}^5_{14} \pm i \, \mathcal{J}^5_{24}) \tag{4.14}$$

and they satisfy

$$[\mathcal{I}^+(t), \mathcal{I}^-(t)] = 2 \, I^3 \quad . \tag{4.15}$$

The matrix element of the right-hand side of (4.15) between single proton states is

$$<p(k)|2 \, I^3|p(k')> = (2\pi)^3 \, \delta(k - k') \quad . \tag{4.16}$$

A complete set of states is inserted in the matrix element of the commutator, separating out the one-nucleon state (only the neutron contributes):

$$<p(k)|[\mathcal{I}^+(t), \mathcal{I}^-(t)]|p(k')> =$$

$$= (2\pi)^3 \, \delta(k - k') \, r_A^2 \, (1 - M_N^2/k_o^2) + \sum_n <p(k)| \mathcal{I}^+(t)|n> \times$$

$$\times <n| \mathcal{I}^-(t)|p(k')> - (\mathcal{I}^+ \leftrightarrow \mathcal{I}^-) \quad . \tag{4.17}$$

An investigation of the continuum term on the right-hand side of (4.17) reveals that (4.17) is a family of sum rules in terms of the parameter k_o. A sum rule for $1 - r_A^{-2}$ is derived in the limit $k_o \to \infty$. Further manipulations finally yield the desired sum rule

$$1 - 1/r_A^2 = \frac{2M_N^2}{G_{\pi N\pi}^2 [F^{\pi N\pi}(0)]^2} \int \frac{2M_N \, dE^2}{(E^2-M_N^2)^2} \left[G^+(E,0)-G^-(E,0)\right] , \tag{4.18}$$

where $F^{\pi N\pi}$ is the pion form factor of the nucleon, normalized so that $F^{\pi N\pi}(-M_\pi^2) = 1$. The derivation is completed by expressing $G^{\pm}(E,0)$ in terms of cross sections for scattering of zero-mass $\pi^{\pm}$ on protons at C.M. energy E. This yields the sum rule

$$1 - 1/r_A^2 = \frac{2M_N^2}{G_{\pi N\pi}^2 [F^{\pi N\pi}(0)]^2} \frac{1}{\pi} \int_{M_N+M_\pi}^{\infty} \frac{dE^2}{E^2 - M_N^2} \times$$

$$\times [\sigma_0^+(E) - \sigma_0^-(E)] \quad . \qquad (4.19)$$

The calculations of Adler and Weisberger give the theoretical prediction $r_A = 1.24$. The experimental value is $(r_A)^{exp} = 1.18 \pm 0.02$.

In order to get one number $r_A = 1.24$ it is necessary to make three assumptions:

A. The axial vector current is partially conserved.

B. The Pomeranchuk theorem is exactly satisfied at high energies.

C. The analytical continuation to zero mass pions is justified.

If B. does not hold at high energies, then a subtraction constant is required and the sum rule provides no information on the value of r_A. It has been shown by Schroer and Stichel [28] that B. is equivalent to the existence of the equal time commutation relation (4.13). Thus, the problem of subtraction constants that plagues the dispersion theoretical approach, plays an important role in this type of calculation. The fact that the sum rule holds for forward scattering and yields such good agreement with experiment makes it an excellent illustrative example of the consequences of the philosophy of using current algebras within the framework of highly convergent dispersion relations.

V. Vector Current Interactions and Mass Relations

We shall now turn our attention to the mass sum rules which constitute an important check of unitary symmetry.

The electromagnetic field breaks the unitary symmetry in a definite way. The electromagnetic interaction is described by the Lagrangian

$$\mathcal{L}_{em} = eA_\mu\ (\bar{B}\gamma_\mu[B,Q] + \partial_\mu\Pi_i\ [\Pi_i,Q]) - \frac{1}{2}\, e^2 A_\mu A_\mu \ \times$$

$$\times\ ([\Pi_i,Q]\ [\Pi_i,Q]) \quad , \qquad (5.1)$$

where the charge operator Q is defined by

$$Q = \hat{F}_3 + \frac{1}{\sqrt{3}}\,\hat{F}_8 \quad . \qquad (5.2)$$

$\mathcal{L}_{em}$ is a scalar in U-spin space and all members of a U-spin multiplet have the same charge. The situation is not so clear when we consider the hypercharge splitting of SU(3). Is there a field in nature that breaks the unitary symmetry by choosing the preferred direction $F_8 = \sqrt{3}/2\ Y$? We have not as yet observed such interactions and the origin of the "semi-strong" splitting remains a mystery.

We shall study the mass sum rules in SU(3) in the framework of the current algebra and field theory [29][30].

The vector currents of the hadrons interact with the vector mesons, the electromagnetic field and the weak lepton currents. It has been suggested by Sakurai [4] that the ρ is universally coupled to the isotopic spin current and ω to the hypercharge current. The stronger form of universality results when it is assumed that the strange vector mesons K^* are coupled to the strangeness changing currents through the same coupling constant and the whole system is unitary symmetric in the absence of bare mass terms. The ratio of $K^* \to K\pi$ to $\rho \to \pi\pi$, in (4.1), shows that this symmetry appears to be remarkably well satisfied.

The vector mesons are themselves a source of the field as they contribute to the current. The derivation of this type of nonlinear theory was carried out by Yang and Mills [31]. A gauge-invariant Lagrangian can be constructed and has the form

$$\mathcal{L} = -\frac{1}{4} F_{i\alpha\beta} F_{i\alpha\beta} - M_{oB} \bar{B}_i B_i - \bar{B}_i \gamma_\alpha (\partial_\alpha B_i + g_{ov} f_{ijk} V_{j\alpha} B_k) -$$

$$- \frac{1}{2} M^2_{ob} \Pi_i \Pi_i - \frac{1}{2} (\partial_\alpha \Pi_i + g_{ov} f_{ijk} V_{j\alpha} \Pi_k) \times$$

$$\times (\partial_\alpha \Pi_i + g_{ov} f_{ijk} V_{j\alpha} \Pi_k) \; , \qquad (5.3)$$

where the eight components of the vector $V_{i\alpha}$ are given by

$$V_{1\alpha} = \frac{1}{\sqrt{2}} (\rho^+_\alpha + \rho^-_\alpha), \qquad V_{5\alpha} = \frac{i}{\sqrt{2}} (K^{+*}_\alpha - K^{-*}_\alpha),$$

$$V_{2\alpha} = \frac{i}{\sqrt{2}} (\rho^+_\alpha - \rho^-_\alpha), \qquad V_{6\alpha} = \frac{1}{\sqrt{2}} (K^{o*}_\alpha + \bar{K}^{o*}_\alpha),$$

$$V_{3\alpha} = \rho^o_\alpha \; , \qquad V_{7\alpha} = \frac{i}{\sqrt{2}} (K^{o*}_\alpha - \bar{K}^{o*}_\alpha),$$

$$V_{4\alpha} = \frac{1}{\sqrt{2}} (K^{+*}_\alpha + K^{-*}_\alpha), \qquad V_{8\alpha} = \omega^o_\alpha \qquad (5.5)$$

and $F_{\alpha\beta}$ denotes the gauge invariant field strength

$$F_{i\alpha\beta} = \partial_\alpha V_{i\beta} - \partial_\beta V_{i\alpha} + g_{ov} f_{ijk} V_j V_k \; . \qquad (5.6)$$

The constant g_{ov} is the bare coupling constant and a vector meson bare mass is perhaps required in $\mathcal{L}$.

If we consider just the hadron-vector meson coupling, then the Lagrangian for the interaction can be written

$$\mathcal{L}^1 = ig_{ov} \{(\rho^0_\alpha \mathcal{F}^{I^3}_\alpha + \rho^+_\alpha \mathcal{F}^{I^+}_\alpha + \omega^0_\alpha \mathcal{F}^M_\alpha) + + (K^{+*}_\alpha \mathcal{F}^{K^*}_\alpha + K^{0*}_\alpha \mathcal{F}^{L^+}_\alpha)\} + \text{h.c.} \qquad (5.4)$$

The first term on the right-hand side of (5.4) represents the coupling to the conserved part of the current, while the second term describes the coupling to the strangeness changing part of the current. In the limit of exact SU(3) symmetry the total current occurring in (5.4) is conserved, because we have only assumed F-type currents in the Lagrangian (5.3) . Thus it is assumed that the vector current of the hadrons is conserved in analogy with the electromagnetic current. Is there also a D type component of the hadron current coupled to the vector mesons? We shall return to this question later.

The total F-spin currents of the baryons and mesons for $A^+ = I^+$, K^+ and L^+ are

$$\mathcal{F}^{I^+}_\mu = \sqrt{2}\,\bar{p}\,\gamma_\mu n - \sqrt{2}\,\bar{\Xi}^0\,\gamma_\mu\,\Xi^- - 2\,\bar{\Sigma}^+\,\gamma_\mu\,\Sigma^0 + + 2\bar{\Sigma}^0\,\gamma_\mu \Sigma^- + \sqrt{2}\,K^-\partial_\mu K^0 - \sqrt{2}\,K^0\partial_\mu K^- - 2\pi^-\partial_\mu\pi^0 + 2\pi^0\partial_\mu\pi^- , \qquad (5.7)$$

$$\mathcal{F}^{K^+}_\mu = -\sqrt{3}\,\bar{p}\gamma_\mu\Lambda + \sqrt{3}\,\bar{\Lambda}\,\gamma_\mu\,\Xi^- - \bar{p}\,\gamma_\mu\Sigma^0 - \sqrt{2}\,\bar{n}\gamma_\mu\Sigma^- + \bar{\Sigma}^0\gamma_\mu\Xi^- + + \sqrt{2}\,\bar{\Sigma}^+\,\gamma_\mu\,\Xi^0 - \sqrt{3}K^-\,\partial_\mu\eta + \sqrt{3}\,\eta\partial_\mu K^- - K^-\,\partial_\mu\pi^0 + + \pi^0\,\partial_\mu K^- - \sqrt{2}\,\bar{K}^0\,\partial_\mu\pi^- + \sqrt{2}\,\pi^-\,\partial_\mu\bar{K}^0 , \qquad (5.8)$$

and

$$\mathcal{F}^{L^+}_\mu = -\sqrt{3}\,\bar{n}\,\gamma_\mu\Lambda + \sqrt{3}\,\bar{\Lambda}\gamma_\mu\,\Xi^0 + \bar{n}\,\gamma_\alpha\Sigma^0 - \sqrt{2}\,\bar{p}\,\gamma_\mu\Sigma^+ -$$

$$- \overline{\Sigma}^{0} \gamma_{\mu} \Xi^{0} + \sqrt{2}\, \overline{\Sigma}^{-} \gamma_{\mu} \Xi^{-} - \sqrt{3}\, \overline{K}^{0} \partial_{\mu}\eta + \sqrt{3}\, \eta \partial_{\mu} K^{0} +$$
$$+ \overline{K}^{0} \partial_{\mu}\pi^{0} - \pi^{0} \partial_{\mu}\overline{K}^{0} - \sqrt{2}\, K^{-} \partial_{\mu}\pi^{+} + \sqrt{2}\, \pi^{+} \partial_{\mu} K^{-} \tag{5.9}$$

We shall assume that the "bare" masses generate the hypercharge breaking [29]

$$\mathcal{H}_{b} = M_{oB}\overline{B}B \quad , \tag{5.10}$$

where $\mathcal{H}_b$ possesses the property (3.11). In particular, it is assumed that the transformation properties of $\mathcal{H}_b$ are of the most general kind which satisfy (3.11). Let us denote by $\mathcal{F}^{A}_{\mu B}$ and $\mathcal{F}^{A}_{\mu M}$ the baryon and meson parts of (5.7)-(5.9), respectively. From the Heisenberg equations of motion obtained from the Lagrangian (2.4), we get

$$\partial_{\mu} \mathcal{F}^{K^{+}}_{\mu B} = - \sqrt{3}\, (M_{op}-M_{o\Lambda})\overline{p}\Lambda + \sqrt{3}\, (M_{o\Lambda}-M_{o\Xi})\overline{\Lambda}\Xi^{-} -$$
$$- (M_{op}-M_{o\Sigma})\overline{p}\Sigma^{0} - \sqrt{2}\, (M_{on}-M_{o\Sigma})\overline{n}\Sigma^{-} +$$
$$+ (M_{o\Sigma}-M_{o\Xi})\overline{\Sigma}^{0} \Xi^{-} + \sqrt{2}\, (M_{o\Sigma}-M_{o\Xi})\overline{\Sigma}^{+} \Xi^{0} + G_{B} \tag{5.11}$$

and

$$\partial_{\mu} \mathcal{F}^{K^{+}}_{\mu M} = \sqrt{3}\, (M^{2}_{oK}-M^{2}_{o\eta})K^{-}\eta + (M^{2}_{oK}-M^{2}_{o\pi})K^{-} \pi^{o} +$$
$$+ \sqrt{2}\, (M^{2}_{oK}-M^{2}_{o\pi})K^{-} \overline{K}^{0} + G_{M} \quad , \tag{5.12}$$

where G_B and G_M are the interaction terms arising from the equations of motion, and p, Λ, Ξ^- etc. denote the Heisenberg field operators that satisfy the commutation relations obtained from (2.22) and (2.23).

By virtue of (3.11), we have

$$\int d^{3}x d^{3}x' \; [\partial_{\mu} \mathcal{F}^{A^{+}}_{\mu} (\underline{x}, t), \mathcal{F}^{A^{+}}_{4} (\underline{x}', t)] = 0 \quad , \tag{5.13}$$

where $A^+ = K^+, L^+$. The term $G = G_B + G_M$ does not contribute to $\partial_\mu \mathcal{F}_\mu^A$, because the interaction Lagrangian, in (2.4), is symmetric under unitary transformations and therefore

$$G = - i \, [\hat{F}_i , \mathcal{H}_I] = 0 \tag{5.14}$$

The commutator in (5.13) can now be evaluated by using the canonical commutation relations (2.22) and (2.23) at equal times. This gives the two independent relations

$$\int d^3x d^3x' \, [\partial_\mu \mathcal{F}^{K^+}_{\mu B}(\underline{x}, t), \mathcal{F}^{K^+}_{4B}(\underline{x}', t)] =$$

$$= \int d^3x \bar{p}(x) \, \Xi^-(x) \, (6M_{o\Lambda} - 4M_{op} + 2M_{o\Sigma} - 4M_{o\Xi}) = 0 \tag{5.15}$$

and

$$\int d^3x d^3x' \, [\partial_\mu \mathcal{F}^{K^+}_{\mu M}(\underline{x}, t), \mathcal{F}^{K^+}_{4M}(\underline{x}', t)] =$$

$$= \int d^3x K^-(x) K^-(x) \, (4M^2_{oK} - 3M^2_{o\eta} - M^2_{o\pi}) = 0 \; . \tag{5.16}$$

We obtain the Gell-Mann Okubo sum rules [32] for the bare masses of the baryons and pseudoscalar mesons

$$M_{op} + M_{o\Xi} - \frac{3}{2} M_{o\Lambda} - \frac{1}{2} M_{o\Sigma} = 0 \tag{5.17}$$

$$4M^2_{oK} - 3M^2_{o\eta} - M^2_{o\pi} = 0 \tag{5.18}$$

to all orders in the F-spin invariant interactions.

Let us now consider the electromagnetic mass sum rule obtained by Coleman and Glashow [33]:

$$M_{\Xi^-} - M_{\Xi^0} = M_{\Sigma^-} - M_{\Sigma^+} + M_p - M_n \quad , \tag{5.19}$$

which is valid to all orders in the medium strong and electromagnetic breaking provided cross terms between medium strong and electromagnetic breaking are neglected [34]. We find that this sum rule cannot be obtained from the current algebra without postulating the existence of D-type currents $\mathcal{F}^{A}_{\mu}$ (D), as well as the F-type currents $\mathcal{F}^{A}_{\mu}$ (F) used to derive the Gell-Mann Okubo sum rules. The D-type current of the baryons for $A^{+} = K^{+}$ is

$$\mathcal{F}^{K^+}_{\mu B} = -\frac{1}{\sqrt{3}}\,\bar{p}\,\gamma_\mu \Lambda + \bar{p}\,\gamma_\mu \Sigma^{0} + \sqrt{2}\,\bar{n}\,\gamma_\mu \Sigma^{-} - -\frac{1}{\sqrt{3}}\,\bar{\Lambda}\gamma_\mu \Xi^{-} + \bar{\Sigma}^{0}\,\gamma_\mu\,\Xi^{-} + \sqrt{2}\,\bar{\Sigma}^{+}\,\gamma_\mu\,\Xi^{0} \quad .. \qquad (5.20)$$

The triple commutation relation

$$\int d^3x d^3x' d^3x'' \Big[[\partial_\mu \mathcal{F}^{I^+}_{\mu B}(F)(\underline{x}, t), \mathcal{F}^{L^+}_{4B}(F)(\underline{x}', t)] \ , \ \mathcal{F}^{K^+}_{4}(D)(\underline{x}'', t)\Big] = 0 \qquad (5.21)$$

is satisfied in virtue of the commutation relations (3.10). A calculation of (5.21) gives the Coleman-Glashow sum rule for the bare masses

$$(M_{o\Xi^0} - M_{o\Xi^-}) + (M_{op} - M_{on}) - (M_{o\Sigma^+} - M_{o\Sigma^-}) = 0 \quad . \qquad (5.22)$$

Here we are assuming that the bare masses of the baryons give rise to a small, intrinsic, and as yet undetected, violation of SU(2).

The sum rule (5.22) can also be derived from the relation

$$\int d^3x d^3x' d^3x'' \left[[\partial_\mu \mathcal{F}^{I^+}_{\mu B}(D)\ (\underline{x}, t), \mathcal{F}^{L^+}_{4B}(F)\ (\underline{x}', t)] \right.,$$

$$\left. \mathcal{F}^{K^+}_{4B}(F)\ (\underline{x}'', t) \right] = 0 \qquad (5.23)$$

provided we go to the higher symmetry $U(3) \times U(3) \equiv W_3$ in order to maintain a conserved baryon current in the unitary symmetry limit [35], [27].

VI. Field Theory Model of Symmetry Breaking Corrections

We shall now study a model calculation of symmetry breaking corrections based on field theory [29][30]. This model is intended as an illustration of what might be expected from a future theory in which unitary symmetry and some features of field theory are combined.

In renormalization theory $M_{oB} = M_B - \delta M_B$, where δM_B denotes the self-mass of the baryons. Let us substitute this expression for M_{oB} into (5.17):

$$M_p + M_\Xi - 3/2\ M_\Lambda - 1/2\ M_\Sigma - \delta A = 0 \quad , \qquad (6.1)$$

where

$$\delta A = \delta M_p + \delta M_\Xi - 3/2\ \delta M_\Lambda - 1/2\ \delta M_\Sigma \quad . \qquad (6.2)$$

A calculation of δM_B in second order perturbation theory gives

$$\delta M_B = \frac{G^2}{32\pi^2 \hbar c} \sum_i C_i^2 \{ M_B \log \left(\frac{\lambda^2}{M_B{}^2}\right) +$$

$$+ M_B \left[- 1/2 + \left(\frac{m_i}{M_B}\right)^2 + 2 \left(\frac{m_i}{M_B}\right)^2 \left(1 - \frac{m_i^2}{2M_B^2}\right) \log \left(\frac{m_i}{M_B}\right) - \right.$$

$$\left. - 2 \left(\frac{m_i}{M_B}\right)^3 \left(1 - \frac{m_i^2}{2M_B^2}\right)^{1/2} \cos^{-1} \left(\frac{m_i}{2M_B}\right)\right]\Big\} \quad , \qquad (6.3)$$

where λ is the invariant Feynman cutoff, the C_i are Clebsh-Gordan coefficients and the m_i $(i = 1...8)$ denote the masses of the octet of pseudoscalar mesons. The cutoff λ is assumed to be the same for all the baryons in the limit $\lambda \to \infty$, and we have kept only the leading term in an expansion of the difference of the external and internal baryon masses in the self-energy graph.

We now assume that the logarithmic divergences cancel, in the limit $\lambda \to \infty$, in the expression δA in (6.2).

For $G^2/\hbar c \simeq 15$ the evaluation of δA for pure D coupling gives

$$\delta A \simeq - 5 \text{ MeV} \qquad (6.4)$$

and for an F/D ratio $\sim 1/5$, we get

$$\delta A \simeq - 2 \text{ MeV}. \qquad (6.5)$$

Thus the corrected sum rule (6.1) is better satisfied than the zeroth order Gell-Mann Okubo formula.

The dependence on the cutoff λ is not unique, and we cannot justify rigorously the assumed cancellation of the logarithmic infinities in the sum rule. However, the remarkable improvement of the sum rule makes it an interesting example of future possibilities of calculating symmetry breaking corrections.

If we apply the same methods to the Coleman-Glashow sum rule (5.22), treating only the part of the corrections arising from single photon emission and absorption, we get the corrected result for the physical masses

$$(M_{\Xi^0} - M_{\Xi^-}) + (M_p - M_n) - (M_{\Sigma^+} - M_{\Sigma^-}) + \qquad (6.6)$$

$$+ \frac{3\alpha}{8\pi}(M_{\Xi^-} - M_p + M_{\Sigma^+} - M_{\Sigma^-}) = 0 \quad ,$$

where α is the fine structure constant. With the current experimental masses the left-hand side of (6.6) has the value 0.21 MeV, which is within the experimental error. We note that the correction

$$\frac{3\alpha}{8\pi}(M_{\Xi^-} - M_p) \qquad (6.7)$$

can be considered as a cross term between the hypercharge and electromagnetic breaking, because it involves a splitting between particles of different hypercharge.

One pleasing feature of the above results deserves mention. Within the framework of the current algebra and field theory the baryons and mesons automatically satisfy linear and squared mass sum rules, respectively. This result cannot be obtained in a pure group theoretical approach to mass sum rules.

VII.Weak Interactions

There are three characteristic features which should play an important role in a mathematical description of weak decay processes:

1. Universality of weak interactions,
2. Damping of strangeness changing decays,
3. Dominance of the $\Delta I = 1/2$ rule in the decay amplitudes of non-leptonic decays.

A two-parameter theory of 1. and 2. for leptonic decays has been suggested by Cabbibo [26] which appears to be in agreement with the experimental branching ratios.

Most of the data seems to be consistent with 3., since small $\Delta I = 3/2$ amplitudes are required to fit the experiments, and the $\Delta I = 5/2$ amplitudes are completely absent. It is tempting to adopt the symmetry suggested by 3., but the experimental situation is not yet reliable. There is no entirely satisfactory theoretical explanation of the dominance of $\Delta I = 1/2$ decays, although some interesting results have been obtained for parity nonconserving decays from various special models [36] and also from the use of the current algebra and the partially conserved axial vector hypothesis discussed in Section IV. [30]

The hadron weak current is first separated into a part generating $\Delta Y = 0$, $\Delta I = 1$ processes and a part generating $\Delta Y = \Delta Q$, $\Delta I = 1/2$ processes. Secondly, the weak current is separated into the vector current part and the axial vector current part. The time components of the hadron vector current $\mathcal{F}_{i\mu}$ and the axial vector current $\mathcal{F}^5_{i\mu}$, integrated over space, lead to the algebra [6]

$$[\hat{F}_i, \hat{F}_j] = if_{ijk}\,\hat{F}_k \ , \tag{7.1}$$

$$[\hat{F}_i, \hat{F}^5_j] = ij_{ijk}\,\hat{F}^5_k \ , \tag{7.2}$$

$$[\hat{F}^5_i, \hat{F}^5_j] = ij_{ijk}\,\hat{F}_k \ , \tag{7.3}$$

We define

$$2\hat{F}^{\pm}_i(t) = \hat{F}_i(t) \pm i\hat{F}^5_i(t) \quad . \tag{7.4}$$

The $\hat{F}^+_i$ and $\hat{F}^-_i$ are two commuting operators which determine the algebra of $SU(3) \times SU(3)$ [27]. The $\hat{F}^+_i$ and $\hat{F}^-_i$ are connected by the parity transformation

$$P\hat{F}_i^{\pm} P^{-1} = \hat{F}_i^{\pm} \quad , \tag{7.5}$$

and the total F-spin is the sum of left-handed and right-handed parts

$$\hat{F}_i = \hat{F}_i^{+} + \hat{F}_i^{-} \quad . \tag{7.6}$$

The weak hadron current is described by

$$J_\mu = \mathcal{F}^{+}_{1\mu} + i\,\mathcal{F}^{+}_{2\mu} + \mathcal{F}^{+}_{4\mu} + i\,\mathcal{F}^{+}_{5\mu} \quad , \tag{7.7}$$

where the term $\mathcal{F}^{+}_{1\mu} + i\,\mathcal{F}^{+}_{2\mu}$ gives $\Delta Y = 0$ and $\mathcal{F}^{+}_{4\mu} + i\,\mathcal{F}^{+}_{5\mu}$ gives $\Delta Y/\Delta Q = +1$.

The field theory model of the weak interactions is based on the current-current interaction Lagrangian (local approximation):

$$\mathcal{L}_W = \frac{G}{\sqrt{2}} J^{+}_\mu J_\mu \quad , \tag{7.8}$$

where J_μ describes the combined hadron-lepton current. For two distinct neutrinos the lepton current is described by

$$\ell_\mu = 2\bar{\nu}_e \gamma_\mu \frac{(1+\gamma_5)}{2} e + 2\bar{\nu}_\mu \gamma_\mu \frac{(1+\gamma_5)}{2} \mu \tag{7.9}$$

and the weak lepton charge Q_ℓ is determined by

$$Q_\ell = - i \int d^3x \; \ell_4 \quad . \tag{7.10}$$

Universality is imposed on the system by requiring that the

lepton current algebra should be the same as the hadron current algebra. For simplicity, one then demands that this algebra is the minimal one consistent with the correct commutation relations [27]. If we consider the spinors

$$\phi_1 = \begin{bmatrix} \frac{1+\gamma_5}{2} \nu_e \\ \\ \frac{1+\gamma_5}{2} e \end{bmatrix}, \quad \phi_2 = \begin{bmatrix} \frac{1+\gamma_5}{2} \nu_\mu \\ \\ \frac{1+\gamma_5}{2} \mu \end{bmatrix}, \tag{7.11}$$

then the lepton current ℓ_μ can be written

$$\ell_\mu = 2\,\{\bar{\phi}_1\, \tau^+ \gamma_\mu \phi_1 + \bar{\phi}_2 \tau^+ \gamma_\mu \phi_2\} \ . \tag{7.12}$$

The weak lepton charge Q_ℓ has the form $Q_\ell \sim \tau_1 + i\tau_2 \equiv 2\tau^+$, and this leads to the algebra of SU(2).

Following Cabbibo [26], we write the weak hadron charge in the form

$$Q_H = \alpha(\hat{F}^+_1 + i\hat{F}^+_2) \cos\theta + \alpha(\hat{F}^+_4 + i\hat{F}^+_5) \sin\theta \tag{7.13}$$

It follows that $Q_H \sim \alpha(\tau_1 + i\tau_2)$ provided we write

$$\tau_1 = \hat{F}^+_1 \cos\theta + \hat{F}^+_4 \sin\theta \quad ,$$

$$\tau_2 = \hat{F}^+_2 \cos\theta + \hat{F}^+_5 \sin\theta \quad ,$$

$$\tau_3 = \hat{F}^+_3 \cos^2\theta + \left(\frac{\sqrt{3}}{2}\, \hat{F}^+_8 + \frac{1}{2}\, \hat{F}^+_3\right) \sin^2\theta - \hat{F}_6 \sin\theta \cos\theta . \tag{7.14}$$

From the decay of O^{14} and the approximate equality of the vector coupling constants, we find that $\alpha \cos\theta \simeq 2$. Universality requires that $\alpha = 2$ and θ is small, and the weak hadron current becomes

$$J_\mu = 2 \cos\theta \, (\mathcal{F}^+_{1\alpha} + i\mathcal{F}^+_{2\alpha}) + 2 \sin\theta \, (\mathcal{F}^+_{4\alpha} + i\mathcal{F}^+_{5\alpha}) \; . \tag{7.15}$$

Cabbibo calculated the angle θ from the ratio of the rates for $K^+ \to \mu^+ + \nu$ and $\pi^+ \to \mu^+ + \nu$:

$$\Gamma(K^+ \to \mu\nu)/\Gamma(\pi^+ \to \mu\nu) = \tan^2\theta \; M_K(1-M_\mu^2/M_k^2)^2/M_\pi(1-M_\mu^2/M_\pi^2)^2 \; . \tag{7.16}$$

The result is $\theta \simeq 0.26$. An independent determination of θ can be obtained from the rate $K^+ \to \pi^0 + e^+ + \nu$, since the matrix element for this process can be connected to that for $\pi^+ \to \pi^0 + e^+ + \nu$, and the result is $\theta \simeq 0.26$.

Let us now turn to the important question of the F and D coupling of the currents in weak interactions. In order to generate the conserved currents, we shall appeal to the "gauge procedure" of Salam and Ward [35]. Consider the nonet of baryons which we denote by χ_i $(i = 0,1...8)$ [38]. χ_i belongs to the representation $(\underline{3}, \underline{3}^*)$ of $SW_3 \equiv SU(3) \times SU(3)$ and transforms as

$$\chi' = U_1 \chi U_2^{-1} \; . \tag{7.17}$$

Let us study the infinitesimal transformations

$$U_1 = 1 + i\theta, \; U_2 = 1 + i\beta \; , \tag{7.18}$$

where θ and β denote 3×3 Hermitian matrices. We obtain from (7.17) the result

$$\chi' = (1 + i\theta)\chi(1 - i\beta) = \chi + i(\theta\chi - \chi\beta) \tag{7.19}$$

and

$$\partial_\mu \chi' = \partial_\mu \chi + i(\theta_\mu \chi - \chi\beta_\mu) \quad . \tag{7.20}$$

In the free Lagrangian the expression $\bar{\chi}\gamma_\mu \partial_\mu \chi$ transforms as

$$\bar{\chi}\gamma_\mu \partial_\mu \chi + i\bar{\chi}\gamma_\mu (\theta_\mu \chi - \chi\beta_\mu) \quad . \tag{7.21}$$

The latter terms in (7.21) describe the coupling of the baryon currents to the vector spin 1 fields θ_μ and β_μ and can be rewritten as

$$\frac{1}{2}\bar{\chi}\gamma_\mu [\theta_\mu + \beta_\mu , \chi] + \frac{1}{2}\bar{\chi}\gamma_\mu \{\theta_\mu - \beta_\mu , \chi\} \, . \tag{7.22}$$

The first term in (7.22) describes the F-type coupling, while the second term describes the D-type coupling of vector currents.

In order to include the axial vector current, we consider in the limit of zero baryon masses the left-handed and right-handed quantities

$$\chi_L = \frac{1 + \gamma_5}{2}\chi \quad ,$$

$$\chi_R = \frac{1 - \gamma_5}{2}\chi \quad , \tag{7.23}$$

There are now four independent transformations

$$\chi_L' = U_1 \chi_L U_2^{-1}$$

$$\chi_R' = U_3 \chi_R U_4^{-1} \tag{7.24}$$

and a total of 32 currents will now be generated

$$F^V,\ F^A,\ D^V,\ D^A\ .$$

This system of currents describes the algebra $[SU(3) \times SU(3)] \times [SU(3) \times SU(3)]$. There are sixteen vector currents, in two octets, and sixteen axial vector currents, also in two octets [39]. If we choose $U_1 = U_4$ and $U_2 = U_3$, then the above reduces to F^V and D^A currents corresponding to $SU(3) \times SU(3)$. The pattern for the hadron weak current, in the limit of zero baryon masses, is then

$$- if_{ijk}\, \bar{\chi}_j\, \gamma_\mu \chi_k + d_{ijk}\, \bar{\chi}_j\, \gamma_\mu\, \gamma_5\, \chi_k \quad (i,j,k = 0,1...8). \tag{7.25}$$

This appears to correspond to the experimental situation for the baryon octet, the vector current is coupled through F and the axial vector current through D, with equal coefficients. The actual experimental ratio $F/D \sim 1/3$ [26]. All these considerations correspond, of course, to the limit of exact unitary symmetry.

The baryon nonet belongs to $(\underline{3}, \underline{3}^*)_L + (\underline{3}^*, \underline{3})_R$ and the ninth baryon has a positive mass with $J = 1/2^-$. There exists an octet of $J = 1^-$ and an octet of $J = 1^+$ particles, belonging to $(\underline{1},\underline{8}) \pm (\underline{8},\underline{1})$ representations. There should also exist $J = 0^+$ and $J = 0^-$ mesons that belong to the representations $(\underline{3},\underline{3}^*) \pm$ $\pm\ (\underline{3}^*,\underline{3})$ (or the representations $(\underline{1},\underline{8}) \pm (\underline{8},\underline{1})$).

The ninth baryon and the octet and singlet scalar mesons are an unpleasant feature of this theory as these particles have not been observed. The $\Lambda(1405)$ is now believed to possess $J = 1/2^-$ and could perhaps fit the bill. The nuclear forces produced by the scalar mesons would correspond to unphysically large cross sections.

Thus, the assignment of a minimal algebraic structure to the weak currents, in the form of the chiral $SU(3) \times SU(3)$, has not met with real success. However, it is possible that the equal time commutation relations among the time components of the vector and axial vector currents are not in fact satisfied in nature. But the most natural scheme to adopt, which leads to universality and a minimal algebra of the weak vector and axial vector currents, is the algebra of SU(3) × SU(3) and, therefore, as a future goal, we should investigate the veracity of the equal time commutation relations, and inquire as to whether scalar mesons really exist in nature.

References

References

1. E. Fermi and C. N. Yang, Phys. Rev. 76, 1739 (1949); E. Teller, Proc. of the Sixth Ann. Rochester Conference on High Energy Nuclear Physics, 1956 (Interscience Pub. Inc., N. Y. 1956).
2. S. Sakata, Progr. Theor. Phys. (Kyoto) 16, 686 (1956).
3. M. Gell-Mann, Phys. Letters 8, 214 (1964). G. Zweig, CERN Report TH. 401, Jan. 1964; and TH. 412, Feb. 1964.
4. J. J. Sakurai, Ann. Phys. 11, 1 (1960)
5. M. Gell-Mann, Calf. Inst. of Techn. Report CTSL-20, March, 1961 (published in the Eightfold Way, W.A. Benjamin, Inc. (1964)).
6. M. Gell-Mann, Phys. Rev. 125, 1067 (1962).
7. Y. Ne'eman, Nucl. Phys. 26, 222 (1961).
8. J. Schwinger, Phys. Rev. Letters 3, 296 (1959).
9. J. J. de Swart, Rev. Mod. Phys. 35, 916 (1963).
10. S. Glashow and A. H. Rosenfeld, Phys. Rev. Letters, 10, 192 (1963).
11. F. Gürsey and L. Radicati, Phys. Rev. Letters 13, 299 (1964); A. Pais, Phys. Rev. Letters 13, 175 (1964); B. Sakita, Phys. Rev. 136, B1756 (1964).
12. R. P. Feynman, M. Gell-Mann, and G. Zweig, Phys. Rev. Letters 13, 678 (1964); R. F. Dashen and M. Gell-Mann, Phys. Letters 17, 142 (1965); Phys. Letters 17, 145 (1965); B.

W. Lee, Phys. Rev. Letters 14, 676 (1965); S. Okubo, Phys. Letters 17, 174 (1965).

13. A. Salam, R. Delbourgo and J. Strathdee, Proc. Roy. Soc. (London), A284, 146 (1965); M. A. Bég and A. Pais, Phys. Rev. Letters 14, 267 (1965); B. Sakita and K. Wali, Phys. Rev. Letters 14, 404 (1965).
14. J. Schwinger, Theor. Physics, Int. Atomic Agency, Vienna p. 122 (1963).
15. S. Coleman, Phys. Letters 19, 144 (1965)
16. I. S. Gerstein and B. W. Lee, preprint (1966).
17. I. S. Gerstein, Phys. Rev. Letters 16, 144 (1966).
18. S. Bergia and F. G. Lannoy, preprint (1966)
19. A. P. Balachandran and H. Pietschmann, Nucl. Phys. 43, 321, (1963); H. Biritz and H. Pietschmann, Proc. Siena, Int. Conf. II. p. 293 (1964).
20. S. Fubini and G. Furlan, Phys. 1, 229 (1965).
21. S. Fubini, G. Furlan and C. Rossetti, CERN Report TH.578 (1965).
22. M. Goldberger and S. Treiman, Phys. Rev. 110, 1478 (1958); J. Bernstein, S. Fubini, M. Gell-Mann, and W. Thirring, Nuovo Cim. 17, 757 (1960); Y. Nambu, Phys. Rev. Letters 4, 380 (1960).
23. M. Gell-Mann and M. Lévy, Nuovo Cim. 16, 705 (1960).
24. S. L. Adler, Phys. Rev. Letters, 14, 1051 (1965); W. I. Weisberger, Phys. Rev. Letters, 14, 1047 (1965); S. L. Adler, Phys. Rev. 140, (1965); W. I. Weisberger, Stanford preprint (1965).
25. S. Okubo, Rochester Report (1965).
26. N. Cabibbo, Phys. Rev. Letters, 10, 531 (1963).
27. M. Gell-Mann, Physics 1, 63 (1964).
28. B. Schroer and P. Stichel, preprint (1966).
29. J. W. Moffat, Toronto, Report (1965) (in the press, Phys. Rev.).
30. J. W. Moffat, and P. J. O'Donnell, Toronto Report, November (1965).
31. C. N. Yang and R. Mills, Phys. Rev. 96, 191 (1954); R.

Shaw, unpublished.

32. S. Okubo, Progr. Theor. Phys. (Kyoto), 27, 949 (1962).
33. S. Coleman and S. Glashow, Phys. Rev. Letters, 6, 423 (1961)
34. P. T. Matthews and G. Feldman, Ann. Phys. 31, 469 (1964); S. Okubo, Phys. Letters, 8, 214 (1964).
35. A. Salam and J. C. Ward, Phys. Rev. 136, B736 (1964); R. E. Marshak and S. Okubo, Nuovo Cim. 19, 1226 (1961); R. E. Marshak, N. Mukunda and S. Okubo, Phys. Letters 13, 180 (1964).
36. B. W. Lee, Phys. Rev. Letters 12, 83 (1964); H. Sugawara, Progr. Theor. Phys. (Kyoto) 31, 213 (1964); S. Okubo, Phys. Letters 8, 362 (1964); B. Sakita, Phys. Rev. Letters 12, 279 (1964); S. P. Rosen, Phys. Rev. Letters 12, 408 (1964); S. P. Rosen, preprint (1965).
37. M. Suzuki, Phys. Rev. Letters 15, 986 (1965); H. Sugawara, Phys. Rev. Letters 15, 870 (1965).
38. J. Schwinger, Phys. Rev. Letters 12, 237 (1964); Phys. Rev. 135, B816 (1964).
39. Y. Nambu and P. Freund, Phys. Rev. Letters 12, 714 (1964).

AXIAL FORM FACTORS BY THE ALGEBRA OF CURRENTS

By

R. JENGO

Istituto di Fisica Teorica dell' Università
Trieste

In this note I shall present the results obtained in three works [1] , in which the matrix elements between nucleon states of the equal time commutation relations of the algebra of SU(2) × SU(2)

$$[\overline{Q}^{(3)}, V_\mu^{(3)}] = 0 \qquad (1)$$

$$[\overline{Q}^{(-)}, V_\mu^{(3)}] = A_\mu^{(-)} \qquad (2)$$

$$[\overline{Q}^{(3)}, A_\mu^{(3)}] = 0 \qquad (3)$$

$$[\overline{Q}^{(+)}, A_\mu^{(-)}] = 2\ V_\mu^{(3)} \qquad (4)$$

have been considered. The $\overline{Q}$ are the axial charges, i.e. the space integral of the fourth component of the axial current, V_μ is the vector current, A_μ is the axial current, the subcripts (-), (3) refer to the isotopic spin.

From these commutators we can obtain relations among the matrix elements of vector and axial currents.

We study the matrix elements of the left hand side of the

---

† Lecture given at the V. Internationalen Universitätswochen für Kernphysik, Schladming, 24 February - 9 March 1966.

commutators by means of the covariant method proposed by Fubini, Furlan and Rossetti [2], i.e., we write for the various commutators

$$\lim_{q \to o} B_\mu = \text{r.h.s.}$$

where

$$B_\mu = \int d^4x \; \theta(-x_o) \; e^{-iqx} \langle N_2 | [\bar{D}(x), (V_\mu \text{ or } A_\mu)] | N_1 \rangle \;, \qquad (5)$$

and

$$\bar{D}(x) = \partial_\mu \; A_\mu(x) \; .$$

This way of approach to the calculation follows from writing formally

$$\bar{Q}(0) = \int d^3x \; A_o(\vec{x},o) = \int d^4x \; \theta(-x_o) \; \partial_\mu \; A_\mu \qquad (6)$$

Then we insert intermediate states in the commutator inside the integral in eq. (5). The eq. (6) gives trouble when one has to deal with the nucleon as intermediate state. To avoid ambiguities we ascribe a mass difference to the nucleons and at the end of the calculation we put it equal to zero. This procedure has the advantage of allowing to write in a compact manner the various contributions to B_μ in eq. (5). However, if we want, we can treat the nucleon intermediate state in a separate way, and consider eq. (5) only when we insert higher lying states, so to avoid any possible trouble connected with the shorthand eq. (6). Let us introduce some kinematics by putting

p_1 = momentum of ingoing nucleon

p_2 = momentum of outgoing nucleon

$$\Delta_\mu = (p_2 - p_1)_\mu \qquad K = p_2 + q - p_1$$

and fix $q^2 = 0$, $q\Delta = 0$, so to define $m\nu = -q.p_1 = -q.p_2$. In this way we are led to dispersion relations at fixed Δ^2 in the "ν-channel".

Decomposing B_μ in some set of invariants, which we specify in the single case,

$$B_\mu(q) = \sum_s \bar{u}(p_2)\, M_\mu^s\, \alpha_s\, (\Delta^2, \nu)\, u(p_1)$$

we can try unsubtracted dispersion relations

$$\alpha_s(\Delta^2, \nu) = \frac{1}{\pi} \int \frac{\alpha_s^I(\Delta^2, \nu')d\nu'}{\nu' - \nu} - \frac{1}{\pi} \int \frac{\alpha_s^{II}(\Delta^2, \nu')d\nu'}{\nu' - \nu}$$

The spectral functions α^I, α^{II}, can be deduced from the general quantities

$$A^I = \frac{i}{2} \sum_\rho (2\pi)^4\, \delta(p_2 + q - p_\rho) <N_2|\bar{D}|\rho><\rho|(V_\mu, A_\mu)|N_1>$$

$$A^{II} = \frac{i}{2} \sum_\rho (2\pi)^4\, \delta(p_1 - q - p_\rho) <N_2|(V_\mu, A_\mu)|\rho><\rho|\bar{D}|N_1>$$

by the same decomposition in invariants. We are interested only in the α_s which refer to the invariants M_s remaining different from zero as $q \to o$:

$$\lim_{q \to o} \sum_s \bar{u}\,(p_2) M_\mu^s\, \alpha_s(\Delta^2, 0)\, u(p_1) = \text{r.h.s of the commutator considered}$$

We start by considering the commutators (1) and (2). We know that the axial vector current is not conserved $\partial_\mu A_\mu \neq o$, so we can decompose $<N_2|A_\mu|N_1>$ into longitudinal and transverse parts. We write

$$a_\mu \equiv <N_2|A_\mu|N_1> =$$

$$= \bar{u}\,(p_2)[i\, r_A\, \gamma_5\, \gamma_\mu\, G(\Delta^2) + \gamma_5\, \Delta_\mu\, \beta(\Delta^2)]u(p_1)$$

$r_A = -1.18$, so $G(\Delta^2)$ is normalized $G(0) = 1$, and then

$$a_\mu = a_\mu^L + a_\mu^T = i\, r_A\, G(\Delta^2)\, \bar{u}_2\, \gamma_5 (\gamma_\mu + i\, 2m \frac{\Delta_\mu}{\Delta^2})\, u_1 +$$

$$+ \frac{\Delta_\mu}{\Delta^2} [2m\, r_A\, G(\Delta^2) + \Delta^2 \beta(\Delta^2)]\, \bar{u}_2\, \gamma_5\, u_1 . \qquad (7)$$

Our aim is to investigate the Δ^2 dependence of form factors (we can say nothing on r_A, because it appears in the l.h.s. and r.h.s.of our equations).

We know from dispersion calculation in the Δ^2 channel that a_μ^L has a contribution from the pion pole:

$$2m\, r_A\, G(\Delta^2) + \Delta^2 \beta(\Delta^2) = \frac{a}{\Delta^2 + m_\pi^2} + \int_{(3m_\pi)^2}^{\infty} \frac{\sigma(\alpha) d\alpha}{\alpha - \Delta^2} \qquad (8)$$

and because m_π is so small we can assume that the pion pole will dominate and obtain for $\Delta^2 = 0$ the Goldberger-Treiman relation

$$\frac{a}{m_\pi^2} = 2m\, r_A$$

We know that our dispersion rules are not valid in the Δ^2 time-like region when Δ^2 approaches singularities (i.e. they will diverge). Thus, because of the smallness of m_π, we can expect a subtraction in the dispersion integral of a_μ^L, or, at least, we can expect that the integrals involved will converge very slowly so to make very poor any approximate calculation.

On the other hand we have a reasonable estimate of a_μ^L by means of (8), so we will treat with our method only the transversal part a_μ^T

Therefore, as

$$k_\mu \xrightarrow[q \to 0]{} \Delta_\mu$$

we write

$$\lim_{q\to o} B^{(-)}(\Delta,q) = \langle N_2|A_\nu|N_1\rangle(\delta_{\nu\mu} - \frac{\Delta_\nu\Delta_\mu}{\Delta^2}), \text{ for eq. (2)}$$

$$B_\mu^{(-)} = (\delta_{\nu\mu} - \frac{k_\nu k_\mu}{k^2})\int d^4x\ \theta(-x_o)e^{-iqx}\langle N_2|[\overline{D}^{(-)}(x),V_\nu^{(3)}(0)]|N_1\rangle \tag{9}$$

and also for deeling with the same formalism,

$$\lim_{q\to o} B_\mu^{(3)}(\Delta,q) = 0, \quad \text{for eq. (1)}$$

$$B_\mu^{(3)}(\Delta,q) =$$

$$= (\delta_{\nu\mu} - \frac{k_\nu k_\mu}{k^2})\ \int d^4x\ \theta(-x_o)e^{-qxi}\langle N_2|[\overline{D}^{(3)}(x),V_\nu^{(3)}(0)]|N_1\rangle \tag{10}$$

Our quantities B_μ are formally gauge invariant in the sense $k_\mu B_\mu = 0$, so our problem looks quite similar to the problem of electroproduction of "$\overline{D}$" particles. Thus we can use the decomposition in invariants of the electroproduction problem [3]

$$B_\mu(\Delta,q) = \sum_{s=1}^{6} \alpha_i(\Delta^2,\nu)\overline{u}_2\ M_\mu^i\ u_1$$

$$M_1 = \gamma_5\ \gamma_\mu\ \gamma_\nu\ F_{\mu\nu} \qquad\qquad P = \frac{p_1+p_2}{2}$$

$$M_2 = \gamma_5\ P_\mu\ k_\nu\ F_{\mu\nu} - \frac{k^2}{4}\ M_1 \qquad\qquad F_{\mu\nu} = k_\mu\ \varepsilon_\nu - k_\nu\ \varepsilon_\mu$$

$$M_3 = \gamma_5\ \gamma_\mu\ q_\nu\ F_{\mu\nu}$$

$$M_4 = \gamma_5\ \gamma_\mu\ P_\nu\ F_{\mu\nu} - i\ \frac{m}{2}\ M_1$$

$$M_5 = \gamma_5\, q_\mu\, k_\nu\, F_{\mu\nu}$$

$$M_6 = \gamma_5\, \gamma_\mu\, k_\nu\, F_{\mu\nu}$$

Only M_1 and M_6 survive as $q \to 0$, (the subtraction represented by the photoelectric term contributes to M_5).

Calculating the nucleon pole we obtain

$$0 = F_2^V(\Delta^2) + \text{continuum, from (1)}$$

$$G(\Delta^2) = F_1^V(\Delta^2) + \text{continuum, from (2)}$$

F_1^V, F_2^V are the isovector Pauli electromagnetic form factors. To evaluate the continuum contribution we can approximate it with few resonant states. From the isobaric model for the pion photo- and electroproduction we know that the $N^*_{33}(1236)$ resonance provides with a rather good approximation of the continuum contribution in a wide energy range [4].

It is resonable to think that the same holds in our case too (we are interested in "low energy theorems" because $\nu \to 0$).

Writing

$$<N_{33}|\overline{D}|N> = -\frac{\lambda'}{m_\pi} q_\mu \overline{u}_\mu u \quad , \qquad q = p^* - p$$

and retaining at the electromagnetic vertex the M1 coupling which is known to be the dominant one

$$<N_{33}|V_\mu^{(3)}|N> = -\frac{A_1}{m_\pi} \overline{u}_\mu \gamma_\nu \gamma_5\, u\, F_{\mu\nu}$$

we obtain

$$G(\Delta^2) = F_1^V(\Delta^2) - \Delta^2 \frac{\lambda' A_1}{r_A m_\pi^2 \sqrt{3}} \frac{M+m}{3M^2} \tag{11}$$

$$0 = F_2^V(\Delta^2) + 4m^2 \frac{\lambda' A_1}{r_A m_\pi^2 \sqrt{3}} \frac{M+m}{3M^2} \tag{12}$$

m = nucleon mass. M = resonance mass.

Therefore we can write the axial form factor in the nice form, substituting eq. (12) into eq. (11),

$$G(\Delta^2) = F_1^V(\Delta^2) + \frac{\Delta^2}{4m^2} F_2^V(\Delta^2) \qquad (13)$$

(to avoid possible ambiguities we remind readers that in our notation the Sachs electric form factor is

$$G_E^{Sachs}(\Delta^2) = F_1^V(\Delta^2) - \frac{\Delta^2}{4m^2} F_2^V(\Delta^2) \;)$$

If we consider eq.(12) we can evaluate λ' by means of a Goldberger Treiman relation, i.e. we assume dominance of the pion pole

$$<\alpha|\overline{D}^{(-)}|\beta> = - \frac{\sqrt{2}\; m\; m_\pi^2}{(p_\alpha - p_\beta)^2 + m_\pi^2} \; \frac{r_A}{g} <\alpha|J_{\pi(-)}|\beta>$$

In this way we write $<N^*|D|N>$ in terms of the coupling constant λ for the vertex $N^*N\pi$, and obtain

$$\frac{F_2^V(\Delta^2)}{2} = - \frac{G\; m^2}{m_\pi^2} \sqrt{2/3}\; \frac{\lambda}{g} A_1(\Delta^2) \; \frac{m(M+m)}{6M^2}$$

From the isobaric model of photoproduction [4] we know at

$\Delta^2 = 0 \qquad \lambda A_1 = - 0.058\ g$

and thus we obtain

$$(F_2^V(0) = 3.7)$$

$$\exp(\frac{F_2^V(0)}{2}) \equiv 1.85 = 1.99 \equiv \text{th}\ (\frac{F_2^V(0)}{2})\upsilon \;,$$

i.e. an evaluation of the anomalous isovector magnetic moment which is right in the 5 % limit.

This gives also an indication of the validity limits of a formula like (13), If we suppose the existence of a 1^+ axial vector meson we can look for the polar form

$$G(\Delta^2) = \frac{M_A^2}{\Delta^2+M_A^2}$$

Assuming the dominance of ρ-mesons in the isovector electromagnetic form factors i.e.

$$F_1^V(\Delta^2) = \frac{F_2^V(\Delta^2)}{3.7} = \frac{M_\rho^2}{\Delta^2+M_\rho^2}$$

and calculating the slope at $\Delta^2 = 0$ of the r.h.s. and l.h.s. of eq. (13) we obtain

$$\frac{M_A}{M_\rho} = 1{,}6$$

which, with the actual mass for the ρ, gives

$$M_A = 1.24 \text{ GeV} .$$

A preliminary fit of $G(\Delta^2)$ from the neutrino experiment gives [5]

$$M_A = 0.84 \pm 0.30 \text{ GeV} .$$

(Actually we know that the polar form of e.m. form factors fits the experiment with a mass somewhat less than the ρ-mass)

We can look for a correction of our results allowing all the possible e.m. vertices $N_{33}- N$, and inserting the higher lying state $N_{13}(1512)$. Then it is no longer possible to use eq. (12) for substituting in eq. (11). We get for the isovector anomalous magnetic moment a better result

$$\text{th}\left(\frac{F_2^V(0)}{2}\right) = 1.90 \quad , \quad \exp\left(\frac{F_2^V(0)}{2}\right) = 1.85 .$$

For the slope of $G(\Delta^2)$, assuming the real ρ mass, in the polar term of the electromagnetic form factors we get $M_A \simeq 1.1$. (The main corrections to eq. (12) come from the N_{13} state, while for (11) come from the other forms of the N_{33} $N\gamma$ vertex, which are not so well established). The corrections remain in the limit of 5 - 10 % (in contrast to the Adler-Weisberger [6] case, which is very different from ours, because of the integral on total cross sections).

We turn now our attention to the other commutators of the algebra

$$[\overline{Q}^{(3)}, A_\mu^{(3)}] = 0 \tag{3}$$

$$[\overline{Q}^{(+)}, A_\mu^{(-)}] = 2V_\mu^{(3)} \tag{4}$$

and take the matrix element between nucleon states $|N_1\rangle$ and $|N_2\rangle$. We cut off the sum over the intermediate states at the N_{33} resonance in order to evaluate the vertex

$$\langle N|A_\mu^{(-)}|N_{33}\rangle = i\overline{u}[\,-H_1\delta_{\mu\nu} - \frac{i}{m_\pi} H_2 p_\nu \gamma_\mu + \frac{1}{m_\pi^2} H_3 p_\nu (p^* + p)_\mu +$$

$$+ \frac{1}{m_\pi^2} H_4 p_\nu (p - p^*)_\mu] u_\nu \quad .$$

The form factors for $\langle N|V_\mu|N_{33}\rangle$ are given by CVC theory, and both vertices allow to calculate the amplitude for the process

$$\nu + N \to \mu + N^*$$
$$\qquad\qquad\qquad \hookrightarrow N + \pi$$

We use the technique described before, introducing the B_μ quantities and decomposing into invariants

$$B_\mu(q, \Delta^2) = \sum_s \overline{u}\, N_\mu^s\, u\, b^s(\Delta^s, \nu)$$

The surviving invariants N_μ in the $q \to o$ limit are now

$$(p_1 + p_2)_\mu \;,\; (p_2 - p_1)_\mu \;,\; \gamma_\mu$$

We obtain an equation from the commutator (3) (the notation is the same as before), going from the $\langle N|\overline{D}|N_{33}\rangle$ to the $\langle N|J_{\overline{V}}|N_{33}\rangle$ vertex by means of a Goldberger-Treiman relation

$$\frac{m\lambda}{\sqrt{6}\, m_\pi g} \frac{M+m}{M} \left[\frac{1}{3} H_1 - \frac{M}{3m_\pi} H_2 + \frac{M}{m_\pi^2} \rho_+ H_4\right] = \frac{1}{4} M \beta \qquad (14)$$

and two equations from the commutator (4)

$$r_A \frac{m\lambda}{\sqrt{6}\, m_\pi g} \frac{M+m}{M} \left[+ \frac{2}{3} H_1 - \frac{2M}{m_\pi} \rho_- H_2\right] = r_A^2 G - F_1^V - F_2^V \qquad (15)$$

$$r_A \frac{m\lambda}{\sqrt{6}\, m_\pi g} \frac{M+m}{M} \left[- \frac{1}{3} H_1 - \frac{M}{3m_\pi} H_2 + \frac{M^2}{m_\pi^2} \rho_+ H_3\right] = \frac{M}{4m} F_2^V \qquad (16)$$

where

$$\rho_\pm = 1 \pm \frac{m}{3M} - \frac{M^2+m^2}{3M^2} - \frac{\Delta^2}{3M^2} \quad .$$

We can add another equation looking at the dispersion relation in Δ^2 channel for $<N|\overline{D}|N_{33}>$ and assuming the pion dominance

$$-H_1 + \frac{M-m}{m_\pi} H_2 - \frac{M^2-m^2}{m_\pi^2} H_3 - \frac{\Delta^2}{m_\pi^2} H_4 = -r_A \frac{\sqrt{2}\ \lambda m}{\sqrt{3}\ g m_\pi} \frac{m_\pi^2}{\Delta^2+m_\pi^2}$$

Thus we have at our disposal four equations from which in principle we can evaluate the form factors $H_i(\Delta^2)$ in terms of F_1^V, F_2^V, G and β.

We know that the pion pole is present in $\beta(\Delta^2)$ and $H_4(\Delta^2)$ only (as can be easily checked), and due to the smallness of m_π we can extrapolate our equations until $\Delta^2 = -m_\pi^2$ and equate the residua of $H_4(\Delta^2)$ as given by eq. (14) and (17). We obtain

$$\frac{\lambda^2}{g^2} = \frac{g m_\pi^2 M^2}{2m(M+m)^2(2M-m)}$$

this gives $\lambda = 1.90$, while the experimental value [7] turns out to be 2.1 . Again the agreement is in the 10 % limit.

Solving the equations for $\Delta^2 = 0$ we get

$$H_1(0) = -0.41 \quad , \qquad H_3(0) = -0.088 \quad ,$$
$$H_2(0) = -1.13 \quad , \qquad H_4(0) = 0.86 \quad .$$

If we suppose that an axial vector meson, if it exists, is coupled to the $\langle N|A_\mu|N_{33}\rangle$ vertex as well, we can get an independent estimate of its mass by looking at the slope of $H(\Delta^2)$ at $\Delta^2 = 0$. Only eq.(15) and (16) are useful for numerical calculation (in the others the pion pole dominates) and describing the e.m. form factors F_1, F_2 by means of a pole located at the ρ mass we obtain

$M_A = 1.07$ GeV from eq. (15) and

$M_A = 1.18$ GeV from eq. (16).

The agreement among these various estimates seems to be rather encouraging.

References

1. S. Fubini, G. Furlan and C. Rossetti, Nuovo Cim. (to be published); G. Furlan, R. Jengo and E. Remiddi, Istituto di Fisica Teorica di Trieste, preprint (submitted to Nuovo Cim.); G. Furlan, R. Jengo and E. Remiddi, Phys. Lett. (to be published).
2. S. Fubini, G. Furlan and C. Rossetti, Nuovo Cim. A40, 1171 (1965).
3. S. Fubini, Y. Nambu and V. Wataghin, Phys. Rev. 111, 329 (1958); Ph. Dennery, Phys. Rev. 124, 2000(1961); N. Dombey, Nuovo Cim. 31, 591 (1965).
4. M. Gourdin and Ph. Salin, Nuovo Cim. 27, 193 (1963); J. P. Loubaton, Nuovo Cim. 39, 591 (1965).
5. A. Zichichi, private communication.
6. W. I. Weisberger, Phys. Rev. Lett., 14, 1047 (1965); S. L. Adler, Phys. Rev. Lett., 14, 1051 (1965).

7. A. H. Rosenfeld, A. Barbaro-Galtieri, W. H. Barkas, P. L. Bastien, J. Kirz and M. Roos, Rev. of mod. Phys. 37, 633 (1965).

COLLINEAR GROUPS AND DYNAMICAL APPROACH TO SYMMETRY†

By

F. BUCCELLA

Istituto di Fisica dell' Università
Firenze

The success of SU(6), as long as concern multiplet structure of elementary particles, mass formulas, ω-ϕ mixing, μ_p/μ_N ratio, D/F for meson baryon couplings, Johnson-Treiman relations and s-wave non-leptonic hyperon decays, lead people to attempt to understand the origin of this symmetry more deeply. From the other side some disappointing features as the selection rule against the p-wave processes $\rho \to \pi+\pi$ and $N^* \to N+\pi$ and more generally the theoretical absence in this scheme of any $\vec{S}.\vec{L}$ interaction made soon clear the static character of this symmetry. So there were many attempts to give a relativistic extension of this rest group. Without going in historical details, I shall follow the approach suggested by Dashen and Gell-Mann [1]. They consider the sixteen covariants one can construct from two Dirac spinor fields, we can call for brevity:

$$1, \vec{\sigma}, \beta, \beta\vec{\sigma}$$

$$\gamma_5, \gamma_5\vec{\sigma}, \beta\gamma_5, \beta\gamma_5\vec{\sigma}$$

where $\vec{\sigma} = -i\beta\gamma_5\vec{\gamma}$.

The symbols written in the first two columns correspond to vector and axial vector currents, which have already a direct

---

†Lecture given at the V. Internationalen Universitätswochen für Kernphysik, Schladming, 24 February - 9 March 1966.

physical interpretation. The 144 charges one obtains from the direct product of the Dirac matrices and SU(3) matrices are supposed to follow the commutation relations one can deduce from the simple quark model:

$$[D(\gamma_i,\lambda_j),D(\gamma_h,\lambda_k)] = D(\{\gamma_i,\gamma_h\},[\lambda_j,\lambda_k]) + D([\gamma_i,\gamma_h],\{\lambda_j,\lambda_k\}) .$$

A less restrictive assumption is to suppose the simple quark model commutation relation only for the fourth components of the vector and axial vector octet charges (they form the SU(3) × SU(3) chiral algebra). Adler and Weisberger [2] deduced, from the SU(2) × SU(2) algebra of isospin and "chiral isospin", the usual connection of the divergence of the axial vector current to the pion field. a good value for G_A/G_V assuming an unsubtracted dispersion relation, correcting the SU(6) value - 1,67 down to - 1,20. An interesting possibility, which has been recently investigated, is the extension of the commutation relations of the charges to the currents. If

$$D(\gamma_i,\lambda_j) = \int A(\gamma_i,\lambda_j,x)\, d^3x$$

one can suppose that

$$[A(\gamma_i,\lambda_j,x),\ A(\gamma_h,\lambda_k,y)] = \delta(\vec{x}-\vec{y})[A(\{\gamma_i,\gamma_h\},[\lambda_j,\lambda_k],x) + A([\gamma_i,\gamma_h],\{\lambda_j,\lambda_k\},x)]$$

this statement is in contradiction with Schwinger's proof [3] of the necessity of the introduction of terms proportional to the gradient of the delta function in the commutator of the spatial and the fourth component of the electric current. However one can obtain sum rules free of Schwinger terms making integrations in such a way they disappear or, supposing they are symmetric in the internal symmetry group indices, multiplying both sides of the sum rules by a tensor antisymmetric in the same indices.

In the charge commutation relation approach one has not to assume that nature is symmetric for transformations associated with the group generated from these charges to get physical predictions: one can connect the matrix elements of the charges to observable quantities in such a way to test the sum rules derived from the commutation relations of the charges.

From the other side, the invariance for a group of transformations acting on physical states gives constraint on nature also if the generators of the group have no physical meaning. In this framework people [4] supposed the existence of a rest group symmetry spanned by the 72 generators of U(12) which commute with β : such a group contains SU_{6S} as a subgroup; for collinear processes one should have SU_{6W} [5] and the coplanar $SU_3 \times SU_3$ spanned by the generators $\lambda_i(1 \pm \sigma_y)$ for events lying in the zx-plane.

Alles and Amati [6] showed that $SU_3 \times SU_3$ coplanar is in contradiction with crossing and unitarity studying quark-quark and quark-antiquark scattering: such inconsistency does not appear to depend from the example considered.

The restriction of the rest symmetry to SU_{6S} which implies that the collinear group is $SU_3 \times SU_3$ [7] and nothing more than usual SU_3 symmetry for coplanar processes has not yet been shown to give contradictions. In the comparison of collinear groups predictions with experiment one has also to keep in account that SU_3 is not an exact theory and so some bad results may be explained in terms of breaking effects at the level of the internal symmetry group; restriction of SU_3 to $SU_{2I} \times U_1(Y)$ sends SU_{6W} in $[SU_4(T) \times SU_2(X) \times V(Y)]_W$ [8] and $SU_3 \times SU_3$ in $SU_2 \times SU_2 \times W(\sigma_2 Y)$ [7].

The inconsistency of the chain $U(6) \times U(6) \to SU_{6W} \to SU_3 \times SU_3$ does not keep off sense to a comparison of W-spin predictions with experiment: in fact also pole models violate unitarity but may also be in agreement with experiment, if their contributions are dominant. In a pole model for electromagnetic form factors with vector meson dominance the imposition of W-spin conservation does not add any contradiction and one finds from SU_{6W} in fair agreement with experiment:

$$G_M^P(q^2) = -\frac{3}{2}\,G_M^N(q^2) \quad , \quad G_E^N(q^2) = 0$$

Now I shall briefly discuss the assignment of the quarks, the pseudoscalar and vector mesons, the octet baryons and the decuplet resonances to the representations of the various collinear groups; let us begin from $SU_2 \times SU_2 \times V(\sigma_2 Y)$ where the first SU_2 corresponds to $\tau_i(1+\sigma_2)$ and the second to $\tau_i(1-\sigma_2)$; we shall classify only states of positive helicity (to obtain the representation corresponding to negative helicity one has simply to exchange the first two numbers used to classify the representations and to change the sign of the third):

$p_0\uparrow$, $n_0\uparrow \epsilon (2,1,0)$ (first SU_2, second SU_2, strange helicity)

$\Lambda_0\uparrow \quad \epsilon\ (1,1,\frac{1}{2})$

$N^*\uparrow, N\uparrow \quad \epsilon\ (3,2,0)$

$N^*\uparrow\uparrow \quad \epsilon\ (4,1,0)$

$\Lambda\uparrow \quad \epsilon\ (2,2,\frac{1}{2})$

$\Sigma\uparrow, Y^*\uparrow \quad \epsilon\ (2,2,\frac{1}{2})$ and $(3,1,-\frac{1}{2})$

$Y^*\uparrow\uparrow \quad \epsilon\ (3,1,\frac{1}{2})$

$\Xi\uparrow, \Xi^*\uparrow \quad \epsilon\ (2,1,0)$ and $(1,2,1)$

$\Omega\uparrow \quad \epsilon\ (0,0,\frac{1}{2})$

$\Omega\uparrow\uparrow \quad \epsilon\ (0,0,\frac{3}{2})$

$\pi \quad \epsilon\ (3,1,0)$ and $(1,3,0)$

$\eta \quad \epsilon\ (1,1,0)$

$X_0 \quad \epsilon\ (1,1,0)$

$K \quad \epsilon\ (2,1,-\frac{1}{2})$ and $(1,2,\frac{1}{2})$

ω_o ϵ (1,1,0)

ϕ^o ϵ (1,1,0)

ρ^o ϵ (1,3,0) and (3,1,0)

$\omega\uparrow, \rho\uparrow$ ϵ (2,2,0)

$\phi\uparrow$ ϵ (0,0,1)

K^{*o} ϵ $(2,1,-\frac{1}{2})$ and $(1,2,\frac{1}{2})$

$K^{*}\uparrow$ ϵ $(2,1,\frac{1}{2})$

When we pass from $SU_2 \times SU_2 \times W(\sigma_2 Y)$ to $[SU_4(T) \times SU_2(X) \times V(Y)]_W$ particles belonging to different representations in the first group collaps in the same multiplet of the bigger one:

p_o, n_o ϵ (4,1) (representation of $SU_4(T)$, representation of $SU_2(X)$)

Λ_o ϵ (1,2)

N, N^* ϵ (20,4)

Λ, Σ, Y^* ϵ (10,2)

Ξ, Ξ^* ϵ (4,3)

Ω ϵ (1,4)

$\pi, \rho, \omega\uparrow$ ϵ (15,1)

η, X_o ϵ (15,1) + (1,3)

$\phi\uparrow$ ϵ (1,3)

ϕ^o, ω^o ϵ (1,1)

K, K^* ϵ (4,2)

$\overline{K}, \overline{K}^*$ ϵ $(\overline{4},2)$

in SU_{6W} we have:

p_o, n_o, Λ_o ϵ 6

baryon octet and resonance decuplet ϵ56 ,

pseudoscalar and vector mesons ϵ 35 .

if we exclude V_1^o ϵ 1 .

Finally the classification in $SU_3 \times SU_3$ is the following:

quark $\uparrow$ $\in (3,1)$ (first SU_3, second SU_3)

$8^+\uparrow$, $10^+\uparrow \in (6,3)$

$10^+\uparrow\uparrow$ $\in (10,1)$

M_8, V_8^o $\in (8,1)$ and $(1,8)$

M_1, V_1^o $\in (1,1)$

$V_8\uparrow, V_1\uparrow$ $\in (3,\bar{3})$

To studv the magnetic moments or more generally the magnetic form factors we have to give an assignment for the photon: the success of the pole models suggests to choose the group property of the combination $\rho_o\uparrow + \omega\uparrow/3 - \sqrt{2}/3\ \phi\uparrow$, in $SU_2\times SU_2\times W(\sigma_2 Y)$ this means to put it in a combination of the representations $(2,2,0)$ and $(0,0,1)$. So we have only one parameter for the nucleon magnetic moments and we find the famous SU_{6S} result $\mu_p/\mu_n = -3/2$: We have also a selection rule against ϕ meson pole contribution to the isoscalar magnetic form factor.

From similar considerations we deduce together with the SU_2 prediction:

$$\mu(\Sigma^+) + \mu(\Sigma^-) = 2\mu(\Sigma^o) \tag{1}$$

the further constraints:

$$\mu(\Lambda) = -\{\mu(\Sigma^+) + 2\mu(\Sigma^-)\}\ ,\ \mu(\Sigma^o\Lambda) = \frac{\sqrt{3}}{4}\{\mu(\Sigma^+) - \mu(\Sigma^-)\} \tag{2}$$

to be compared with the different SU_3 statements:

$$\mu(\Lambda) = -\frac{1}{2}\{\mu(\Sigma^+) + \mu(\Sigma^-)\}\ ,\ \mu(\Sigma^o\Lambda) = \frac{\sqrt{3}}{2}\{\mu(\Sigma^+) + \mu(\Sigma^-)\} \tag{3}$$

The consistency there is only for $\mu(\Sigma^+) = -3\mu(\Sigma^-)$; $SU_3\times SU_3$ which contains both SU_3 and $SU_2\times SU_2\times W(\sigma_2 Y)$, predictes in terms of the magnetic moment of the proton all the others:

$$\mu(p) = \mu(\Sigma^+) = -\frac{3}{2}\mu(n) = -\frac{3}{2}\mu(\Xi^0) = -3\mu(\Xi^-) =$$

$$= -3\mu(\Sigma^-) = \sqrt{3}\mu(\Sigma^0\Lambda) = 3\mu(\Sigma^0) = -3\mu(\Lambda) \qquad (4)$$

$\mu(\Sigma^0\Lambda)$ is not known in terms of the decay rate $\Sigma^0 \to \Lambda + \gamma$, which is too short to make measurements, but can be connected through charge independence to the energy asymmetry of the final leptons in the weak decay $\Sigma^\pm \to \Lambda + e^\pm + \nu(\bar{\nu})$.

If we impose W-spin conservation and make the natural assumption of W-spin 1 for the photon, we find that the electromagnetic transition octet-decuplet are pure M_{1+} in good agreement with the photoproduction results; besides SU_{4W} gives:

$$\mu^* = \frac{2\sqrt{2}}{3}\mu_p \qquad (5)$$

which is not to far from the experimental value which is $\sim\frac{5}{4}$ bigger.

Let us now discuss the axial vector couplings and baryon-meson vertices which are related each other by PCAC hypothesis as long as we are concerned with the strangeness conserving current; if we assume that the generator $\tau^+\sigma_2$ of the collinear groups is just the integral of the corresponding β-decay axial vector current (which is the usual assumption) we obtain again SU_6 predictions:

$$-\frac{G_A}{G_V} = \frac{5}{3} \quad , \quad G_A(n \to N^{*+}) = \frac{2\sqrt{2}}{3}$$

$$G_A(\Sigma^+ \to \Lambda) = \frac{2}{\sqrt{3}} \qquad (6)$$

For strangeness changing axial vector current the situation is different between $SU_3\times SU_3$ and $SU_2\times SU_2\times W(\sigma_2 Y)$: for the first group one finds the SU_6 predictions in the same way, while $V^+\sigma_2$ is not a generator for the other group, if one makes the natural assumption that the raising spin part of the strangeness changing axial vector current belongs to the representation $(2,1,\frac{1}{2})$, the result is the branching ratio:

$$\frac{A(\Sigma^- \to n)}{A(\Lambda \to p)} = - \frac{\sqrt{2}}{3\sqrt{3}} \tag{7}$$

which is in good agreement with experiment and

$$\frac{A(Y^{*-} \to n)}{A(\Lambda \to p)} = \frac{4}{3\sqrt{3}} \tag{8}$$

to be tested in neutrino experiments. $SU_2 \times SU_2 \times W(\sigma_2 Y)$ does not say anything beyond SU_2 about the couplings of the vector current in the static limit.

$SU_3 \times SU_3$ predicts all the static weak current coupling and so nothing new can be said assuming SU_{6W} symmetry. On the contrary from SU_{4W} we find some new prediction with respect to $SU_2 \times SU_2$:

$$\frac{V(\Lambda \to p)}{V(\Sigma^- \to n)} = \frac{V(\Xi^- \to \Lambda)}{V(\Xi^0 \to \Sigma^+)} = \sqrt{3/2} \tag{9}$$

$$\frac{A(\Xi^- \to \Lambda)}{A(\Xi^0 \to \Sigma^+)} = \frac{1}{5}\sqrt{3/2} \tag{10}$$

If one believes in generalized Goldberger-Treiman relations the results found in $SU_2 \times SU_2$ and $SU_3 \times SU_3$ for axial vector couplings are still valid for f couplings of baryon-meson vertex; so we find in $SU_2 \times SU_2$:

$$\frac{f_{\Sigma^+ \Lambda \pi^+}}{f_{pn\pi^+}} = \frac{\sqrt{6}}{5} \quad , \quad \frac{f_{n\Sigma^- K^+}}{f_{p\Lambda K^+}} = \frac{-1}{3}\sqrt{2/3} \tag{11}$$

$$\frac{f_{pN^{*++}\pi^-}}{f_{pN^{*0}\pi^+}} = \frac{2\sqrt{6}}{5} \quad , \quad \frac{f_{nY^{*-}K^+}}{f_{p\Lambda K^+}} = \frac{4}{3\sqrt{3}} \tag{12}$$

These results are obtained in SU_{4W} without assuming Goldberger-Treiman relations because pseudoscalar mesons have W = 1 as axial currents. From $SU_3 \times SU_3$ one deduces for pseudoscalar couplings

$$\frac{f^*}{f} = \frac{2\sqrt{2}}{\sqrt{3}} \frac{D}{D+F} \tag{13}$$

SU_{6W} or Goldberger-Treiman relations give D/F = 3/2.

I want to consider the consequence of the relation $\frac{f^*}{f} = \frac{2\sqrt{6}}{5}$ following from SU_{4W} or $SU_2 \times SU_2 \times W(\sigma_2 Y)$ and the additional hypothesis of PCAC, first one has to define the strength f^* keeping into account that N^* has a width. Gatto and Veneziano [9] proposed to define f^{*2} by the following integral

$$f^{*2} = \frac{\mu^2}{\pi} \int_{\mu}^{\infty} \frac{\mathrm{Im}\, f_{33}}{q^2} d\omega \tag{14}$$

where μ is the pion mass, q and ω are c.m. pion momentum and energy. One can insert for Im f_{33} the Chew-Low formula: in such a way we have f^{*2} expressed in terms of f^2 and of the position of the resonance ω_r;if we put $f^{*2} = \frac{24}{25} f^2$ according to symmetries we find

$$\frac{24}{25} f^2 = \frac{\mu^2}{\pi} \int_{\mu}^{\infty} \frac{\mathrm{Im}\, f_{33}}{q^2} d\omega \tag{15}$$

the relation between f^2 and ω_r so obtained is in perfect agreement with their known values. An exact cancellation of the lowest order s-wave terms (proportional to mf^2/μ^2 and to f^2/μ) occurs with the assumed f^{*2}/ f^2 branching and may give a theoretical explanation of the smallness of this wave.

It is possible to obtain eq. (15) with a different procedure; if we consider only the 3-3 term in the Adler-Weisberger [2] sum rule, we have

$$1 - (G_V/G_A)^2 = \frac{2}{3}\,\frac{\mu^2}{\pi\, f^2}\int_\mu^\infty \frac{\text{Im } f_{33}}{q^2}\, d\omega \tag{16}$$

the limitation to the N^* contribution should reproduce the SU_6 result $(G_V/G_A)^2 = 9/25$; so we have

$$\frac{\mu^2}{\pi}\int_\mu^\infty \frac{\text{Im } f_{33}}{q^2}\, d\omega = \frac{3}{2}\, f^2(1 - 9/25) = \frac{24}{25}\, f^2 \quad . \tag{17}$$

This is a good check of the validity of the approach followed.

These considerations may be extended to resonant photoproduction and Compton scattering on nucleon.

From W-spin invariance we deduce the suppression of electric contribution to the 3-3 resonance in good agreement with experiment. We can then define

$$\sqrt{3/4\pi}\;\; \frac{f^*\mu^*}{2}\, \frac{e}{2m} = \frac{\mu^2}{\pi}\int_\mu^\infty \frac{\text{Im } M_{1+}^{(3)}}{kq}\, d\omega \tag{18}$$

where μ^* is the N-N^* transition magnetic moment in units $\frac{e}{2m}$ and $M_{1+}^{(3)}$ is the $T = 3/2$ magnetic dipole.

The integral in the right side of eqs. (14) and (18) are found to be proportional by unitarity; then

$$\mu^* = \frac{\sqrt{3}}{3}(\mu_p - \mu_n)\, \frac{f^*}{f} \tag{19}$$

which follows also from W-spin invariance from collinear N-N^* vertex (the incident pion and the photon have $W = 1$ with $W_z = 0$ and ± 1 respectively). For the symmetric value of f^* the SU_6 result $\mu^* = 2\sqrt{2}/5\ (\mu_p - \mu_n)$ is reproduced.

A quite similar analysis for resonant Compton scattering leads to the relation:

$$\frac{3\mu^{*2}}{(\mu_p - \mu_n)^2} = \frac{f^{*2}}{f^2} \tag{20}$$

which is exactly the square of eq. (19). The resonant electroproduction amplitude $M_{1+}^{(3)}(k^2,\omega)$ depends from the additional variable k^2, the squared virtual photon momentum. Again one can connect the electroproduction amplitude (at fixed q^2) to the scattering amplitude, by means of unitarity. One then verifies that eq. (19) must hold at each k^2 , among the k^2 -dependent $\mu^*(k^2)$, $\mu_p(k^2)$ and $\mu_n(k^2)$, confirming one of the most important consequences of the collinear group. The result holds for both signs of k^2 (from consideration of $\pi + N \to N + e^+ + e^-$).
The axial vector amplitude for neutrino-production of pions, analyzed along similar lines and by means of the Goldberger-Treiman relation, leads, from application of the unitarity condition to

$$\frac{f^{*2}(k^2)}{f^2(k^2)} = \frac{24}{25}$$

I want to use the definition (14) of the couplings of the pseudoscalar mesons between octet baryons and decuplet resonances to establish a connection between Johnson-Treiman [10] relations and other symmetry predictions, using sum rules deduced by chiral U(3)×U(3) algebra. To this extent I shall evaluate between proton states the commutators:

$$[Q_A^{\pi^+} , Q_A^{\pi^-}] = 2I_3$$

$$[Q_A^{K^+} , Q_A^{K^-}] = 3Q - 2I_3$$

$$[Q_A^{K^0} , Q_A^{\overline{K}^0}] = 3\,Q - 4I_3 \qquad (21)$$

finding:

$$\frac{r_A^2}{f^2}\, I_{\pi^-} = 1$$

$$\frac{r_A'^2}{f'^2}\, I_{\overline{K}^0} = 1$$

$$\frac{r_A''^2}{f''^2} I_{K^-} = 2 \qquad (22)$$

where

$$I_{\pi^-} = \frac{\mu^2}{2\pi} \int_\mu^\infty \frac{d\omega}{q^2} (\mathrm{Im}\, f_{\pi^- p} - \mathrm{Im}\, f_{\pi^+ p}) \; .$$

$f_{\pi p}$ are forward amplitudes, and similarly for $I_{\overline{K}^0}$ and I_{K^-} ; also

$$r_A^2 = (r_D + r_F)^2 \; , \; r_A'^2 = (r_D - r_F)^2 \; , \; r_A''^2 = 2(r_F^2 + \frac{1}{3} r_D^2) \; .$$

Using

$$\frac{r_A^2}{f^2} = \frac{r_A'^2}{f'^2} = \frac{r_A''^2}{f''^2}$$

we get integrated Johnson-Treiman relations:

$$I_{\pi^-} = I_{\overline{K}^0} = \frac{1}{2} I_{K^-} \qquad (23)$$

If we suppose, in agreement with experiment, that Johnson-Treiman relations are valid for energies above decuplet zone (decuplet dominance breaks strongly J.T. relations:

$$\sigma_{\pi^- p} - \sigma_{\pi^+ p} = 0 \quad \text{while} \quad \sigma_{\overline{K}^0} - \sigma_{K^0 p} > 0),$$

we obtain as a consequence that octet and decuplet contributions must compensate each other, which gives the condition:

$$(f^2 - \frac{2}{3} f^{*2}):(f'^2 + \frac{1}{3} f^{*2}):(f''^2 + \frac{1}{6} f^{*2}) = 1:1:2 \qquad (24)$$

from which we can deduce [11]

$$\frac{D}{F} = \frac{3}{2} \quad \text{and} \quad \frac{f^{*2}}{f^2} = \frac{24}{25} \quad . \tag{25}$$

This means that, while G_A/G_V discrepances from SU_6 value are due to contributions to the integrals I_M by states different from octet and decuplet (which really seem not to be negligible) the eqs. (25) are only brought by assumption of J.T. in the region above decuplet. The relations we find are better consistent with experiment then $G_A/G_V = 5/3$. If we believe only on the J.T. relation one can deduce by U(3)×U(3) collinear.

$$\sigma_{\pi^- p} + \sigma_{K^+ p} + \sigma_{\bar{K}^0 p} = \sigma_{\pi^+ p} + \sigma_{K^- p} + \sigma_{K^0 p} \tag{26}$$

one finds:

$$\frac{f^*}{f} = 2\sqrt{2/3}\ \frac{D}{D+F} \tag{27}$$

which is the same relation one obtains in U(3) U(3) collinear without assuming G.T. relations.

I want to present a theoretical analysis of sum rules deduced by the following prescription for the local commutators of the fourth components of the vector current:

$$[j_4^{(i)}(x),\ j_4^{(j)}(y)]_{x_o=y_o} = - f_{ijk}\ j_4^{(k)}(x)\ \delta(\vec{x} - \vec{y}) \ . \tag{28}$$

Performing some simple modifications to the sum rule obtained by Adler |11'| for the contribution of the vector weak current of strong interacting particles to the neutrino-production, one obtains the following sum rule for electroproduction:

$$1 = \frac{(G_E^V(q^2))^2}{1+q^2/4m^2} + \frac{q^2}{4m^2}\ \frac{(G_M^V(q^2))^2}{1+q^2/4m^2} +$$

$$+ \frac{q^2}{2\pi^2\alpha} \int_{m+\mu}^{\infty} \frac{dW}{|\vec{Q}|} [2\sigma_T^{(1/2)}(q^2,W) + 2\sigma_L^{(1/2)}(q^2,W) -$$

$$- \sigma_T^{(3/2)}(q^2,W) - \sigma_L^{(3/2)}(q^2,W)] \tag{29}$$

where $G_E^V(q^2)$ and $G_M^V(q^2)$ are Sachs electric and magnetic isovector form factors of the nucleon and the superscripts (1/2) and (3/2) indicate isovector transition to T = 1/2 and T = = 3/2 respectively. Dividing Eq. (29) by q^2 and passing to $q^2 \to 0$ one finds the sum rule:

$$\frac{1}{3} (\langle r_p^2\rangle - \langle r_n^2\rangle) = (\frac{\mu_p - \mu_n}{2m})^2 - \frac{1}{2\pi^2\alpha} \int_{(m+\mu)^2}^{\infty} \frac{dW^2}{W^2-m^2} [\sigma^{(3/2)}(W) -$$

$$- 2\sigma^{(1/2)}(W) \tag{30}$$

already derived by Cabibbo and Radicati[12]; $\langle r_p^2\rangle$ and $\langle r_n^2\rangle$ are the protons' and neutron's charge radii, defined according to

$$- \frac{1}{6} \langle r^2\rangle = (\frac{d\ G_E(q^2)}{dq^2})_{q^2=0} - \frac{1}{8m^2} \quad . \tag{31}$$

μ_p and μ_n are the proton's and the neutron's magnetic moments; $\sigma^{(3/2)}(W)$ and $\sigma^{(1/2)}(W)$ are the total cross-section for photoproduction on protons of states with T = 3/2 and 1/2 respectively through isovector transition.

We now use the optical theorem to write:

$$\sigma^{(3/2)}(W) - 2\sigma^{(1/2)}(W) = \frac{4\pi}{|\vec{Q}|} \frac{2}{3} [\mathrm{Im}\ C^{(3/2)}(0,\omega) -$$

$$- \mathrm{Im}\ C^{(1/2)}(0,\omega)] \tag{32}$$

where $C^{(3/2)}(0,\omega)$ and $C^{(1/2)}(0,\omega)$ are forward Compton-scattering amplitudes at photon energy ω = W-m for isovector photons, in the pure states T = 3/2 and T = 1/2 respectively. Let us first limit the evaluation of the integral in the sum rule

(30) to the resonant N^* contribution, taking the dominating magnetic moment term M_{1+}. If we keep only the $C^{(3/2)}$ amplitude approximated by its resonant part $C^{(3/2)}_{M,M}$

$$\mathrm{Im}\, C^{(3/2)}(o,\omega) \simeq 2\, \mathrm{Im}\, C^{(3/2)}_{M,M} \,. \tag{33}$$

We find

$$\frac{1}{3}\,(<r_p^2> - <r_n^2>) = (\frac{\mu_p - \mu_n}{2m}) - \frac{8}{3\alpha\pi} \int \frac{d\omega}{|\vec{Q}|^2}\, \mathrm{Im}\, C^{(3/2)}_{M\,M} \,. \tag{34}$$

The integral in the r.h.s. may be related to the square of the N-N^* transition magnetic moment μ^* [9]

$$\frac{e^2}{4\pi}\,(1/2m)^2\, \frac{3\mu^{*2}}{4} = \frac{1}{\pi} \int \frac{d\omega}{|\vec{Q}|^2}\, \mathrm{Im}\, C^{(3/2)}_{M,M} \,. \tag{35}$$

So we obtain:

$$\frac{1}{3}(<r_p^2> - <r_n^2>) = (\frac{\mu_p - \mu_n}{2m})^2 - 2(\frac{\mu^*}{2m})^2 \tag{36}$$

which differs from the relation derived by Lee [13] and by Dashen and Gell-Mann [1] starting from assumed commutation relations between magnetic moments

$$\frac{1}{6}\,(<r_p^2> - <r_n^2>) = (\frac{\mu_p - \mu_n}{2m})^2 - 2(\frac{\mu^*}{2m})^2 \tag{37}$$

only by a factor two on the left hand side.
Suppose $\mu^{*2} = 8/25\,(\mu_p - \mu_n)^2$ which is the SU_6 result by may also be obtained from the relation

$$\frac{3\mu^{*2}}{(\mu_p - \mu_n)^2} = \frac{f^{*2}}{f^2} \tag{20}$$

gained on the basis of purely dynamical argument and $f^{*2}/f^2 = 24/25$ which follows from assuming chiral $SU_3 \times SU_3$ commutation rule, PCAC and validity of integrated J.T. relations above decuplet zone; then eq. (36) predicts $\sqrt{\langle r_p^2 \rangle} = 0{,}52$ fermi while eq. (37) gives the better value 0,73 fermi.

However it would be surprising if (36) gave a better comparison with the data, considering that the integral in the sum rule (30) has a quite similar structure as the integral appearing in the Adler-Weisberger sum rule for G_A/G_V, where it is known that keeping only the N^* contribution results in the SU_6 prediction $G_A/G_V \equiv 5/3$ not well satisfied by experiment. To bring in a quantitative form the above considerations on the relation to the Adler-Weisberger sum rule, we compare the sum rule (30) with inserted (32)

$$\frac{1}{3}\left(\langle r_p^2 \rangle - \langle r_n^2 \rangle\right) = \left(\frac{\mu_p - \mu_n}{2m}\right)^2 - \frac{4}{3\pi\alpha} \int \frac{d\omega}{|\vec{Q}|^2} \{\mathrm{Im}\, C^{(3/2)}(o,\omega) - \mathrm{Im}\, C^{(1/2)}(o,\omega)\} \tag{38}$$

to the Adler-Weisberger sum sule, written in the form

$$(G_V/G_A)^2 = 1 - \frac{1}{3\pi f^2} \int \frac{d\omega}{|\vec{p}|^2} \{\mathrm{Im}\, f^{(3/2)}(o,\omega) - \mathrm{Im}\, f^{(1/2)}(o,\omega)\} \tag{39}$$

where f is the pion-nucleon coupling constant, $|\vec{p}|$ is the c.m. pion momentum and $f^{(T)}(0,\omega)$ is the forward scattering amplitude with isotopic spin T; eq. (39) is obtained from the definition of a covariant amplitude, we call $\phi(\nu)$, satisfying an unsubtracted dispersion relation

$$\phi(\nu) = \frac{1}{\pi} \int \frac{\mathrm{Im}\phi(\nu')}{\nu' - \nu} d\nu' \quad . \tag{40}$$

Eq. (39) then expresses the equality of $(G_V/G_A)^2$ to the limit for $\nu \to o$ of the derivative of $\phi(\nu)$

$$\lim_{\nu\to 0} \frac{\partial\phi(\nu)}{\partial\nu} = \frac{1}{\pi} \int \frac{\mathrm{Im}\phi(\nu')}{\nu'^2} d\nu'^2 \tag{41}$$

with the nucleon pole included in the integral. A corresponding amplitude $\psi(\nu)$, satisfying an unsubtracted dispersion relation, will allow us to express the sum rule (38) as the equality of the limit for $\nu \to 0$ of the derivative $\partial\psi(\nu)/\partial\nu$ to the squared isovector radius $1/3(\langle r_p^2\rangle - \langle r_n^2\rangle)$. At $\nu = 0$ the Born terms are however dominating, suggesting that we approximate the ratio of the two derivatives in terms of the pole contributions. We thus find:

$$\frac{1}{3}(\langle r_p^2\rangle - \langle r_n^2\rangle) \simeq \left(\frac{\mu_p-\mu_n}{2m}\right)^2 (G_V/G_A)^2 \tag{42}$$

We note that if one inserts into (42) the SU_6 relations (dominance of N^*)

$$(G_V/G_A)^2 = 9/25 \quad , \quad \left(\frac{\mu_p-\mu_n}{2m}\right)^2 = 25/9 \frac{\mu_p^2}{4m^2} \tag{43}$$

one obtains

$$\frac{1}{3} (\langle r_p^2\rangle - \langle r_n^2\rangle) = (\mu_p/2m)^2 \tag{44}$$

giving a too large value for μ_p. One sees from eq. (42) that the inclusion of additional contributions besides N^* produces a lowering of the prediction for μ_p^2 in the same way as they lower the G_A/G_V value from that predicted with N^* dominance.

Before coming to the comparison with data we extend our analysis to the sum rules analogous to (29) that hold for currents with strangeness.

Following the same line of reasoning used to derive eq. (42) we find, supposing the validity of SU_3:

$$\frac{1}{3} \langle r_p^2\rangle \simeq \frac{1}{4m^2} (\mu_p^2+\mu_p\mu_n+\mu_n^2)\frac{3}{4r^2-2r+1} (G_V/G_A)^2 \tag{45}$$

and

$$\frac{1}{3}(\langle r_p^2\rangle+2\langle r_n^2\rangle) \simeq \frac{1}{4m^2}(\mu_p+2\mu_n)^2(1-2r)^{-2}(G_V/G_A)^2 \tag{46}$$

where r is related to the F/D ratio of the axial currents

$$r = \frac{F}{F+D} \tag{47}$$

We can look at eqs. (42), (45) and (46) as at a system of equations in the three unknown quantities r, $\langle r_p^2\rangle$ and $\langle r_n^2\rangle$ with μ_p, μ_n and G_V/G_A inserted as known parameters. The system has two real solutions for r: the first gives, independently of μ_p/μ_n ratio

$$\left(\frac{D}{F}\right)_{\text{axial current}} = \left(\frac{D}{F}\right)_{\text{magnetic moments}} \tag{48}$$

in agreement with data [14] . The second solution gives for the axial current (D/F) $\simeq$ - 11, with the experimental value of μ_p/μ_n , and is therefore physically unacceptable. We must therefore choose the first solution, yielding the result (48) and also

$$\langle r_n^2\rangle = 0 \tag{49}$$

$$\frac{1}{3}\langle r_p^2\rangle = \left(\frac{\mu_p-\mu_n}{2m}\right)^2(G_V/G_A)^2 \tag{50}$$

equs. (49) and (50), the latter giving $\sqrt{\langle r_p^2\rangle} = 0{,}75$ f , are in fair agreement with experiment.

To conlude it seems to me worth to be remarked that the discrepancy between eq. (36) and (37) may be understood in terms of some incompatibility of the quark model commutation rules between the spatial components of the vector current and between the time components of the same currents.

References

1. R. F. Dashen and M. Gell-Mann, Phys. Lett. 17, 142 and 145 (1965).
2. S. L. Adler, Phys. Rev. Lett. 14, 1047 (1965); W. I. Weisberger, Phys. Rev. Lett. 14, 1051 (1965).
3. J. Schwinger, Phys. Rev. Lett. 3, 296 (1959).
4. K. Bardakci, J. M. Cornwall, P. G. O. Freund and B. W. Lee Phys. Rev. Lett. 13, 698 (1964); 14, 48 and 264 (1965).
5. H. J. Lipkin and S. Meshkov, Phys. Rev. Lett. 14, 670 (1965) K. J. Barnes, Phys. Rev. Lett. 14, 798 (1965).
6. W. Alles and D. Amati, Nuovo Cim. 39, 758 (1965).
7. D. V. Volkoff, JETP Lett. 1, 129 (1965); F. Buccella and R. Gatto, Nuovo Cim. 40, 684 (1965).
8. F. Buccella, Nuovo Cim. 41, 613 (1966).
9. R. Gatto and G. Veneziano, Phys. Lett. 19, 512 (1965) and 20, 439 (1966).
10. K. Johnson and S. B. Treiman, Phys. Rev. Lett. 14, 189 (1965).
11. G. Veneziano, Florence TH 66/5 (to be published)

11'.S. L. Adler, Phys. Rev. 143, B1144 (1966).

12. N. Cabibbo and L. Radicati, Phys. Lett. 19, 697 (1966).
13. B. W. Lee, Phys. Rev. Lett, 14, 676 (1965).
14. W. Willis et al., Phys. Rev. Lett. 13, 291 (1964).

RADIATIVE CORRECTIONS IN NUCLEON-β-DECAY AND ELECTROMAGNETIC FORM FACTORS†

By

G. KÄLLEN

Department of Theoretical Physics, University of Lund
Lund, Sweden

Summary

The calculation of the radiative corrections in ordinary β-decay is reviewed with particular emphasize on the modifications necessary because of the fact that the nucleons are not point particles but have a finite extension. The calculations are conveniently performed in a particular gauge where the field operator renormalization constants are all finite (to order e^2). In this gauge one obtains a finite result when one uses a dispersion theoretic approach and neglects all intermediate states with mesons. The result contains the strong interactions only through the conventional nucleon form factors and one new set of form factors which are not experimentally known today. However, in principle they can be observed from independent experiments.

1. Introduction

The radiative corrections to ordinary β-decay were worked out many years ago by several authors [1]. In these calcu-

† Lecture given at the V. Internationalen Universitätswochen für Kernphysik, Schladming, 24 February - 9 March 1966

lations one usually treats all particles taking part in the process as point particles. Using standard techniques one finds that the diagrams in Fig. 1 contribute to the process.

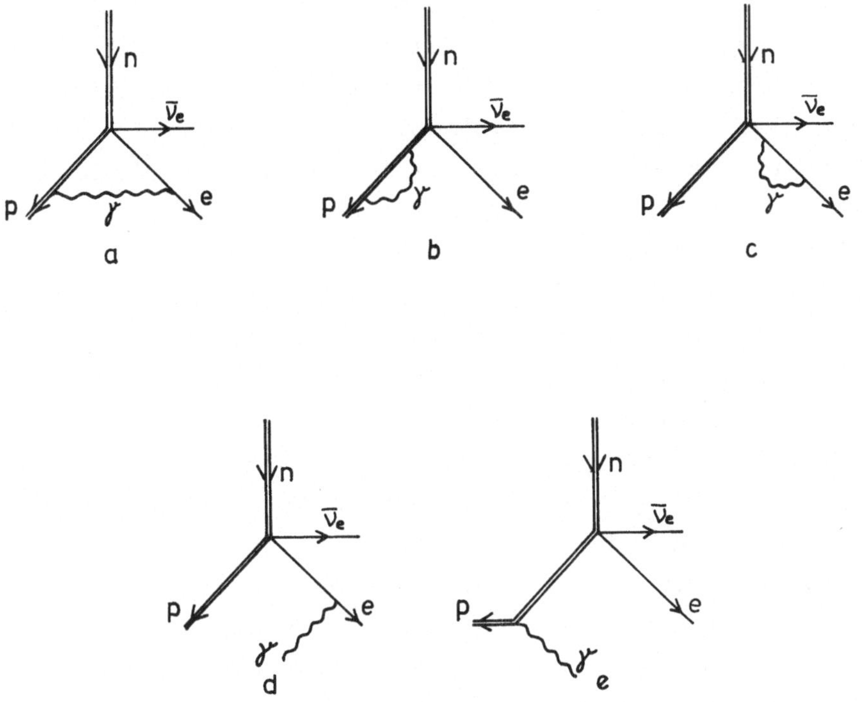

Fig. 1

Diagrams contributing to radiative corrections in β-decay.

The first three of these diagrams correspond to virtual photon processes while the two last diagrams d and e correspond to the emission of real photons. As is well-known, various infinities appear in these calculations and one usually tries to handle them with standard renormalization technique. The two diagrams b and c give rise to electromagnetic self energies of the proton and of the electron as well as to infinite renormalizations of the field operators. These last quantities are both ultraviolet and infrared divergent. Diagram a also gives rise to an ultraviolet and an infrared singularity. Diagrams d and e, finally, contain only infrared singularities. When the contributions from all the diagrams in Fig. 1 are added, one finds, as usual, that the infrared divergences cancel. However, in contradistinction to the experience from other electrodynamical calculations one finds that the contributions from the three diagrams a, b and c together give rise to a logarithmic singularity. Therefore, an arbitrary cut off quantity has to be introduced. Evidently, this situation is very unsatisfactory. The physical reason which is usually given for the logarithmic divergence of the radiative correction to β-decay that the calculations are conventionally done disregarding all problems connected with the structure of the nucleons. However, one knows experimentally that the proton and the neutron have a structure, and this structure is expected to introduce a modification in the calculation when the virtual energy of the photon becomes of the order of magnitude of the inverse of the charge radius of the nucleon. In practice, this means that one expects an important modification of the calculation based on point particles at energies somewhere between a few π-meson masses and a few nucleon masses. Consequently, the cut off energy which is introduced by brute force in the calculation is normally taken to be of the order of magnitude of the nucleon mass. As the result depends logarithmically on this cut off energy, the exact value of this energy is not too important. Further, the actual radiative corrections which one calculates in this way turn out to be rather small - of the order of magnitude one per cent. However, the experimental accuracy in the determination of the coupling constant in β-decay etc. is of the order of magnitude 0.1 per cent and, therefore, the radiative corrections are not insignificant. Further, they are mainly of interest as far as a comparison between the coupling constant appearing in μ-particle decay and in the vector part of the nuclear β-interaction is concerned. According to the conserved vector current theory, these two coupling constants should be equal [2]. However, more recent speculations in this field based on an application of the SU_3 symmetry to weak interactions, would allow for a slight difference in these two coupling constants and even relate this difference to the occurrence of other decay processes. [3] In any case, it is important for these arguments that one is able to make as precise a determination of the vector part of the coupling constant in ordinary β-decay as possible [4]. From the point of view of principle, it is therefore of interest to try to modify the theory for the radiative corrections in β-decay in such a way as to be able to take into account nucleon structure.

Experimentally it is known that both the neutron and the proton have a certain finite extension usually described with the aid of electric and magnetic form factors. These form factors have been determined from electron-nucleon scattering experiments and are, essentially, functions of the momentum transfer in the scattering process. However, both the final and the initial particles in these scattering experiments are on their mass shells. In the radiative processes described by the diagrams in Fig. 1 intermediate nucleons off the mass shells also appear. Therefore, the known electromagnetic form factors do not seem to be enough to calculate the modifications of the diagrams in Fig. 1 because of nucleon structure.

Another problem appears here because the virtual photon in diagram c interacts only with the electron which does not have any structure (as far as we know). Consequently, the field operator renormalization from that diagram is infinite while all the other contributions from the diagrams in Fig.1 are, at least in principle, influenced by the nucleon structure and, therefore, probably finite. Consequently, it appears that first of all we do not know enough about the structure of the nucleon to be able to calculate in a reliable way all the diagrams in Fig. 1 and, second, that even if enough information about off shell form factors were available, we could still never hope to get a total finite contribution from our calculation because of diagram c.

The main point which we should like to make here is that the situation is perhaps not quite so hopeless as it might appear at the first moment. The field operator renormalization constants coming from diagrams b and c are gauge dependent quantities. Therefore, any contribution of this kind does not have an absolute significance but can be changed considerably with the aid of gauge transformation. In an exact calculation such a gauge transformation would be of no significance except for transferring some of the cancellations between infinities which occur between diagrams a, b and c to a cancellation of infinities within diagram a itself. However, the calculation which we are going to make below is not exact but based on various approximations. Therefore, the question of gauge becomes important. We shall prefer to work in a gauge which is non-covariant but such that no actual infinities appear in the field operator renormalizations from diagrams b and c even for point particles. Therefore, only diagram a is (ultraviolet) divergent and this divergence is effectively cut off by the nucleon form factors. The non-covariance of the gauge which we are going to use causes no serious practical difficulty in the calculations. In an exactly gauge invariant formalism the non-covariant terms have to drop out in the final result. Because of the approximate character of our scheme, this cancellation is not quite trivial in our calculation. Nevertheless, the non-covariant terms do drop out apart, possibly, from terms which are very small when the effective cut off becomes very large.

Next comes the question about the contribution from form factors off the mass shell. Neglecting finer details like γ-matrices etc. it is well-known that diagram a in Fig. 1

gives a contribution essentially proportional to the following integral

$$I = \int \frac{dk}{(k^2+\mu^2)_F[(q_e-k)^2+m^2]_F[(q_p+k)^2+M^2]_F} \quad (1)$$

Here, k is the four vector of the virtual photon, q_e is the energy momentum vector of the external electron and q_p is the corresponding quantity for the external proton. The symbol M stands for the mass of the proton, m denotes the mass of the electron and μ denotes the small photon mass which is introduced to handle the infrared problem. The index F on the denominators in Eq.(1) denotes, as usual, the Feynman prescription for the integration over the four momentum vector k. Off hand, this integral appears to get contributions from essentially everywhere in momentum space when the integrations over the vector k is carried out. However, in reality we integrate over an algebraic denominator and it is well-known that the actual contributions to the integral can be written as a sum of residues of the poles of the integrand, i.e., the zeros of the denominator. Consequently, the integral gets contributions only from points where either $k^2 = 0$, $(q_e - k)^2 + m^2 = 0$ or $(q_p + k)^2 + M^2 = 0$. In another way of thinking we can say that these contributions occur only when at least on of the particles corresponding to the internal lines is on its mass shell. Such a way of describing the situation is also closely related to more old fashioned perturbation theory where one sums over intermediate states with virtual particles, all of which are on their mass shells, but where energy conservation does not hold at each vertex. This line of reasoning suggests that only form factors where at least one of the intermediate particles is on its mass shell is going to contribute effectively. One of our main tasks below will be to give a reasonable motivation for this conclusion. For this purpose we shall not base our computations on the standard

Feynman diagram technique, but use other methods more directly emphasizing the importance of the various intermediate states which appear in old fashioned perturbation theory. Alternatively, our technique can be described as a dispersion theoretic method taking into account some intermediate states but not all. At the end of the discussion it turns out that we obtain a final result which depends on the strong interaction effects partly through the standard electromagnetic form factors of the nucleon and partly through a similar set of other form factors not available from experiments so far. However, in principle these new form factors can also be measured. Because this second set of form factors is not available today, we shall not be able to arrive at definite numerical conclusions.

2. Choice of Gauge

As has already been mentioned in the introduction, we want to perform our calculations in a particular gauge where the field operator renormalization coming from diagrams b and c are finite also for point particles. As is well-known, this is not the case for the standard gauge of Gupta and Bleuler which is otherwise very convenient to use in practical calculations. The main advantage of this gauge is the fact that we have, simultaneously, the two equations

$$[A_\mu(x), A_\nu(x')] = -i\delta_{\mu\nu} D(x'-x) \quad , \tag{2a}$$

$$\langle 0|\{A_\mu(x), A_\nu(x')\}|0\rangle = \delta_{\mu\nu} D^{(1)}(x'-x) \quad . \tag{2b}$$

Here, $A_\mu(x)$ denotes, as usual, the electromagnetic four potential while $D(x'-x)$ and $D^{(1)}(x'-x)$ denote conventional singular functions [5]. To obtain Eq. (2b) we have used the particular case of the Gupta-Bleuler formalism where the vacuum contains no photons, neither transverse nor longitudinal or scalar. However, the Gupta-Bleuler formalism with this particular vacuum is not the only possible choice. In general, we can replace the operator of the electromagnetic field by any expression which differs from the Gupta-Bleuler $A_\mu^{GB}(x)$ by a term proportional to a gradient

$$A_\mu(x) = A_\mu^{GB}(x) + \frac{\partial\Lambda(x)}{\partial x_\mu} \quad . \tag{3}$$

In what follows we shall find it convenient to use the following special function $\Lambda(x)$

$$\Lambda(x) = \frac{-i}{\sqrt{V}} \sum_k \sqrt{\omega/2}\ \{e^{ikx}\ C(k)[a^{(3)}(\bar{k}) + i\ a^{(4)}(\bar{k})] -$$

$$- e^{-ikx}\ C^*(k)\ [a^{(3)^*}(\bar{k}) + i\ a^{(4)^*}(\bar{k})]\} \quad . \tag{4}$$

The function C(k) in Eq. (4) is an arbitrary, complex c-number depending on k, in general in a non-covariant way. However, we note that $\Lambda(x)$ is always a self joint operator in the Gupta Bleuler metric

$$\eta\Lambda^*(x)\eta = \Lambda^+(x) = \Lambda(x) \quad , \tag{5}$$

where η is the metric operator in the Gupta-Bleuler formalism. [5] By direct and straight forward calculations one also finds the relations

$$\langle 0|\Lambda(x)\ \Lambda(x')|0\rangle = 0 \quad , \tag{6a}$$

$$\langle 0|A_\mu^{GB}(x)\ \Lambda(x')|0\rangle = \frac{i}{(2\pi)^3} \int dk\ e^{ik(x-x')}\delta(k^2)\theta(k)C^*(k)k_\mu . \tag{6b}$$

These two results imply that the vacuum expectation value of a product of our two new operators can be written in the following form

$$<0|A_\mu(x)\ A_\mu(x')|0> =$$

$$= \frac{1}{(2\pi)^3} \int dk\ e^{ik(x-x')} \delta(k^2)\theta(k)[\delta_{\mu\nu} + 2k_\mu k_\nu \mathrm{Re}\ C(k)] \quad . \quad (7)$$

Equation (7) replaces Eq. (2a) and (2b) in our new gauge.

It is now a straight forward matter to calculate, e.g., the field operator renormalization of the electron operator $\psi_e(x)$. It can be done by considering the matrix element between the vacuum and a one-particle state of the unrenormalized field operator itself. We start from the basic Dirac equation ($e > 0$)

$$(\gamma \frac{\partial}{\partial x} + m)\ \psi_e(x) = -ie\ \gamma_\lambda\ \psi_e(x)\ A_\lambda(x) + \delta m\ \psi_e(x) \equiv f(x) \qquad (8)$$

and find to order e^2

$$<0|\psi_e(x)|q> = -\int dx'\ S_R(x-x')\ <0|f(x')|q>\ e^{\varepsilon x'_0} \ , \qquad (9)$$

$$<0|f(x)|q> = -\ ie\ \gamma_\lambda\ <0|\psi_e^{(1)}(x)\ A_\lambda^{(in)}(x)|q> -$$

$$-\ ie\ \gamma_\lambda\ <0|\psi_e^{(in)}(x)\ A_\lambda^{(1)}(x)|q> + \delta m^{(2)} <0|\psi_e^{(in)}(x)|q> \ , \qquad (9a)$$

$$\psi_e^{(1)}(x) = ie \int dx'\ S_R(x-x')\gamma_\lambda\ \psi_e^{(in)}(x')\ A_\lambda^{(in)}(x') e^{\varepsilon x'_0}, \qquad (9b)$$

$$A_\mu^{(1)}(x) = -\ ie \int dx'\ D_R(x-x')\ :\ \overline{\psi}_e^{(in)}(x')\gamma_\mu\ \psi_e^{(in)}(x'):\ e^{\varepsilon x'_0} \qquad (9c)$$

In these equations we have introduced a factor $e^{\varepsilon x'_o}$ to make all integrations properly convergent at $x'_o = -\infty$. In the literature, this damping factor is normally referred to as an "adiabatic switching off of the interaction" at infinity. The parameter ε is a small positive number, which is supposed to be put equal to zero after all integrations have been performed. This careful handling of the limit $x'_o \to -\infty$ is rather essential for our calculation. If we did not take it into account, Eq. (9) would give an indeterminate ratio 0/0.

Substituting in the formula above and grinding through all integrations one finds

$$<0|\psi_e^{unren}(x)|q> = (1 + \delta) <0|\psi_e^{(in)}(x)|q> \quad , \tag{10}$$

$$\delta = \frac{\alpha}{4\pi^2} \int \frac{dk}{m^2}\{[\delta(k^2-2qk)\theta(q-k)-\delta(k^2)\theta(-k)](1-\frac{m^2}{qk}) +$$

$$+ \frac{m^4}{(qk)^2}\delta(k^2+\mu^2)\theta(-k) - 4m^2 \operatorname{Re} C(-k)\delta(k^2)\theta(-k)\} \quad . \tag{10a}$$

The term involving $\delta(k^2-2qk)$ in Eq. (10a) comes from the second term in Eq. (9a) while the remaining terms involving $\delta(k^2)$ come from the first term on the right-hand side of Eq. (9a). Further, a small artificial photon mass μ has been introduced in one of the terms in Eq.(10a) to avoid infrared divergences. Evidently, the last term involving Re(C(k)) comes from the new gauge dependent term in Eq. (7). The terms which appear in Eq. (10a) and which do not contain this function C(k) are the conventional field operator renormalization terms in the Gupta-Bleuler gauge. We note in passing that if the self mass δm is calculated by this same method one finds a result which does not contain the function C(k) or, rather, where the terms containing this function can be put equal to zero for symmetry reasons [6]. In reality, they are strongly divergent but we can put them equal to zero by the physical requirement that the self mass must be gauge independent.

The terms exhibited in Eq. (10a) and independent of the function C(k) are divergent when the integration over the vector k is performed. We handle this divergence by the regularization technique, i.e., we introduce a set of heavy, auxiliary photons with masses μ_i and coupling constants C_i adjusted in such a way that the sum of all the constants C_i vanishes. Further, $C_1 = 1$ and $\mu_1 = \mu$ is the mass of the ordinary photon, which is very small, while all the other μ_i are very large. In this way one finds [7]

$$\delta = \frac{\alpha}{2\pi} \sum_i C_i \int_0^1 \frac{tdt}{1-t} \left\{\frac{1-2t+3t^3/4}{t^2+(1-t)\mu_i^2/m^2} + \frac{1}{4}\right\} -$$

$$- \frac{\alpha}{\pi^2} \sum_i C_i \int dk\ \delta(k^2+\mu_i^2)\ \theta(-k)\ \mathrm{Re}\ C(-k) =$$

$$= \frac{\alpha}{4\pi} \sum_i C_i \log \frac{\mu_i}{m} + \frac{3\alpha}{4\pi} \left(\log \frac{m}{\mu} - \frac{3}{4}\right) -$$

$$- \frac{\alpha}{\pi^2} \sum_i C_i \int dk\ \delta(k^2+\mu_i^2)\ \theta(-k)\ \mathrm{Re}\ C(-k) \quad . \qquad (11)$$

Next, consider functions C(k) of the following form

$$C(k) = -\frac{1}{8} \frac{1}{(nk)^2+a^2} \quad , \qquad (12)$$

where a is some constant to be determined later and n is a unit vector in some time-like direction

$$n^2 = \bar{n}^2 - n_0^2 = -1 \quad . \qquad (12a)$$

The last integral in Eq. (11) can now be evaluated with the same cut off convention as before. We find

$$- \frac{\alpha}{\pi^2} \sum_i C_i \int dk\, \delta(k^2 + \mu_i^2)\, \theta(-k)\, \mathrm{Re}\, C(-k) =$$

$$= \frac{\alpha}{4\pi} \{ \sum_i C_i \log \frac{m}{\mu_i} + \log \frac{\mu}{2a} + 1 \} \quad . \tag{13}$$

Adding the two results (11) and (13) we find that the cut off quantities C_i and μ_i ($i \neq 1$) disappear (to order e^2) and we obtain

$$\delta = \frac{\alpha}{4\pi} (\log \frac{m^3}{2a\mu^2} - \frac{5}{4}) \ . \tag{14}$$

We note that this result has a completely covariant form in spite of the fact that our gauge function (12) depends on an arbitrary time-like vector. The formal reason for this is that, with our covariant cut off technique, the integral in Eq. (13) can only depend on the square of n and this quantity is a pure number as shown in Eq. (12a). Consequently, there is no dependence on, e.g., the angle between n and q.

The particular choice

$$a = \frac{m^3}{2\mu^2} e^{-5/4} \tag{15}$$

is of special interest as Eq. (10) then simplifies to

$$\langle 0 | \psi_e^{unren}(x) | q \rangle = \langle 0 | \psi_e^{(in)}(x) | q \rangle \quad . \tag{16}$$

Consequently, our gauge is chosen in such a way that there is no field operator renormalization for the electron field at all.

Before we close this section we want to say a few words about the corresponding renormalization constant for the proton. If the proton had been a point particle, the calculation of the corresponding renormalization constant could have been done in a way completely analogous to the calculation of this quantity for the electron. The only difference would have been that Eq. (14) would appear with the mass m of the electron replaced by the mass M of the proton. However, the constant a has been determined once and for all in Eq. (15) and contains the mass of the electron. Therefore, we get instead of Eq. (16)

$$<0|\psi_p^{unren}(x)|q> = (1 + \frac{3\alpha}{4\pi} \log \frac{M}{m})<0|\psi_p^{(in)}(x)|q> \quad . \qquad (17)$$

We see that also this constant is finite although it is not equal to one. Probably, this result is not very accurate because we know that the proton has an electromagnetic structure which is possible important for the large virtual momenta considered here. However, we do not want to discuss this problem for the moment.

3. Radiative Corrections for the vector part of the β-interaction

We now proceed to a discussion of the proper radiative corrections in nucleon-β-decay and thereby restrict ourselves to the vector interaction. We have already mentioned in the introduction that we do not want to make use of standard Feynman techniques but rather of a more dispersion theoretic approach to be able to introduce form factors afterwards in a straight forward way. Consequently, we consider a Lagrangian containing two interaction parts corresponding to the electromagnetic interaction and the vector part of the β-interaction

$$\mathcal{L}^{int} = \mathcal{L}_W + \mathcal{L}_{EM} \quad , \qquad (18)$$

$$\mathcal{L}_W = -\frac{g}{\sqrt{2}}\, \bar{\psi}_e(x)\gamma_\lambda(1+\gamma_5)\ \psi_\nu(x) \times \bar{\psi}_p(x)\gamma_\lambda\ \psi_n(x) + \text{h.c.}, \quad (18a)$$

$$\mathcal{L}_{EM} = A_\lambda(x)\{\frac{ie}{2}[\bar{\psi}_p(x),\gamma_\lambda\ \psi_p(x)] - \frac{ie}{2}[\bar{\psi}_e(x),\gamma_\lambda\ \psi_e(x)]\}\ . \quad (18b)$$

Here, the electromagnetic interaction is much stronger than the weak interaction. Consequently, we imagine that we first have solved the equations of motion with all electromagnetic effects exactly included. The solution of these equations is then used in a perturbation theory approach and we calculate the transition rate for the β-decay using the weak interaction Lagrangian in Eq. (18a)

$$\frac{\delta W}{\delta \tau} = 2\pi \sum_{\substack{\text{final}\\ \text{states}}} |\int d^3x\ \langle p,e,\bar{\nu}|\mathcal{L}_W|n\rangle|^2\ \delta(E) \quad . \quad (19)$$

In Eq. (19) we have considered the particular case that a neutron is decaying into a proton, an electron, and an antineutrino. The neutron is not necessarily a free particle but may very well be bound inside a nucleus.

In calculating the matrix element of the interaction Lagrangian appearing in Eq. (19) we first remember that the neutrino field and the neutron field are not modified because of the electromagnetic interactions as they involve only neutral particles [8]. Consequently, we have

$$\langle p,e,\bar{\nu}|\mathcal{L}_W|n\rangle = \frac{-g}{V\sqrt{2}} \sum_{|z\rangle} \langle p,e|\bar{\psi}_e(x)|z\rangle\gamma_\lambda(1+\gamma_5)\ u_\nu^{(-)}(-\bar{q}_\nu)\ \times$$

$$\times\ \langle z|\bar{\psi}_p(x)|0\rangle\gamma_\lambda\ u_n^{(+)}(\bar{q}_n)\ e^{i(q_n-q_\nu)x} \quad . \quad (20)$$

The notation of Eq. (20) is essentially self explanatory. The symbols $u(\overline{q})$ denote, as usual, the plane wave solutions of the Dirac equation. The symbol V stands for the volume of quantization and the summation over $|z\rangle$ goes over all intermediate states introduced because of the electromagnetic interaction for our point particles. Further, the various momenta of the particles have been denoted by q_i where the index i indicates which particle is being considered.

We now take the electromagnetic part of the Lagrangian (18) into consideration by expanding the operators ψ_e and ψ_p in powers of the electric charge e. Up to and including order e^2 we have

$$\psi_e(x) = \psi_e^{(in)}(x) + ie \int S_R(x-x')\gamma_\lambda\, \psi_e^{(in)}(x')\, A_\lambda^{(in)}(x')\, dx' -$$

$$- e^2 \iint S_R(x-x')\gamma_\lambda\, \psi_e^{(in)}(x')\, D_R(x'-x'')[:\overline{\psi}_p^{(in)}(x'')\gamma_\lambda \times$$

$$\times\, \psi_p^{(in)}(x''): - :\overline{\psi}_e^{(in)}(x'')\gamma_\lambda\, \psi_e^{(in)}(x''):]\, dx'\, dx'' -$$

$$- e^2 \iint S_R(x-x')\gamma_\lambda\, S_R(x'-x'')\gamma_\rho\, \psi_e^{(in)}(x'')\, A_\lambda^{(in)}(x') \times$$

$$\times\, A_\rho^{(in)}(x'')\, dx'\, dx'' - \delta m^{(2)} \int S_R(x-x')\, \psi_e^{(in)}(x')dx', \quad (21)$$

together with a similar expression for the operator $\psi_p(x)$ which we do not write down explicitly. By inspection we see from Eq. (21) that the states $|z\rangle$ which contribute in Eq. (20) are of three kinds, viz. states with one proton, either the same as the proton in the final state or a different one, states with one proton and one photon and states with one proton together with one electron-positron pair. Therefore, we have

$$\langle p,e,\bar{\nu}|\mathcal{L}_W|n\rangle = \frac{-g}{V\sqrt{2}}\{\langle e|\bar{\psi}_e(x)|0\rangle\gamma_\lambda(1+\gamma_5)\ u_\nu^{(-)}(-\bar{q}_\nu)\ \times$$

$$\times\ \langle p|\bar{\psi}_p(x)|0\rangle\gamma_\lambda\ u_n^{(+)}(\bar{q}_n)\ +$$

$$+\sum_{\bar{k}}\ \langle e|\bar{\psi}_e(x)|\gamma\rangle\gamma_\lambda(1+\gamma_5)\ \ u_\nu^{(-)}(-\bar{q}_\nu)\langle p,\gamma|\bar{\psi}_p(x)|0\rangle\gamma_\lambda u_n^{(+)}(\bar{q}_n)+$$

$$+\sum_{\bar{q}_{p'}}\ \langle p,e|\bar{\psi}_e(x)|p'\rangle\gamma_\lambda(1+\gamma_5)\ u_\nu^{(-)}(-\bar{q}_\nu)\langle p'|\bar{\psi}_p(x)|0\rangle\gamma_\lambda$$

$$\times\ u_n^{(+)}(\bar{q}_n)\ +\sum_{\bar{q}_{e'}}\langle 0|\bar{\psi}_e(x)|e'\rangle\gamma_\lambda(1+\gamma_5)\ u_\nu^{(-)}(-\bar{q}_\nu)\ \times$$

$$\times\ \langle p,e,e'|\bar{\psi}_p(x)|0\rangle\gamma_\lambda\ u_n^{(+)}(\bar{q}_n)\}\quad . \tag{22}$$

We note in passing that only the first term on the right-hand side of Eq. (22) contributes in the case when the radiative corrections are neglected.

The various matrix elements appearing in Eq. (22) can be worked out directly from the expression (21) and a similar formula for the proton field. Using the result of the previous section for each term we find

$$\langle e|\bar{\psi}_e(x)|0\rangle\gamma_\lambda(1+\gamma_5)\ u_\nu^{(-)}(-\bar{q}_\nu)\langle p|\bar{\psi}_p(x)|0\rangle\gamma_\lambda\ u_n^{(+)}(\bar{q}_n)\ =$$

$$=\ [1\ +\frac{3\alpha}{4\pi}\log\frac{M}{m}]\ \bar{u}_e(\bar{q}_e)\gamma_\lambda(1+\gamma_5)\ u_\nu^{(-)}(-\bar{q}_\nu)\ \times$$

$$\times\ \bar{u}_p^{(+)}(\bar{q}_p)\gamma_\lambda\ u_n^{(+)}(\bar{q}_n)\ \frac{e^{-i(q_p+q_e)x}}{V} \tag{23a}$$

$$\sum_{\bar{k}} \langle e|\bar{\psi}_e(x)|\gamma\rangle \gamma_\lambda (1+\gamma_5)\; u_\nu^{(-)}(-\bar{q}_\nu)\langle p,\gamma|\bar{\psi}_p(x)|0\rangle \gamma_\lambda \; u_n^{(+)}(\bar{q}_n) =$$

$$= \frac{e^2}{V} \iint dx'\, dx''\; \bar{u}_e^{(+)}(\bar{q}_e)\gamma_\rho \; S_A(x''-x)\gamma_\lambda \; u_\nu^{(-)}(-\bar{q}_\nu) \times$$

$$\times \; \bar{u}_p^{(+)}(\bar{q}_p)\gamma_\tau \; S_A(x'-x)\gamma_\lambda \; u_n^{(+)}(\bar{q}_n)\; e^{-i(x''q_e + x'q_p)} \times$$

$$\times \; \langle 0|A_\rho^{(in)}(x'')\; A_\tau^{(in)}(x')|0\rangle =$$

$$= \frac{e^2}{V} \frac{e^{-i(q_e+q_p)x}}{(2\pi)^3} \int dk\; \delta(k^2+\mu^2)\theta(k)\; \bar{u}_e^{(+)}(\bar{q}_e)\gamma_\rho \; \frac{i\gamma(q_e-k)-m}{[(q_e-k)^2+m^2]_A} \times$$

$$\times \; \gamma_\lambda(1+\gamma_5)\; u_\nu^{(-)}(-\bar{q}_\nu)\; \bar{u}_p(\bar{q}_p)\gamma_\tau \; \frac{i\gamma(q_p+k)-M}{[(q_p+k)^2+M^2]_A}\gamma_\lambda \; u_n^{(+)}(\bar{q}_n) \times$$

$$\times \; \left[\delta_{\rho\tau} - \frac{k_\rho k_\tau}{4} \frac{1}{(nk)^2+a^2}\right] \quad , \tag{23b}$$

$$\sum_{\bar{q}_{p'}} \; \langle p,e|\bar{\psi}_e(x)|p'\rangle \gamma_\lambda(1+\gamma_5)\; u_\nu^{(-)}(-\bar{q}_\nu)\langle p'|\bar{\psi}_p(x)|0\rangle \gamma_\lambda \; u_n^{(+)}(\bar{q}_n) =$$

$$= \frac{e^2}{V} \frac{e^{-i(q_e+q_p)x}}{(2\pi)^3} \times$$

$$\times \int dq_{p'}\; \bar{u}_e^{(+)}(\bar{q}_e)\gamma_\rho \; \frac{i\gamma(q_p+q_e-q_{p'})-m}{[(q_p+q_e-q_{p'})^2+m^2]_A} \; \gamma_\lambda(1+\gamma_5)\; u_\nu^{(-)}(-\bar{q}_\nu) \times$$

$$\times \; \frac{1}{[(q_{p'}-q_p)^2+\mu^2]_R} \; \bar{u}_p^{(+)}(\bar{q}_p)\gamma_\rho \; (i\gamma q_{p'}-M)\gamma_\lambda \; u_n^{(+)}(\bar{q}_n) \times$$

$$< \delta(q_{p'}^2 + M^2)\theta(q_{p'}) = \frac{e^2}{V} \frac{e^{-i(q_e+q_p)x}}{(2\pi)^3} \int dk \frac{\delta((q_p+k)^2+M^2)\theta(q_p+k)}{[k^2+\mu^2]_R} \times$$

$$\times \bar{u}_e^{(+)}(\bar{q}_e)\gamma_\rho \frac{i\gamma(q_e-k)-m}{[(q_e-k)^2+m^2]_A} \gamma_\lambda(1+\gamma_5)\, u_\nu^{(-)}(-\bar{q}_\nu) \times$$

$$\times \bar{u}_p^{(+)}(\bar{q}_p)\gamma_\rho\, [i\gamma(q_p+k) - M]\gamma_\lambda\, u_n^{(+)}(\bar{q}_n) \quad , \qquad (23c)$$

$$\sum_{\bar{q}_{e'}} <0|\bar{\psi}_e(x)|e'>\gamma_\lambda(1+\gamma_5)\, u_\nu^{(-)}(-\bar{q}_\nu)<p,e,e'|\bar{\psi}_p(x)|0> =$$

$$= \frac{e^2}{V} \frac{e^{-i(q_e+q_p)x}}{(2\pi)^3} \int dk \frac{\delta((q_e-k)^2+m^2)\theta(k-q_e)}{[k^2+\mu^2]_A} \times$$

$$\times \bar{u}_e^{(+)}(\bar{q}_e)\gamma_\rho\, [i\gamma(q_e-k)-m]\gamma_\lambda(1+\gamma_5)\, u_\nu^{(-)}(-\bar{q}_\nu) \times$$

$$\times \bar{u}_p^{(+)}(\bar{q}_p)\gamma_\rho \frac{i\gamma(q_p+k)-M}{[(q_p+k)^2+M^2]_A} \gamma_\lambda\, u_n^{(+)}(\bar{q}_n) \quad . \qquad (23d)$$

We note here, in particular, that the gauge function C(k) from Eq. (12) appears only in Eq. (23b) and has its origin in the second term on the right-hand side of Eq. (22), where we sum over virtual photons. Otherwise, the expressions given above are essentially standard. From here it is a straight forward matter to substitute the expressions from Eqs. (23) in Eqs. (22) and (19) and calculate the transition probability including the radiative corrections. After some algebraic rearrangements one finds

$$\left(\frac{\delta W}{\delta \tau}\right)_{total} = \frac{1}{\tau} = \frac{1}{\tau_0} \left[1 + \frac{3\alpha}{4\pi} \log \frac{M}{m} - \frac{\alpha}{4\pi^2} \sum_i C_i \int dk \frac{\delta(k^2+\mu_i^2)\theta(k)}{(nk)^2+a^2}\right] +$$

$$+ \frac{g^2m^5}{2\pi^3} \int_0^{\varepsilon} dx\ (\varepsilon-x)^2\sqrt{x^2+2x}\ \ \frac{\alpha}{2\pi^2}\ f(x) \quad , \tag{24}$$

$$\frac{1}{\tau_0} = \frac{g^2m^5}{2\pi^3}\left[\frac{\eta^5}{30} - \frac{\eta^3}{12} - \frac{\eta}{4} + \frac{1+\varepsilon}{4}\log\ (1+\varepsilon+\eta)\right] \quad , \tag{24a}$$

$$\eta = \sqrt{\varepsilon^2+2\varepsilon} \quad , \tag{24b}$$

$$f(x) = \frac{-1}{32m}\int dk\ \Delta\ \mathrm{Sp}\Big[(i\gamma q_e-m)\gamma_\rho\,[i\gamma(q_e-k)-m]\gamma_\lambda\ \times$$

$$(1+\gamma_5)\gamma_4\gamma_{\lambda'}(1+\gamma_5)\Big]\ \mathrm{Sp}\Big[(1+\gamma_4)\gamma_\rho\ \ [i\gamma k-M(1+\gamma_4)]\gamma_\lambda(1+\gamma_4)\gamma_{\lambda'}\Big] \ , \tag{24c}$$

$$\Delta = \frac{\delta(k^2+\mu^2)\theta(k)}{[(q_e-k)^2+m^2]_A[(q_p+k)^2+M^2]_A} + \frac{\delta((q_e-k)^2+m^2)\theta(k-q_e)}{[k^2+\mu^2]_A[(q_p+k)^2+M^2]_A} +$$

$$+ \frac{\delta((q_p+k)^2+M^2)\theta(q_p+k)}{[k^2+\mu^2]_R[(q_e-k)^2+m^2]_A} \ . \tag{24d}$$

In the formula above τ_0 denotes the lifetime without any radiative corrections. The symbol ε stands for the maximum kinetic energy of the electron expressed in units of the electron rest mass while η is the corresponding momentum. Further, the nucleons have been considered non-relativistic and their momenta have been put exactly equal to zero in Eq. (24c).

Equation (24) has been written in a form where a comparison with the standard diagram technique is possible. The divergent term in the coefficient for $1/\tau_0$ would, in a normal calculation, correspond to the contributions of the two diagrams b and c in Fig. 1. However, in our gauge it actually occurs as one of the contributions from diagram a. The expression involving the function f(x) corresponds to the rest of diagram a in Fig. 1. The integral over k in Eq.(24c) contains the three residue contributions exhibited in Eq. (24d) instead of the product of Feynman denominators in Eq. (1). Because of the relation

$$\int \frac{dk}{[k^2+\mu^2]_F[(q_e-k)^2+m_e^2]_F[(q_p+k)^2+M^2]_F} = 2i\pi \int dk\ \Delta \quad , \qquad (25)$$

the two expressions are equivalent.

The integration over the vector k in Eq. (24c) is straight forward although somewhat time consuming. Most of the terms converge, the only exception being the expression

$$I = \frac{1}{32m} \int dk\ \Delta\ Sp[(i\gamma q_e - m_e)\gamma_\rho i\gamma k\ \gamma_\lambda(1+\gamma_5)\gamma_4\gamma_{\lambda'}(1+\gamma_5)] \times$$

$$\times\ Sp[(1+\gamma_4)\gamma_\rho i\gamma k\ \gamma_\lambda(1+\gamma_4)\gamma_{\lambda'}] = \frac{4}{m} \int dk\ \Delta(E_e k^2 + \omega q_e k) \quad . \qquad (26)$$

We calculate this integral by using the same regularization technique as before. This gives

$$I = \pi(1+x)[\log \frac{\mu^2}{M^2} - 5 \sum_i C_i \log \frac{\mu_i^2}{M^2} - 5] \quad . \qquad (27)$$

In this same way we find that the remaining divergent integral in Eq. (24) is given by Eq. (13) above

$$\sum_i C_i \int \frac{dk\ \delta(k^2+\mu_i)^2\theta(k)}{(nk)^2+a^2} = \frac{\pi}{4} \{\sum_i C_i \log \frac{\mu_i}{m} - \log \frac{\mu}{2a} - 1\} \quad . \qquad (28)$$

Collecting all the formulae given above and adding the finite terms not shown explicitly one finds

$$\frac{1}{\tau} = \frac{1}{\tau_0} \left[1 - \frac{3\alpha}{4\pi} \left(\sum_i C_i \log \frac{\mu_i^2}{M^2} + 1\right) + \frac{\alpha}{2\pi} \left(\log \frac{m}{\mu} - 5\right) \right. +$$

$$+ \frac{g^2 m^5}{2\pi^3} \int_0^\varepsilon dx (\varepsilon - x)^2 \sqrt{x^2+2x}\, (1+x) \frac{\alpha}{\pi} f_0(x) \quad , \tag{29}$$

$$f_0(x) = \frac{1+x}{\sqrt{x^2+2x}} \left[\ell\left(-\log \frac{m^2}{\mu^2} + 1 + \frac{x^2+2x}{(1+x)^2}\right) - \Phi(\zeta) + \frac{11}{12} \pi^2\right] , \tag{29a}$$

$$\ell = \log \left(1 + x + \sqrt{x^2 + 2x}\right) \quad , \tag{29b}$$

$$\zeta = 2\sqrt{x^2 + 2x} \left(1 + x + \sqrt{x^2 + 2x}\right) \quad , \tag{29c}$$

$$\Phi(\zeta) = \int_1^\zeta \frac{dt}{t} \log |1 + t| \quad . \tag{29d}$$

The small photon mass μ appearing in Eq. (29) disappears if the contributions from the two diagrams d and e in Fig. 1 are added. The evaluation of these terms offers no difficulty in principle and one finds

$$\frac{1}{\tau} = \frac{1}{\tau_0} \left[1 - \frac{3\alpha}{4\pi} \left(\sum_{i \neq 1} C_i \log \frac{\mu_i^2}{M^2} + 1\right) + \frac{3\alpha}{2\pi} \left(\log \frac{M}{m} - 1\right)\right] +$$

$$+ \frac{g^2 m^5}{2\pi^3} \int_0^\varepsilon dx \sqrt{x^2+2x}\, (\varepsilon - x)^2 (1+x) \frac{\alpha}{\pi} g(x) \quad , \tag{30}$$

$$g(x) = \frac{1+x}{\sqrt{x^2+2x}} \{\ell[2\ell + 1 + \frac{x^2+2x}{(x+1)^2}] - 2\Phi(\zeta) + \frac{5}{6}\pi^2\} +$$

$$+ 2[\frac{1+x}{\sqrt{x^2+2x}} \ell - 1] [\frac{\varepsilon - x}{3(1+x)^2} - \frac{3}{2} + \log [2(\varepsilon - x)]] \quad . \qquad (30a)$$

Equation (30) is essentially the standard result for the radiative corrections in nucleon-β-decay specialized for the particular case of a pure vector interaction for the weak interaction Lagrangian part. Further, a term normally already included in the Coulomb corrections has been subtracted from the radiative corrections. As has already been mentioned in the introduction our result is not convergent but depends on the cut off parameters μ_i and C_i $(i \neq 1)$.

4. Influence of Nucleon Structure on the β-Decay Rate

The result exhibited in Eqs. (30) shows an explicit dependence on the cut off quantities C_i and μ_i. In principle, the masses μ_i are much larger than both the electron mass and the proton mass. Mathematically, this cut off has entered the result through Eqs. (27) and (28). Using, e.g., the first of these expressions we have

$$\int k^2 \Delta \, dk = \int dk \, [\frac{\delta((q_e-k)^2+m^2)\theta(k-q_e)}{(k+q_p)^2+M^2} +$$

$$+ \frac{\delta((q_p+k)^2+M^2)\theta(k+q_p)}{(q_e-k)^2+m^2}] = - \pi \sum_{i \neq 1} C_i \log \frac{\mu_i}{M^2} \quad . \qquad (31)$$

By tracing the origin of the two terms appearing in Eq. (31) one finds that they come from those terms in Eq. (22) which contain, respectively, a sum over intermediate states involving one extra positron or one extra proton. The characteristic matrix elements appearing in the first of these terms are the products

$$<0|\overline{\psi}_e(x)|e'>\gamma_\lambda(1+\gamma_5)\ u_\nu^{(-)}(-\overline{q}_\nu)<p,e,e'|j_\lambda^\beta(x)|n> \qquad (32)$$

while the second term contains the expression

$$<p,e|\overline{\psi}_e(x)|p'>\gamma_\lambda(1+\gamma_5)\ u_\nu^{(-)}(-\overline{q}_\nu)<p'|j_\lambda^\beta(x)|n> \qquad (33)$$

Here, the quantity $j_\lambda^\beta(x)$ is the vector current entering into the nucleon part of the weak interaction Lagrangian. Until now we have approximated matrix elements of this quantity by treating all particles, also the nucleons, as point particles. The fact that by doing this we encounter divergent integrals shows that this approximation is not good enough. In reality, we know that those particles which have strong interactions, i.e., the proton and the neutron in Eqs (32) and (33) have a structure and it is to be expected that this structure is going to introduce an effective cut off in the formalism. We now want to discuss this aspect of the theory in some detail. Although we shall not be able to arrive at definite numerical results at this point, we still hope to be able to indicate the general idea according to which a treatment of this feature of the theory becomes possible.

According to the conserved vector current hypothesis [2] the quantity $j_\lambda^\beta(x)$ should be proportional to the electromagnetic current. Consequently, we have, e.g., for the last factor which enters in Eq. (33)

$$<p'|j_\lambda^\beta(x)|n> = \frac{2g}{V}\, e^{ix(q_n-q_{p'})}\ \overline{u}_p^{(+)}(\overline{q}_{p'})[\gamma_\lambda\ F_1^V((q_n-q_{p'})^2) +$$

$$+ \frac{\mu_V}{2M}(q_n-q_{p'})_\tau\ \sigma_{\lambda\tau}\ F_2^V(q_n-q_{p'})^2)]\ u_n^{(+)}(\overline{q}_n) \qquad , \qquad (34)$$

$$F_1^V(Q^2) = \frac{1}{2}\,[F_1^{proton}(Q^2) - F_1^{neutron}(Q^2)] \quad , \tag{34a}$$

$$F_2^V(Q^2) = \frac{1}{\mu_V}\,[\mu_p\, F_1^{proton}(Q^2) - \mu_n\, F_1^{neutron}(Q^2)] \quad , \tag{34b}$$

$$\mu_V = \frac{1}{2}(\mu_p - \mu_n) \quad . \tag{34c}$$

Here, $F_i^{proton}(Q^2)$ and $F_i^{neutron}(Q^2)$ are the conventional Pauli and Dirac form factors for the proton and neutron, respectively. Experimentally, these quantities are known for space like values of Q from nucleon-electron scattering. With the exception of $F_1^{neutron}$ they are all normalized to 1 for $Q^2 = 0$ while $F_1^{neutron}(0)$ is equal to zero. All the form factors are known to decrease for space like values of Q^2 rather rapidly. They do not differ appreciably from their values at the origin unless Q^2 is of the order of magnitude of the square of the π-meson mass. Nowadays, the experimental results are very often given not in terms of the Pauli and Dirac form factors but rather in terms of the electric and magnetic form factors $G(Q^2)$

$$G_E(Q^2) = F_1(Q^2) - \frac{Q^2}{4M^2}\,\mu\, F_2(Q^2) \tag{35a}$$

$$G_M(Q^2) = F_1(Q^2) + \mu\, F_2(Q^2) \tag{35b}$$

Recent experimental data [9] can be summarized in formulae

$$G_E^{proton}(Q^2) \approx \frac{G_M^{proton}(Q^2)}{1 + \mu_p} \approx \frac{G_M^{neutron}(Q^2)}{-\mu_n} \cong$$

$$\cong \frac{1}{(1+\xi Q^2/(4M^2))^2} \tag{36a}$$

$$G_E^{neutron}(Q^2) \approx 0 \tag{36b}$$

Here, ξ is a dimensionless phenomenological constant approximately given by

$$\xi \approx 5 \tag{36c}$$

Clearly, the form factors $F_i(Q^2)$ when substituted in Eq. (31) are going to introduce an effective cut off thereby giving $\sum_i C_i \log (\mu_i^2/M^2)$ a well defined value.

Evidently, the last factor in Eq. (33) is not the only place where the nucleon form factors enter. Even more directly, the first factor in Eq. (33) is going to contain the proton form factors. A comparison with Eq. (23c) or Eq. (21) gives

$$<p,e|\overline{\psi}_e(x)|p'> =$$

$$= \frac{ie^2}{\sqrt{V}} e^{i(q_{p'}-q_p-q_e)x} \overline{u}_e^{(+)}(\overline{q}_e)\gamma_\sigma \frac{i\gamma(q_p+q_e-q_{p'})-m}{[(q_p+q_e-q_{p'})^2+m^2]_A} \times$$

$$\times \frac{1}{(q_{p'}-q_p)_R^2} <p|j_\sigma^{EM}|p'> \quad , \tag{37}$$

$$<p|j_\sigma^{EM}|p'> = \frac{ie}{V} \overline{u}_p^{(+)}(\overline{q}_p)[\gamma_\sigma F_1^{proton}((q_{p'}-q_p)^2) +$$

$$+ \frac{\mu_p}{2M}(q_{p'}-q_p)_\tau \sigma_{\tau\sigma} F_2^{proton}((q_{p'}-q_p)^2)] u_p^{(+)}(\overline{q}_{p'}) \quad . \tag{37a}$$

If, for the moment, we neglect the terms involving the anomalous magnetic moment we find that the last term appearing in Eq. (31) should be replaced by

$$I_1 = \int dk \frac{\delta((q_p+k)^2+M^2)\theta(k+q_p)}{(q_e-k)^2+m^2} \rightarrow$$

$$\rightarrow 2\int dk\, F_1^V((k+q_p-q_n)^2)\, F_1^{proton}(k^2) \frac{\delta(k^2+2q_pk)\theta(k+q_p)}{-2Qk} , \tag{38}$$

$$Q = q_p + q_e \tag{38a}$$

Within the approximation which we have made here both the vectors q_p and q_n have only a time component. Therefore, we can write the argument of the form factor F_1^V in Eq. (38) in the following way (making use of the δ-functions appearing in the integrand)

$$(k + q_p - q_n)^2 \approx (k + (q - \frac{M_n}{M_p})q_p)^2 =$$

$$= k^2 \frac{M_n}{M_p} - (M_p - M_n)^2 \approx k^2 \qquad (39)$$

Therefore, the argument of the first form factor in Eq.(38) is, within our approximation scheme, the same as the argument of the second form factor in the same integrand.

To estimate what value we are going to get for the cut off parameters we imagine the expressions (35) substituted for the form factors on the right-hand side of Eq. (38). As we have already neglected terms involving the anomalous magnetic moment, it is consistent with our present calculation to drop also all terms involving μ in Eqs. (35) and hence we have

$$I_1 \approx \int dk\ [F_1^{proton}(k^2)]^2 \frac{\delta(k^2+2q_p k)\theta(k+q_p)}{-2Qk} \approx$$

$$\approx \int dk \frac{\delta(k^2+2q_p k)\theta(k+q_p)}{[1+\xi k^2/(4M^2)]^4(-2Qk)} \qquad (40)$$

The integration over the vector k in Eq. (40) is conveniently done by introducing an auxiliary δ-function in the following way

$$I_1 = \int_0^\infty \frac{d\alpha}{(1+\xi\alpha/(4M^2))^4} J(\alpha) \qquad (41)$$

$$J(\alpha) = \int dk \frac{\delta(k^2+2q_p k)\theta(k+q_p)}{-2Qk} \delta(k^2 - \alpha) =$$

$$= \frac{\pi}{2\sqrt{\lambda}} \log \left|\frac{\sqrt{\alpha}(Q^2-M^2-m^2)-\sqrt{\lambda}\ \sqrt{\alpha+4M^2}}{\sqrt{\alpha}(Q^2-M^2-m^2)+\sqrt{\lambda}\ \sqrt{\alpha+4M^2}}\right| , \qquad (41a)$$

$$\lambda = \lambda(-Q^2,M^2,m^2) = (Q^2+M^2+m^2)^2 - 4M^2m^2 \tag{41b}$$

Apart from a very small neighbourhood of the origin, the first term in the numerator and denominator in the logarithm in Eq. (41a) dominates over the second term and I_1 is approximately given by

$$I_1 \approx \frac{\pi}{2M^2}\int_0^\infty \frac{d\alpha}{\sqrt{\alpha}} \frac{\sqrt{\alpha+4M^2}}{[1+\xi\alpha/(4M^2)]^4} = 2\pi\int_0^\infty \frac{dx}{\sqrt{x}} \frac{\sqrt{x+1}}{(1+\xi x)^4} \approx \frac{5\pi^2}{4}\frac{1}{\sqrt{\xi}} . \tag{42}$$

The last approximation in Eq. (42) implies $\xi >> 1$ and is probably good enough for our present purposes. Consequently, Eq. (31) becomes

$$I = \frac{5\pi^2}{4}\frac{1}{\sqrt{\xi}} + I_2 = \frac{5\pi^2}{4}\frac{1}{\sqrt{\xi}} + \int dk \frac{\delta((q_e-k)^2+m^2)\theta(k-q_e)}{2Qk}. \tag{43}$$

The remaining integral I_2 in Eq. (43) is also divergent. However, according to the discussion above it should be modified by a factor corresponding to the ratio between the matrix element actually appearing in Eq. (32) and the approximation of this expression corresponding to point like particles which has been used up till now. In principle, this ratio depends on several invariant quantities. If we use only general arguments of invariance, a very large number of form factors appears. However, in the discussion of the integral I_1 we had only kept form factors which are proportional to those terms which appear already in the approximation where all particles are point like. Even if we adopt the same general idea here and simply modify the matrix elements in Eq. (32) by scalar factors, these new form factors can depend on two independent invariants. Therefore, our "Ansatz" would be

$$\langle p,e,e'|j^{\beta}_{\lambda}(x)|n\rangle = \frac{2eg}{V^2} e^{i(q_n-q_p-q_e-q_{e'})x} \frac{\bar{u}^{(+)}_e(\bar{q}_e)\gamma_\sigma u^{(-)}_e(-\bar{q}_{e'})}{[q_e+q_{e'}]^2{}_A} \times$$

$$\times \bar{u}^{(+)}_p(\bar{q}_p) \frac{i\gamma(q_p+q_e+q_{e'})-M}{(q_p+q_e+q_{e'})^2+M^2}\gamma_\lambda\, u^{(+)}_n(\bar{q}_n) \times$$

$$\times \mathcal{F}^{\text{proton}}_1((q_e+q_{e'})^2,(q_p+q_e+q_{e'})^2) \times$$

$$\times \mathcal{F}^{V}_1((q_n-q_p-q_e-q_{e'})^2,(q_p+q_e+q_{e'})^2) \tag{44}$$

The two new form factors $\mathcal{F}^{\text{proton}}_1(k^2,q^2)$ and $\mathcal{F}^{V}_1(k^2,q^2)$ have been introduced in analogy with the form factors in Eq. (38). The first of them is supposed to have its origin in the appearance of the electromagnetic interaction while the second one is caused by the weak interaction current. The two new form factors are not identical with the previous ones and formally correspond to one of the nucleons being off the mass shell. In principle, and if the conserved vector current hypothesis is accepted, both these new form factors can be determined from electromagnetic experiments where an incoming nucleon impinges upon a very heavy nucleus and is allowed to create an electron-positron pair through electromagnetic processes. However, such form factors have not yet been experimentally measured with an accuracy which is sufficient for our purpose. Anyhow, it appears reasonable to assume that also these form factors are of the same general form as those discussed previously. Consequently, we assume

$$\mathcal{F}^{i}_1(k^2,q^2) \approx \frac{C_i}{(1+\xi_i k^2/(4M^2))^2(1+\eta_i(q^2+M^2)/(4M^2))^2} , \tag{45}$$

$$C_{\text{proton}} = 1, \tag{45a}$$

$$C_V = \frac{1}{2} , \tag{45b}$$

where ξ_i and η_i are dimensionless constants assumed to be of the same order of magnitude as ξ in Eq. (36c). Therefore, the integral I_2 becomes

$$I_2 = \int \frac{dk}{Qk} \frac{\delta(k^2-2q_e k)\theta(k-q_e)}{[1+\xi_p k^2/(4M^2)]^2[1+\xi_V(q_n-q_p-k)^2/4M^2)]^2} \times$$

$$\times \frac{1}{[1+\eta_p[(q_p+k)^2+M^2]/(4M^2)]^2[1+\eta_V[(q_p+k)^2+M^2]/(4M^2)]^2} . \quad (46)$$

At this stage it is convenient to make the same approximation which we made before, viz. to replace the argument of the third factor in the denominator by k^2

$$(q_n - q_p - k)^2 \approx k^2 \quad (47)$$

Consequently, both the form factors in Eq. (46) depend only on the variables k^2 and $q_p k$. Using the same technique as before we have

$$I_2 = -2 \iint_{-\infty}^{+\infty} \frac{du dv}{u+2v} \frac{1}{(1-\xi_p u/(4M^2))^2} \frac{1}{(1-\xi_V u/(4M^2))^2} \times$$

$$\times \frac{1}{[1-\eta_p(u+2v)/(4M^2)]^2[1-\eta_V(u+2v)/(4M^2)]^2} J(u,v) \quad , \quad (48)$$

$$J(u,v) = \int dk\ \delta(k^2-2q_e k)\ \delta(k^2+u)\ \delta(q_p k+v)\ \theta(k-q_e) \approx$$

$$\approx \frac{\pi}{2Mm\sqrt{x^2+2x}} \theta(u-4m^2)\theta(-4v^2m^2+ 4uvMm(1+x)-u^2M^2-4uM^2m^2(x^2+2x)), \quad (48a)$$

$$x = \frac{E_e}{m} - 1 \quad (48b)$$

In deriving the result (48) we have neglected terms of order of magnitude m/M and powers of this number everywhere where this can be safely done. At this stage it is convenient to introduce dimensionless variables and write

$$u = 4m^2 s \quad , \tag{49a}$$

$$v = Mmt \quad . \tag{49b}$$

Using these definitions and substituting in Eq. (48) we find

$$I_2 = -\frac{\pi}{\sqrt{x^2+2x}} \frac{m}{M} \int_1^\infty ds \int_{t_2}^{t_1} \frac{dt}{t+2sm/M} \frac{1}{(1-\xi_p sm^2/M^2)^2} \frac{1}{(1-\xi_V sm^2/M^2)^2}$$

$$\frac{1}{[1-\eta_p(s+tM/(2m))m^2/M^2]^2[1-\eta_V(s+tM/(2m))m^2/M^2]^2} \quad , \tag{50}$$

$$t_{1,2} = 2\left[s(1+x) \pm \sqrt{x^2+2x}\,\sqrt{s^2-s}\,\right] . \tag{50a}$$

We here note that the first argument in both of the form factors is of the order of magnitude m^2/M^2 as long as s and t both are of the order of magnitude 1. The second variable, however, is of the order of magnitude m/M. Consequently, both the form factors assume essentially their point particle values as long as s and therefore also t is of the order of magnitude 1. Because of the explicit factor m/M in front of the integral, the corresponding contribution to I_2 can be neglected. The situation changes only when x becomes of the order of magnitude M/m. In that case, the first argument in the two form factors is still very small and can be replaced by 0 while the second argument effectively cuts the integral off. Therefore, we have

$$I_2 \approx \frac{-\pi}{\sqrt{x^2+2x}} \frac{m}{M} \int_1^\infty ds \int_{t_2}^{t_1} dt \left[1 - \eta_p mt/(2M)\right]^{-2} \left[1-\eta_V mt/(2M)\right]^{-2} \approx$$

$$\approx -\pi \frac{m}{M} \int_0^\infty \frac{dt}{\left[1-\eta_p mt/(2M)\right]^2 \left[1-\eta_V mt/(2M)\right]^2} =$$

$$= \frac{2\pi}{\eta_p} \int_0^\infty \frac{dx}{(1+x)^2(1+(\eta_V/\eta_p)x)^2} \qquad (51)$$

In writing down Eq. (51) we have assumed for definiteness that both the η's are negative. If we further put the two η's equal we find

$$I_2 = \frac{2\pi}{3} \frac{1}{\eta} \qquad (52)$$

Collecting Eqs. (52), (43) and (31) we find

$$\log \frac{2\lambda}{M} + 1 = \frac{5\pi}{4} \frac{1}{\sqrt{\xi}} + \frac{2}{3} \frac{1}{\eta} \approx \frac{5\pi}{4} \frac{1}{\sqrt{\xi}} \approx \frac{\pi\sqrt{5}}{4} \qquad (53)$$

The last approximation in Eq. (53) assumes that η and ξ are of the same order of magnitude and both about equal to five. Clearly, this estimate is more or less a guess, but it still indicates that the exact shape of the off shell form factors entering in Eq. (44) is not very significant as long as they are assumed to be of the same general form as the known electromagnetic form factors. If the result indicated in Eq. (53) is accepted, we get a finite result for the radiative corrections in Eq. (30). Indeed, we find

$$\frac{1}{\tau} = \frac{1}{\tau_0}\left[1 + \frac{3\alpha}{4\pi}\left(\frac{\pi\sqrt{5}}{4} + \frac{1}{2}\right) + \frac{3\alpha}{2\pi}\left(\log\frac{M}{m} - 1\right) + \frac{\alpha}{\pi}h(\varepsilon)\right], \quad (54)$$

$$h(\varepsilon) = \int_0^\varepsilon dx\,\sqrt{x^2 + 2x}\,(\varepsilon - x)^2(1+x)g(x)\,\times$$

$$\times\left[\int_0^\varepsilon dx\,\sqrt{x^2 + 2x}\,(\varepsilon - x)^2(1+x)\right]^{-1} \quad (54a)$$

The function $h(\varepsilon)$ in Eq. (54a) can be calculated numerically as a function of ε. Asymptotically, it is given by

$$h(\varepsilon) = -3\log(2\varepsilon) + \text{const.} \quad (55)$$

The result exhibited in Eq. (54) can be compared with the formula given by Kinoshita and Sirlin [1]. Using our notation $h(\varepsilon)$ their result for a pure vector interaction reads

$$\frac{1}{\tau} = \frac{1}{\tau_0}\left[1 + \frac{3\alpha}{2\pi}\left(\log\frac{\lambda^{KS}}{M} + \frac{3}{4}\right) + \frac{3\alpha}{2\pi}\left(\log\frac{M}{m} - 1\right) + \frac{\alpha}{\pi}h(\varepsilon)\right], \quad (56)$$

where λ^{KS} is the cut off introduced by Kinoshita and Sirlin. A comparison between Eqs. (56) and (54) gives

$$\log\frac{\lambda^{KS}}{M} = -\frac{1}{2} + \frac{\pi\sqrt{5}}{8} = 0.378 \quad (57)$$

In most numerical applications of the formula by Kinoshita and Sirlin λ^{KS} has been put equal to M. Our result (57) would provide a justification for this practice. Actually, the change in the radiative corrections implied by Eq. (57) as compared to the standard procedure would be only

$$\frac{\delta\tau}{\tau} = \frac{1}{\tau}\left[\tau(\lambda^{KS} = M) - \tau\right] = \frac{3\alpha}{2\pi}\,0.378 = 0.13\ \% \quad (58)$$

This difference is of the same order of magnitude as the experimental error involved in the determination of the β-decay coupling constant.

5. Discussion

Evidently, the numerical result arrived at in the last section and exhibited in Eq. (58) is not very reliable. Our treatment is incomplete in several respects. First of all, we have neglected the term involving the unknown $1/\eta$ in Eq. (53). If we assume that η is of the same order of magnitude as ξ, i.e., about 5, this introduces an error in the result (57) by less than 20 %. Because of the smallness of the result shown in Eq. (58) such an error is admissable. However, if η should actually turn out to be considerably smaller than ξ, the approximation made in Eq. (53) may be more questionable. Probably a more serious defect in our calculation lies in the fact that we have considered the influence of the electromagnetic form factors only in the divergent terms. Actually, the result shown, e.g., in Eq. (54) contains a term proportional to $\log(M/m)$. Such a term has its origin in contributions from virtual states with relatively high momenta of the particles involved. Therefore, it is probable that the nucleon form factors are going to influence this term too. As this term gives the largest numerical contribution to the radiative corrections in ordinary β-decay, this effect may be more significant than the change indicated by Eq. (58). Further, we have so far neglected all terms involving anomalous magnetic moment contributions. The only reason for doing this was to simplify the actual calculations as much as possible and quickly arrive at a preliminary numerical result. Evidently, this question must be reinvestigated. Before the work reported on here can be considered complete, these two effects must be taken into account. Even so, there remains the question of the magnitude of the parameters η in the new form factors $\mathcal{F}$. We can here only repeat what has been said above, viz. that this parameter is, in principle, possible to determine from independent experiments. If this can be done also in practice, we should arrive at a more definite numerical value for the Kinoshita-Sirlin cut off than the number given in Eq. (57).

Finally we should like to say a few words about the general approximation scheme which has been used here. It can be roughly so described that we have made a dispersion theoretical approach to our problem and considered those intermediate states which do contribute in first non-trivial order of perturbation theory. In principle, and in an exact calculation, there would also be other virtual states appearing containing one or more mesons. We have neglected all such contributions. First of all, they are evidently somewhat difficult to calculate. However, this is hardly enough as a motivation for not considering them, but we should like to argue that the contributions from these virtual states are probably smaller than both terms which we have considered because all such states correspond to rather high virtual energies. In this connection it is perhaps of interest to note that the terms which we neglected in Eq. (53) correspond to virtual states in Eq. (22) which have a slightly higher mass than those states which we considered. Actually, the mass difference between the various states in Eq. (22) is only given by twice the electron mass while the meson states which we have neglected have a virtual mass which is larger than those considered by at least one π-meson mass. Consequently, the numerical indications we have so far from our formalism at least do not contradict the idea that states involving virtual mesons in Eq. (22) can be neglected. This is, of course, also the main reason why only comparatively special cases of the general off shell form factors enter into our result.

References and Footnotes

1. Cf., e.g., T. Kinoshita and A. Sirlin, Phys. Rev. 113, 1652 (1959); S. M. Berman, Phys. Rev. 112, 267 (1958); S. M. Berman and A. Sirlin, Ann. Phys. 20, 20 (1962); L. Durand, L. F. Landowitz, R. B. Marr, Phys. Rev. 130, 1188 (1963). These papers contain references to earlier work by various authors.
2. S. S. Gershtein, I. B. Zeldovich, Soviet Phys. JETP, 2, 576 (1956); R. P. Feynman, M. Gell-Mann, Phys. Rev. 109, 193 (1958).
3. N. Cabibbo, Phys. Rev. Lett. 10, 531 (1963).
4. The corresponding problem in μ-particle decay offers no serious difficulty as the radiative corrections there turn out to be finite in the limit of point particles.
5. The notation used here is the same as in G. Källén, Quantenelektrodynamik, Handbuch der Physik V_1, Springer (1958).
6. To obtain this result formally one also has to symmetrize the right hand side of Eq. (18) and replace $\psi_e(x)\ A_\lambda(x)$ by $\frac{1}{2}\{\psi_e(x), A_\lambda(x)\}$.
7. Cf. ref. [5], esp. p. 303, Eq. (34,6).
8. At this stage we neglect the magnetic moment of the neutron as well as its charge distribution and consider only the static charge of the particle.
9. F. M. Pipkin, Proc. of the Oxford International Conference on Elementary Particles, September 1965.

DESCRIPTION OF $K^{\pm}p$ INTERACTIONS†

By

H. PILKUHN

University of Lund

1. Introduction

The strong interactions of kaons and other strange particles have been studied for many years. Experimentally, we now have a fairly detailed picture of kaon-hyperon associated production ($\pi p \rightarrow KY$) and of K^+p and K^-p interactions. In particular, K^-p interactions have developed into a new branch of hadron physics [1].

For many years, physicists took a special interest in strange particles. One tried to determine spins, parities and coupling constants. Proposed schemes of higher symmetries (global symmetry, Sakata model and others) had to be tested. These subjects were reviewed by Dalitz [2] four years ago. Last but not least, one expected that the strange particles might show some unusual effects, e.g. isospin, P, CP or T invariance could be broken in the strong interactions of strange particles.

Nowadays we believe that these basic questions are settled, at least if we believe in the eightfold way. The four kaons K^+, K^0, $\overline{K}^0$ and K^- form an octet together with π^+, π^0, π^- and η. There is nothing particularly strange about kaons any more.

Unfortunately, SU(3) symmetry is badly broken and a number of practical problems remain. Compared with the present theoretical understanding of πN elastic scattering, we don't understand $K^{\pm}$ scattering at all. Whereas the pion-nucleon coupling

---

† Lecture given at the V. Internationalen Universitätswochen für Kernphysik, Schladming, 24 February - 9 March 1966.

constant, $G^2_{\pi N}/4\pi = 14.9$ is known within 2 - 3 %, there are no reliable estimates of $G^2_{K\Lambda}/4\pi$ or $G^2_{K\Sigma}/4\pi$. Presumably these parameters have some value between 2 and 15.

Let me first remind you of the main features of total cross sections. At high energies, the total K^+p cross section is about 18 mb, being smaller than the corresponding π^+p cross section (30 mb) and pp-cross section (40 mb).

The total K^-p cross section is larger than the total K^+p cross section. This is due to the pion-hyperon channels $\pi\Lambda$, $\pi\Sigma$, $\pi\pi\Lambda$ etc., which are open for the K^-p system but not for the K^+p system. At small momenta the exothermic K^-p reactions cause a 1/v behaviour of $\sigma_{total}(K^-p)$. At 20 GeV/c, the difference between the total cross sections has decreased to 4 mb. Why it decreases so slowly is of course more difficult to explain. In some respects, the K^- behaves like a light antiproton. The total $\bar{p}p$ cross section is extremely large, due to the many annihilation channels. The above pion-hyperon channels correspond to the annihilation channels.

The production of kaons in pion-nucleon collisions rises to the order of a few per cent at high energies. At small energies, reactions like $\pi^-p \to K^0\Lambda$ or $\pi^-p \to K^+K^-n$ are of course suppressed because of their small phase space.

However, even in statistical models, with phase space properly included, the cross-section for kaon and hyperon production usually is too large by a factor of 5, if equal interaction volumes are taken for pions and kaons. Likewise, the matrix elements for kaon photoproduction are smaller than those of pion photoproduction.

From these facts, the conclusion has been drawn that the KNY coupling constants are smaller than the πNN coupling constant. Such arguments are, however, dangerous. For example, the production cross section for antiprotons in πp collisions is extremely small. This certainly does not imply that the pion-antinucleon coupling constant is small.

2. Low-Energy K^+p Interaction

The K^+p interaction is purely elastic up to kaon momenta of the order of 800 MeV/c. This is called the low-energy region. In this region, K^+p scattering is mainly s-wave. The corresponding phase shift varies linearly with momentum, changing from zero to -45° as the momentum increases from zero to 800 MeV/c. This is very different from π^+p scattering, which is dominantly p-wave, with a resonance at 300 MeV/c.

The one-particle exchanges which may contribute to K^+p scattering are ρ, ω, ϕ, Λ, Σ, Y_0^* and Y_1^* exchanges. From NN scattering and πN scattering, one knows that the exchange of a low-energy s-wave pion pair can also give large contributions to the exchange force. For KN scattering (both for K^+p and K^-p), the two-pion exchange force is in fact the force of longest range. Thus one could imagine that low-energy KN-scattering is dominated by two-pion exchange. This, however, seems not to be the case. Martin and Spearman [3] have analyzed K^+p scattering with the special aim of isolating the 2π-contribution. They used partial wave dispersion relations, which they solved by means of the N/D method. The effects of all one-particle exchanges were parameterized by poles using the method of Balázs. The result is expressed in terms of the s-wave $K\pi$ scattering lengths, which are determined as $-0.07/m_\pi$ and $-0.16/m_\pi$ for isospin 1/2 and 3/2 of the $K\pi$ system, respectively. I hope that Professor Hamilton will tell you more about the extraction of low-energy meson-meson scattering from meson nucleon scattering.

Qualitatively, it is clear that the long-range force must be weak if the p-wave scattering remains small up to 800 MeV/c.

3. Low-Energy $\overline{K}N$ Interactions

At zero kinetic energy, the K^-p system has three inelastic channels open, namely $\Lambda\pi$, $\Sigma\pi$ and $\Lambda\pi\pi$. The three-particle final state is fortunately negligible. In isospace, the K^-p state is a mixture of I = 0 and I = 1. Thus the $\Sigma\pi$ final states will be a mixture of I = 0 and I = 1. The $\Lambda\pi$ state has, of course I = 1. This means that for a given partial wave the S-matrix containing the zero-energy KN matrix elements is a 2 × 2 matrix for I = 0 and a 3 × 3 matrix for I = 1. The s-wave $\overline{K}N$ scattering lengths A^I have been determined by Kim and Sakitt et al. [4]. Recently the charge exchange cross section has been remeasured by Kittel, Otter and Wacek [5]. When combined with the previous measurements, it leads to the scattering lengths

$$A^0 = -\ 1.57 + i\ 0.54\ [\text{fermi}]$$
$$A^1 = -\ 0.24 + i\ 0.43\ [\text{fermi}] \qquad (1)$$

Please forgive me that I do not quote the statistical error. For many years, two sets of scattering lengths were possible. The second set gives a poor description of the recent experiments. In addition, it is ruled out by the s-d interference of the $D_{3/2}\ Y_0^*(1520)$ resonance and by the $\Sigma^-\pi^+/\Sigma^+\pi^-$ ratio in K^-d reactions. Because of the presence of inelastic channels, the scattering lengths are complex. The large negative value of the real part of the scattering length for I = 0 indicates a $\overline{K}N$ bound state not far below threshold. This is the Y_0^* state with a mass of 1405 MeV, which decays into $\Sigma\pi$. In order to discuss this in more detail, let me first remind you of a few basic formulae of multichannel scattering [2].

In the single-channel case, the connection between the scattering length A and the 1 × 1 S-matrix for s-waves is

$$S = \frac{1 + iqA}{1 - iqA} \qquad (2)$$

where q is the cms momentum. Unitarity requires that A is real. In the multi-channel case, A has to be replaced by the matrix K of Wigner and Heitler. The unitarity of S requires Hermiticity for K. Moreover, due to time-reversal invariance, S is symmetric, which means that K is symmetric too. Since a symmetric Hermitian matrix is real, the proper generalization of the single-channel case is not a complex scattering length for each separate channel, but a real matrix of "lengths". This means three real numbers for I = 0, and six for I = 1. The final formula is

$$S = \frac{1 + i\sqrt{q}\,K\sqrt{q}}{1 - i\sqrt{q}\,K\sqrt{q}} \quad ; \quad q = \begin{vmatrix} q_{\bar{K}N} & & 0 \\ & q_{\pi\Lambda} & \\ 0 & & q_{\pi\Sigma} \end{vmatrix} \tag{3}$$

In the following we shall need the partial-wave T-matrix, defined by

$$S = 1 + 2i\sqrt{q}\,T\sqrt{q} \quad . \tag{4}$$

The relation between T and K is

$$T = K(1 - iqK)^{-1} \quad \text{or} \quad T^{-1} = K^{-1} - iq \tag{5}$$

From the extended unitarity equation

$$\mathrm{Im}\,T = T^{+}\,\theta(q^2)\,q\,T \tag{6}$$

it follows that K has no threshold singularities. It is the multichannel generalization of $(1/q)$ tg δ. In the scattering length approximation, it is energy-independent.

We are now able to give the connexion between the complex scattering lengths A and the K-matrix. Let us call the $\bar{K}N$ channel the "closed" channel $\underline{c}$, and the $\Lambda\pi$ and $\Sigma\pi$ channels the "open" channels $\underline{o}$. The $\bar{K}N$ scattering length A_c is defined as follows

$$T_{c \to c} \equiv \frac{A_c}{1 - iq_c A_c} \tag{7}$$

Comparison of this expression with the K-matrix equation leads to

$$A_c = K_{cc} + i\, K_{co}(1 - iq_o K_{oo})^{-1}\, q_o\, K_{oc}\ . \tag{8}$$

Similarly, we can define complex "lengths" for the transitions $c \to o$

$$T_{co} \equiv \frac{A_o}{1 - iq_c A_c}\ , \qquad A_o = K_{co}(1 - iq_o K_{oo})^{-1}\ . \tag{9}$$

The scattering length A_c and the vector of reaction lengths A_o are related by

$$\mathrm{Im}\, A_c = A_o^+\, q_o\, A_o\ . \tag{10}$$

It is clear that an energy-independent K-matrix will give rise to some energy-dependence in the "lengths", due to the matrix

$$q_o = \begin{vmatrix} q_{\Lambda\pi} & 0 \\ 0 & q_{\Sigma\pi} \end{vmatrix} \tag{11}$$

But at least they are independent of q_c, which is decisive for the treatment of Y_o^* as a $\overline{K}N$ bound state, as we shall see below. Of course, it would be nicer to work with the K-matrix elements instead of the complex lengths, which are not even independent of each other. Unfortunately, the experimentalists are unable to measure $\Lambda\pi$ scattering, $\Sigma\pi$ scattering or the reaction $\Lambda\pi \to \Sigma\pi$. They are confined to the $\overline{K}N$ initial state. The cross sections for the transitions to the different final states are then given by

$$\left.\begin{matrix} K^-p \\ \overline{K}^o n \end{matrix}\right\} = \pi \left| \frac{A^1}{1 - iqA^1} \pm \frac{A^o}{1 - iqA^o} \right|^2 \tag{12}$$

$$\Lambda^o \pi^o = 2\pi\, \frac{q_\Lambda}{q}\, \frac{|A_\Lambda^1|^2}{1 + q^2 |A^1|^2} \tag{13}$$

$$\Sigma^0\pi^0 = \frac{2}{3}\pi \frac{q_\Sigma}{q} \frac{|A_\Sigma^0|^2}{1 + q^2|A^0|^2} = \frac{2\pi}{3q} \frac{\text{Im } A^0}{1 + q^2|A^0|^2} \tag{14}$$

$$\left.\begin{matrix}\Sigma^+\pi^- \\ \Sigma^-\pi^+\end{matrix}\right\} = \pi \frac{q_\Sigma}{q} \left|\frac{A_\Sigma^1}{1 - iqA^1} \pm \sqrt{\frac{2}{3}} \frac{A_\Sigma^0}{1 - iqA^0}\right|^2. \tag{15}$$

Here I have dropped the index c on q_c and A_c. The superscript refers the total isospin. These six reactions allow the determination of the two complex scattering lengths and two more parameters. It is customary to choose here the ratio,

$$\varepsilon = \frac{q_\Lambda|A_\Lambda^1|^2}{\text{Im } A^1} = \frac{\sigma^1(\pi\Lambda)}{\sigma^1(\pi\Lambda) + \sigma^1(\pi\Sigma)} =$$

$$= \frac{\sigma(\pi^0\Lambda^0)}{\sigma(\Sigma^+\pi^-) + \sigma(\Sigma^-\pi^+) + \sigma(\pi^0\Lambda) - 2\sigma(\Sigma^0\pi^0)} \tag{16}$$

and the relative phase ϕ between the I = 0 and I = 1 $\Sigma\pi$ amplitudes. With these parameters, the cross sections for the $\Lambda\pi$, $\Sigma^+\pi^-$ and $\Sigma^-\pi^+$ final states are given by

$$\Lambda\pi^0 = \frac{2\pi}{q}\,\varepsilon\,\frac{\text{Im } A^1}{1+q^2|A^1|^2} \tag{17}$$

$$\left.\begin{matrix}\Sigma^+\pi^- \\ \Sigma^-\pi^+\end{matrix}\right\} = \pi \frac{q_\Sigma}{q} \left(|T_\Sigma^1|^2 + \frac{2}{3}|T_\Sigma^0|^2 \pm 2\sqrt{2/3}\,|T_\Sigma^1|\cdot|T_\Sigma^0|\cos\phi\right) \tag{18}$$

For the final comparison with experiment, one has to include the Coulomb interaction in the K^-p state and the mass difference between the K^-p and $\overline{K}^0n$ states. The experimental values are [4]

$$\phi = 53.8^\circ\,, \qquad \varepsilon = 0.32 \tag{19}$$

3. The Y_o^* (1405) as a Virtual $\overline{K}N$ Bound State

Now we turn to the problem of the Y_o^* (1405). Remember that in the I = 0 state only the $\Sigma\pi$ channel is open below the $\overline{K}N$ threshold. The amplitude for $\Sigma\pi$-scattering can be written in a form analogous to (7) and (8)

$$T_{o \to o} = \frac{A_{oo}}{1-iq_o A_o} \qquad A_{oo} = K_{oo} + iq_c \frac{K_{co}^2}{1-iq_c K_{cc}} \quad . \tag{20}$$

Unitarity now requires A_{oo} to be real. This requirement is satisfied automatically since K is real and q_c is imaginary below threshold. A_{oo} is usually called the reduced K-matrix. We see that A_{oo} can have a pole at $iq_c K_{cc} = 1$. This is the location of the resonance. For the sake of calculation, it is assumed that K_{oo} can be neglected. Then K_{cc} is given by the real part of A_c, according to eq. (8). In terms of the "Binding Energy" E_B and the reduced mass μ of $\overline{K}N$ bound state, we have

$$- 2\mu\, E_B = q_c^2 \text{ (res.)} = - (\text{Re } A_c)^{-2} \tag{21}$$

Insertion of Re A_c from (1) gives 21.5 MeV binding energy, which gives a Y_o^* mass of

$$494 + 938 - 21.5 = 1410 \text{ MeV} \tag{22}$$

The width of the Y_o^* can be calculated in the same approximation. Remembering that $q_o A_{oo}$ is $\delta(\pi\Sigma \to \pi\Sigma)$ we have

$$\text{tg } \delta = \frac{\Gamma/2}{m-E} = q_o A_{oo} = \frac{iq_o q_c K_{co}^2}{1-iq_c K_{cc}} \tag{23}$$

With $K_{oo} = 0$, we have $K_{co} = A_o$ according to (9), i.e. A_o should be real. Then we can apply (10)

$$\frac{\Gamma/2}{m-E} = \frac{\text{Im } A_c}{\frac{1}{iq_c} - K_{cc}} \tag{24}$$

Taking Im A_c and K_{cc} as energy-independent, we need only find the variation of $(iq_c)^{-1}$ in the neighbourhood of E = m,

$$(iq)^{-1} = K_{cc} + (E-m) \frac{d}{dE} \left(\frac{1}{iq_c}\right)_{E=m} \quad . \qquad (25)$$

From $(iq)^{-1} = (2\mu E_B)^{-1/2}$, one finds the value $(2\mu E_B)^{-3/2}/\mu$ for the derivative, which implies

$$\frac{\Gamma}{2} = - \frac{\text{Im } A_c}{\mu} (2\mu E_B)^{3/2} = - \frac{\text{Im } A_c}{\mu} (\text{Re } A_c)^3 \quad . \qquad (26)$$

Insertion of the experimental value of A gives Γ = 37 MeV. If we use the experimental binding energy instead of Re A_c, we get Γ = 52 MeV. The experimental value is 35 MeV. This of course is by no means an "explanation" of the Y_o^*. It merely shows that the Y_o^* is strongly coupled to the $\overline{K}N$ state.

4. The Y_1^* (1385)

As you know, the Y_1^*(1385) is produced copiously in the reaction $K^- p \to \pi^+ \pi^- \Lambda$. Recent values for its width and branching ratio are [6]

$$\Gamma = (35 \pm 3) \text{ MeV}, \qquad \frac{\Gamma(Y_1^* \to \Sigma\pi)}{\Gamma(Y_1^* \to \Lambda\pi)} = 0.14 \pm 0.03 \quad . \qquad (27)$$

It is a $3/2^+$ state like the N_{33}^* . There is an attempt by Martin [7] to explain the Y^* by means of partial wave dispersion relations for $\pi\Lambda$-scattering analogous to the N_{33}^* -case. He finds $G_{\Sigma\Lambda\pi}^2/4\pi$ = 11. He also finds that the dynamics of the resonance depends as much on the exchange of a low-energy T = 0 s-wave π-π pair as on the exchange of a Σ hyperon. The absolute magnitude of the π-π exchange contribution is about the same as in the N_{33}^* - case. I wonder how this analysis would change if the $\Sigma\pi$ channel was included. The small branching ratio (27) alone is a poor motivation for neglecting the $\Sigma\pi$ channel. In Y_o^* resonance we have an illustration that even closed channels can be extremely important.

5. K^-p Interactions at Higher Energies

Below 300 MeV/c, the elastic K^-p scattering seems to be s-wave with a large absorptive part. Three-particle final states like $\Lambda\pi\pi$ gain rapidly in importance. In the GeV region, a large number of resonance is produced: ω, ρ, η, η^*, ϕ, K^*. In addition to the Y_0^* and Y_1^* already discussed, there are several other baryonic resonances of zero hypercharge. The Y_0^* (1520) has three decay channels, whereas the $Y_1^*(1660)$ decays mainly into $Y_0^* + \pi$[8,9]. The next three resonances, namely $Y_1^*(1765)$, Y_0^* (2065), show up mainly in elastic scattering.

Reviews covering some or all of these resonances have been given by Dalitz [10], Tripp [11] and Peyrou [12]. On the theoretical side, no models or dynamical calculations of these resonances have been published so far. Therefore all I can do is to recapitulate some of the experimental facts.

The $Y_0^*(1520)$ deserves special attention [11,13]. It is the lightest resonance which may be "formed" in K^-p interactions. The corresponding kaon lab. momentum is 395 MeV/c. It has a width of 16 MeV. The percentages in the $\bar{K}N$, $\Sigma\pi$ and $\Lambda\pi\pi$ channels are 30 %, 55 % and 15 % respectively. The weights of the different charge combinations follow from isospin invariance. K^-p and $\bar{K}^0n$ have equal weights, $\Sigma^0\pi^0$ $\Sigma^+\pi^-$ and $\Sigma^-\pi^+$ have equal weights, and $\Lambda\pi^0\pi^0/\Lambda\pi^+\pi^- = 1/2$.

In some channels, the resonance is superimposed on a large background. In elastic K^-p scattering for example, there is hardly any bump in the integrated cross section. In order to find the resonance, one has to analyze the differential cross section. This requires an analysis of the "background" as well. It turns out that the $Y_0^*(1520)$ is a $D_{3/2}$ resonance. The angular distribution of K^-p scattering at resonance is of the form

$$|S + 2D|^2 \cos^2\theta + |S - D|^2 \sin^2\theta \quad .$$

Terms proportional to $\cos\theta$ and $\cos^3\theta$ are small, which means that the p-waves are small. The coefficient of $\cos^2\theta$ is very large, telling us that there is constructive S-D interference. It drops off symmetrically on both sides of the resonance. This shows that the s-wave amplitude is mainly imaginary. A similar investigation of $K^- p \to \Sigma\pi$ shows that also here the $\cos^2\theta$ term is maximal at resonance energy. This again indicates constructive S-D interference, in good agreement with the scattering length analysis mentioned in section 3. The relevant quantity is the angle ϕ of eq. (19). By the way, the $\Sigma\pi$ decay mode of the Y_o^* (1520) gave the first indication of positive Σ parity. If the intrinsic parities of initial and final state in $K^- p \to \pi\Sigma$ had been opposite, then the $\pi\Sigma$ final state of $Y_o^*(1520)$ would have been p-wave. This is ruled out by the Σ polarization, observed in the decay $\Sigma^+ \to p\pi^o$.

The next hyperon resonance on the mass scale is the Y_1^* (1660). It has a width of 45 MeV. In the tables of resonances you find quite a number of decay modes. However, the recent experiments of Eberhard et al. [8] and London et al. [9] are consistent with 100% $Y_1^*(1660) \to Y_o^*(1405) + \pi$. The parity is as yet uncertain. London et al. favour the assignment $3/2^-$, i.e. the same quantum numbers as for the $Y_o^*(1520)$.

The next two resonances, $Y_1^*(1765)$ and $Y_o^*(1815)$ have both spin 5/2. They have opposite parities. This follows from the presence of a large $\cos^5\theta$ term in the differential cross section. The parity of $Y_1^*(1765)$ was then determined from the decay into $Y_o^*(1520) + \pi$, which turns out to be a p-wave. Thus we have $D_{5/2}$ for Y_1^* and $F_{5/2}$ for Y_o^*

Nowadays it is customary to classify the resonances into SU(3) supermultiplets. The $Y_o^*(1405)$ is a unitary singlet, whereas the $Y_1^*(1385)$ makes a decuplet together with N^*, Ξ^* and Ω^-. A classification of the higher resonances might be useless. To be more specific, let me assume that one day one will be able to calculate all resonances from the masses and coupling constants of the stable hadrons. Since there is a large violation of SU(3) symmetry in these basic quantities, there is no reason why the higher resonances should form complete supermultiplets. This would be analogous to nuclear physics, where only the lighter nuclei contain complete isospin multiplets.

References

1. M. M. Nikolić, Experimental information on negative kaons, in Progress in Elementary Particle and Cosmic Ray Physics, VIII, edited by J. G. Wilson and S. A. Wouthysen, North-Holland Publishing Company, Amsterdam, 1965.
2. R. H. Dalitz, Rev. Mod. Physics 33,471 (1961); Strange particles and strong interactions, Oxford Univ. Press, 1962.
3. A. D. Martin and T. D. Spearman, Phys. Rev. 136, B1480 (1964). See also: D. P. Roy, Phys. Rev. 136, B804 (1964).
4. J. K. Kim, Phys. Rev. Lett. 14, 29 (1965); Sakitt et al., Phys. Rev. 139 B719 (1965).
5. W. Kittel, G. Otter, I. Wacek, Phys. Lett. (to be published).
6. Armenteros et al., Phys. Lett. 19, 75 (1965).
7. B. R. Martin, Phys. Rev. 138, B 1136 (1965).
8. Eberhard et al., Phys. Rev. Lett. 14, 466 (1965).
9. G. London et al., Phys. Rev. 143, 1034 (1966).
10. R. H. Dalitz, Ann. Rev. Nucl. Science 13, 339 (1963).
11. R. D. Tripp, Baryon Resonances, CERN Yellow report 1965-7.
12. Ch. Peyrou, Proc. Oxford Conference on Elementary Particle Physics, 129 (1965)
13. M. Watson, M. Ferro-Luzzi and R. D. Tripp, Phys. Rev. 131, 2248 (1963).

DYNAMICS OF THE π - N SYSTEM

By

J. HAMILTON
NORDITA, Kopenhagen

Introduction

We wish to explain π-N scattering in the energy regions where it is a one-channel problem,or where the one-channel approximation is good. We shall see that this means the energy region up to 600 or 700 MeV (lab. system), and in addition it includes the case of several resonances at higher energies. The reasons why we consider only the one-channel approximation are (a) it is a simple case to treat, (b) we do not really have any good techniques at present for treating the two-channel or several-channel cases.

We wish to explain π-N scattering in terms of the basic interactions which produce the scattering. In this way we can extract physical information from the experimental data on differential cross-sections, phase shifts etc. It is only this physical description in terms of the basic interactions which is of any fundamental significance. Once we have understood this physical description properly, we can then predict any property of π-N scattering (so far as it is useful to use the one-channel approximation). It will be shown below that we now know a great deal about the physical description, but we do not yet know the full story in the case of the S-waves.

Because the interactions are strong, we must use dispersion relation methods. On account of the comparative lightness

Lecture given at the V. Internationalen Universitätswochen für Kernphysik, Schladming, 24 February - 9 March 1966.

of the pion we must of course use relativistic dispersion relations. Moreover, in dynamical studies the centrifugal potential barrier plays a very important part, and the simplest way to study the effects of the angular momentum is by the use of partial wave dispersion relations.

Now we shall start our more detailed discussion by first examining the problem of inelasticity.

I. Inelasticity in $\pi + N \to \pi + N$

1. Low Energies:

The pion has a much smaller mass than the other strongly interacting bosons, and the nucleon is the lightest baryon. A result of this is that over a considerably energy range $\pi + N \to \pi + N$ can be treated as a one-channel process. Let us look at the details.

The π+N threshold is approximately 1079 MeV, and the η + N channel opens at 1487 MeV, i.e. 558 MeV lab. pion kinetic energy. (A simple formula to use here is

$$s = M^2 + \mu^2 + 2M\,\omega_L$$

where s = (total energy in center of mass system)^2, M = nucleon mass, μ = pion mass, ω_L = total lab. pion energy. Also $M/\mu = 6.7$ and μ = 140 MeV). In what follows when we say lab. pion energy, we shall mean lab. pion kinetic energy, unless otherwise indicated. The various thresholds are:

Process	Threshold (lab. pion energy)
η + N	558 MeV
K + Λ	765 MeV
ρ + N	900 MeV

On this basis π + N should be elastic up to about 550 MeV, where the process $\pi + N \to \eta + N$ begins to set in. However we would expect $\eta + N$ to occur first in the state where η and N have orbital angular momentum $l = 0$ relative to each other, provided that the $\eta + N$ amplitude does not have any marked structure near 550 MeV. Now η has isospin $T = 0$ and $J = 0^-$, so $l = 0$ gives a $T = 1/2$, $J = (1/2)^-$ state. This can only be reached from the $\pi + N$ state S_{11} (the notation for the $\pi + N$ states is $L_{2T,\,2J}$ where L is the orbital angular momentum). In fact the η production around 600 MeV is quite large and some authors have suggested that there is an $\eta + N$ antibound state with $l = 0$. It might be however that the structure is in the $l = 1$ state, so that the P_{11} $\pi + N$ state is involved.

On the basis of the thresholds above, $\pi + N$ is elastic up to 550 MeV, and moreover, except for S_{11} (and possibly P_{11}). The $\pi + N$ particle waves are elastic up to well above 600 MeV. However we have forgotten the process $\pi + N \to 2\pi + N$. The $2\pi + N$ threshold is 171 MeV which is much below the other inelastic thresholds. This might be thought to give much inelasticity at low energies but that is not so. The reason is that only when two of the outgoing particles attract each other strongly, do we get any important effects. This can occur in two ways:

a) $(\pi\pi)_0 + N$

b) $\pi + N^*$ $\quad$ ($N^* = (T = 3/2,\ J = 3/2)$ π-N-Resonance).

Here $(\pi\pi)_0$ is the notation for a pair of pions in the $T = 0$, $J = 0^+$ state. It is known (see below) that when the total energy of the two pions lies between 330 MeV and 420 MeV they attract each other strongly.

Consider $\pi + N \to (\pi\pi)_0 + N$, when the orbital angular momentum of $(\pi\pi)_0$ relative to N is $l = 0$, this process should begin to become important just above 300 MeV lab. pion energy. The $\pi + N$ state involved is $T = 1/2$, $J = 1/2^+$ i.e. P_{11}. Analysis of the experimental data shows that P_{11} does become appreciably inelastic just above 300 MeV and the inelasticity increases rapidly up to 500 MeV (see Fig. 1). The state $(\pi\pi)_0$ + N where $(\pi\pi)_0$ has orbital angular momentum l = 1 relative to the nucleon can be reached from the π + N states S_{11} and D_{13} but on account of the centrifugal barrier we do not expect these inelastic states to become important until the pion lab. energy exceeds 500 MeV. In fact S_{11} is almost elastic right up to 550 MeV (Fig. 1).

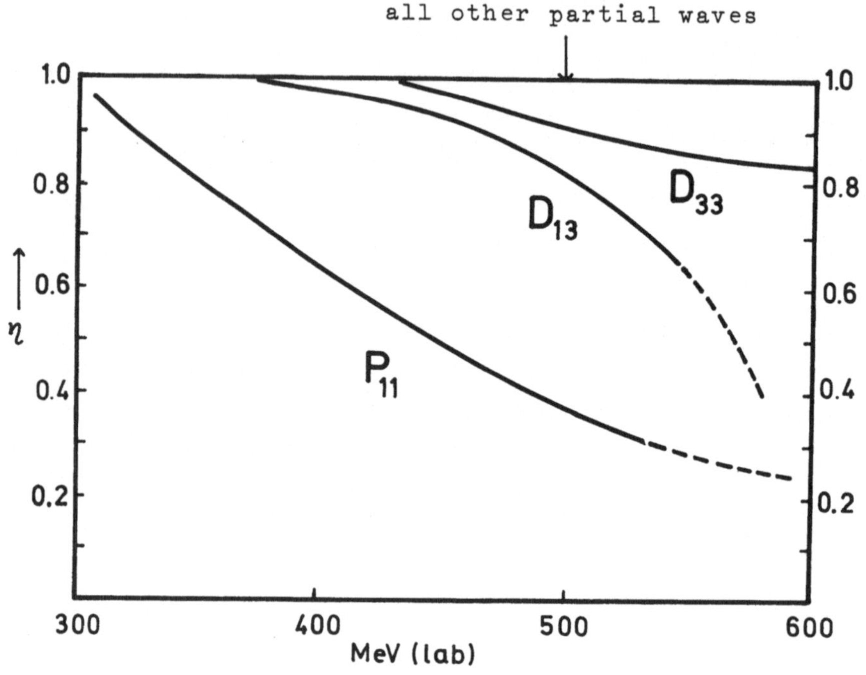

Fig. 1

The threshold for $\pi + N \to \pi + N^*$ is around 330 MeV. Assuming that the extra pion is in an S-wave relative to N^* this process can only occur in the π + N states D_{13} and D_{33} (since N^* is T = 3/2, J = $3/2^+$). Analysis of the experimental data shows that D_{13} and D_{33} do become slightly inelastic above 400 MeV, and by 550 MeV the inelasticity is fairly large in the case of D_{13} (Fig. 1).

The validity of the above discussion is confirmed by the fact that analysis of the experimental data shows that except for D_{11}, D_{13}, D_{33} no other $\pi + N$ amplitude has any appreciable inelasticity up to 550 MeV. We can however treat D_{33} and even D_{13} as approximately one-channel problems but the inelasticity of P_{11} is so strong and is spread over such a range of energy that we cannot at present give a dynamical theory for that amplitude.

2. Inelasticity and Resonances:

Any $\pi + N$ partial wave amplitude can be written

$$f(s) = (e^{2i\delta(s)} - 1)/2iq \qquad (1)$$

where q is the momentum in the center of mass system, and s is the square of the energy in the c.m. system. The phase shift $\delta(s)$ will be complex if there is any inelasticity and then $\mathrm{Im}\delta(s) > 0$. It is customary to write

$$e^{2i\delta(s)} = \eta(s)\, e^{2i\alpha(s)}$$

where $\alpha(s)$ is the real part of the phase shift, and $\eta(s) = e^{-2\mathrm{Im}\delta(s)}$. Clearly $0 < \eta(s) < 1$ and the smaller $\eta(s)$ the greater is the inelasticity. In Fig. 1 experimental values of $\eta(s)$ are shown.

If we consider only one partial wave, then the total elastic cross section is

$$\sigma(el) = \frac{\pi}{q^2}(J + 1/2)\,|\eta e^{2i\alpha} - 1|^2$$

and the inelastic cross section is

$$\sigma(inel) = \frac{\pi}{q^2}(J + 1/2)(1-\eta^2) \qquad (2)$$

where $J = l \pm 1/2$ is the total angular momentum and l is the orbital angular momentum.

We can now evaluate the quantity

$$R = \sigma_{tot}/\sigma(el)$$

where $\sigma_{tot} = \sigma(el) + \sigma(inel)$ is the total cross section for this partial wave. By eq. (1) and (2)

$$R(s) = \frac{1}{q} \frac{\mathrm{Im}\, f(s)}{|f(s)|^2}$$

and this gives the optical theorem for the partial wave

$$\frac{1}{q\, R(s)} \mathrm{Im}\, f(s) = |f(s)|^2 \tag{3}$$

In order to make any progress it is necessary to know something about how the parameter R(s) behaves at and near a resonance. For this purpose we shall make the assumption that the general resonance formulae of nuclear physics- which can be derived by very general methods - are approximately valid for normal elementary particle resonances. Of course there are classes of "resonances" for which this is a poor or bad approximation, but it is reasonable to ignore them in the first instance.

In our notation, the multi-channel Breit-Wigner formulae can be written

$$\sigma_{tot} = \frac{\pi}{q^2} (J + 1/2) \frac{\Gamma_{el} \cdot \Gamma}{(E - E_R)^2 + (\Gamma/2)^2} \tag{4a}$$

$$\sigma(el) = \frac{\pi}{q^2} (J + 1/2) \frac{\Gamma_{el}^2}{(E - E_R)^2 + (\Gamma/2)^2} \tag{4b}$$

$$\sigma(inel.,\beta) = \frac{\pi}{q^2} (J + 1/2) \frac{\Gamma_{el}\, \Gamma_\beta}{(E - E_R)^2 + (\Gamma/2)^2} \quad . \tag{4c}$$

Here Γ, Γ_{el}, Γ_β denote the total width, the elastic width and the width for the inelastic channel β respectively. Also

$$\Gamma = \Gamma_{el} + \sum_\beta \Gamma_\beta$$

We assume that Γ, Γ_{el}, Γ_β only vary slowly with energy in the resonance region.

Now using (4a) and (4b) we get

$$R(s) = \Gamma/\Gamma_{el} \quad , \tag{5}$$

so we expect R(s) to vary slowly with s in the region of the resonance. We complete our analysis by using the Adair plot. This is the plot of Im(2q f(s)) vs. Re(2q f(s)). Writing

$$Z = 2q\ f(s) \ ,$$

eq. (1) gives

$$|Z - i| = \eta \tag{6}$$

So if $\eta = 1$, Z(s) moves on a circle of radius unity whose center is C (Fig. 2). For $\eta < 1$, Z(s) must lie inside this circle.

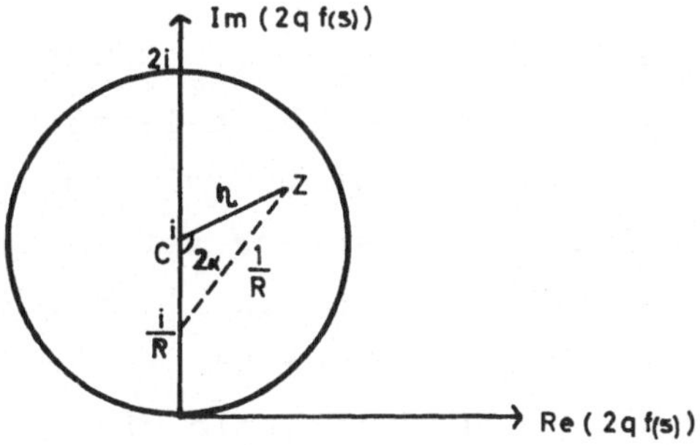

Fig. 2

The phase angle $\alpha(s)$ is given by $2\alpha(s) = \sphericalangle$ PCO. Causality implies that near a resonance the point P describes a loop in the counterclockwise direction as s increases.

Also equation (3) gives

$$|Z - \frac{i}{R}| = \frac{1}{R} \quad . \tag{7}$$

If R is constant P moves around a circle of radius 1/R centred at i/R. Fig. 2 shows these various relations.

Now using Fig. 2 we can classify those resonances for which R(s) varies slowly in the resonance region with two categories: (i) elastic or quasi-elastic resonances, (ii) inelastic resonances. In category (i) P passes above C; α goes through $\pi/2$ and at the resonance $R < 2$. In the category (ii) P describes a counter clockwise motion which passes below C, so $\alpha = 0$ at the resonance and $R > 2$ (Fig. 3a, 3b).

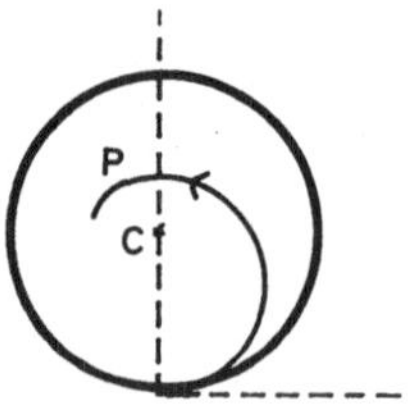

Fig. 3a

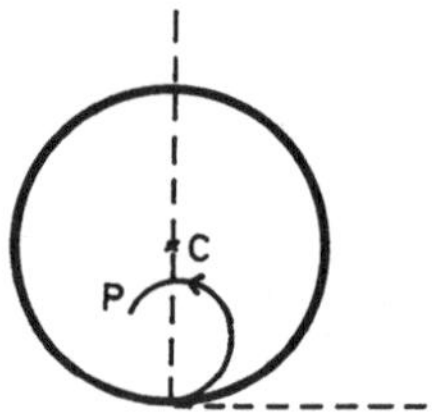

Fig. 3b

In the quasi-elastic type $\Gamma_{el} > 1/2\Gamma$ and in many cases the coupling to the inelastic channels is not important. This is particularly so when P passes well above C. In the quasi-elastic type we look for a strong attraction in the elastic channel as a dynamical explanation. In the inelastic type $\Gamma_{el} < 1/2\ \Gamma$ and the coupling to the inelastic channels is very important. If one inelastic channel β is dominant then $\Gamma_{\beta'} > 1/2\ \Gamma$ and the phenomenon is probably a quasi-elastic resonance in the β'channel. In that case we should look at the β' channel for a dynamical explanation.

It should be remembered that in types (i) and (ii) there may be a background amplitude as well as the simple Breit-

Wigner amplitude. This will move the circle or loop to one side or the other. This effect is more likely to be important in the inelastic type since there the Breit-Wigner amplitude is not so large.

It should be remembered that we have given only a limited discussion above, and there may be many exceptions to our analyses. We shall however show below that the π-N resonances D_{13}, F_{15}, F_{37} can be understood as quasi-elastic resonances driven by strong attractions in the elastic channel.

II. Partial Wave Dispersion Relations

1. The Singularities

The singularities of the π-N partial wave amplitude f(s) can easily be discovered from the Mandelstam representation. Mandelstam deals with scattering amplitudes of the form T(s,t) where $t = -2q^2(1-\cos\theta)$ and θ is the scattering angle in the c.m. system. To obtain $f_l(s)$ (where l is the orbital angular momentum) we use an expression like

$$f_l(s) = \frac{1}{2} \int_{-1}^{+1} dx \; T(s,t(x))$$

where $x \equiv \cos\theta$. We can express s and t in terms of q^2 and find the behaviour of f_l as an analytic function of q^2 and so of s. We now list the singularities of $f_l(s)$ and their origins.

(a) T(s,t) has a Born pole at $s = M^2$. This gives a pole at $s = M^2$ in the P_{11} partial wave amplitude, which represents the nucleon itself.

(b) T(s,t) has also a Born pole at $u = M^2$. From the relation $s + t + u = 2M^2 + 2\mu^2$ we find the effect on $f_l(s)$ using eq. (8). In $f_l(s)$ we get a cut $(M - \mu^2/M)^2 \leqslant s \leqslant M^2 + 2\mu^2$ and a cut $-\infty < s \leqslant 0$ (Fig. 4).

(c) T(s,t) has a cut $(M + \mu)^2 \leqslant s < \infty$ which gives the same cut in $f_1(s)$. This represents rescattering.

(d) $T(s,t)$ has a cut $(M + \mu)^2 \leqslant u < \infty$ which gives the cut $-\infty < s \leqslant (M + \mu)^2$ in $f_1(s)$.

(e) $T(s,t)$ has a cut $4\mu^2 \leqslant t < \infty$ which gives cuts in $f_1(s)$ along the circle $|s| = M^2 - \mu^2$ and along the line $-\infty < s \leqslant 0$. (Fig. 4)

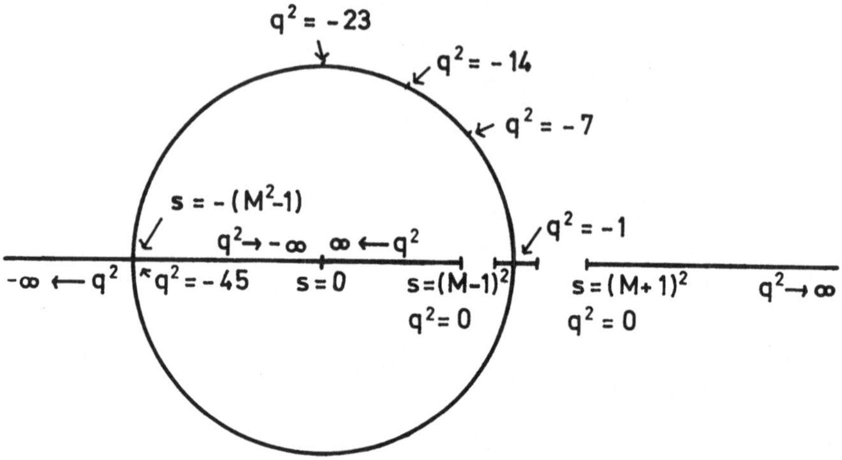

Fig. 4

What is the physical meaning of these singularities?

(b) The Born pole $u = M^2$ comes from the graph in Fig. 5a. This graph has two very different parts, as we can see if we assume time is increasing up the page. Then Fig. 5a corresponds to the simple process of nucleon exchange, while Fig. 5b involves nucleon pair production. The former is associated with the cut $(M - \mu^2/M)^2 \leqslant s \leqslant M^2 + 2\mu^2$ which is not very far from the physical threshold $s_o = (M + \mu)^2$ and therefore gives a fairly long range interaction (range $\sim 10^{-13}$cm). The pair production graph is associated with the cut $-\infty < s \leqslant 0$ and it gives a very short range interation.

(d) The cut $(M + \mu)^2 \leqslant u < \infty$ is associated with processes like Fig. 5c where a nucleon isobar is exchanged. This belongs to cut $0 \leqslant s \leqslant (M - \mu)^2$. There are other parts of this graph which concern anti-particle processes and contribute to the

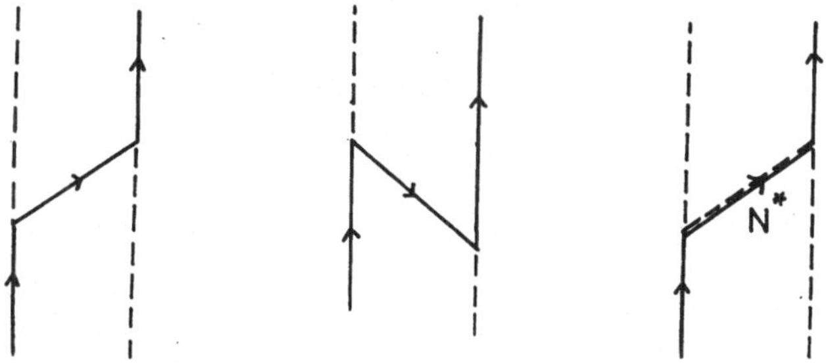

Fig. 5a Fig. 5b Fig. 5c

very short range interactions on $-\infty \leqslant s \leqslant 0$.

(e) The singularities $4\mu^2 \leqslant t \leqslant \infty$ in $T(s,t)$ come from graphs like Fig. 6, i.e. the process $\pi + \pi \to N + \overline{N}$. In $f_1(s)$ singular-

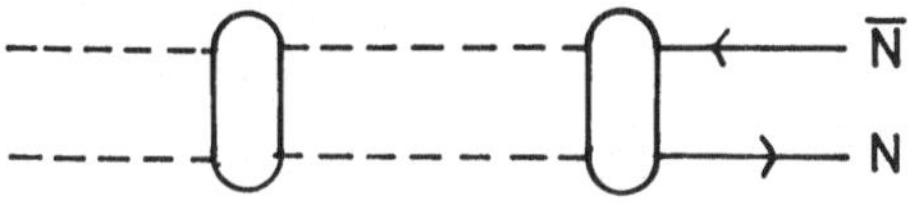

Fig. 6

ities on $|s| = M^2 - \mu^2$ near $s = M^2 - \mu^2$(i.e. singularities from the front of the circle) only arise from the smaller values of t. In turn these only arise from graphs like Fig. 6 where two mesons join the two parts of the figure. This gives the long range part of the effect of the $\pi - \pi$ interaction on π-N scattering. It corresponds to scattering the incident pion on the pion cloud which surrounds the nucleon. The further off parts of the circle and the line $-\infty \leqslant s \leqslant 0$ correspond to the short range part of the effect of $\pi - \pi$ or multi-pion interactions on π- N scattering.

2. Application to S-wave π - N scattering

The first application of this method was to S-wave π-N scattering at low energies. We summarize the method used and the results. The dispersion relation is

$$\mathrm{Re}f(s) = \frac{1}{\pi} P \int_{(M+\mu)^2}^{\infty} \frac{\mathrm{Im}\, f(s')}{s' - s} ds' + \frac{1}{2\pi i} \int_{\substack{\text{unphysical}\\ \text{cuts}}} \frac{\Delta f(s')}{s' - s} ds' \quad (9)$$

Here $\Delta f(s')$ is the discontinuity in $f(s')$ across the cut at s'. We calculate the following parts of $\Delta f(s')$: The contribution from $(M - \mu^2/M)^2 \leqslant s \leqslant M^2 + 2\mu^2$, which is very small in the S-wave case, the contribution from the exchange of the (3/2, 3/2) resonance N^* to the cut $0 \leqslant s \leqslant (M - \mu)^2$. We cannot calculate any of the other contributions to the unphysical cuts because (a) we do not yet know the coupling constants for the graph in Fig. 6, (b) the back of the circle $|\arg s| > 66°$ and the cut $-\infty \leqslant s \leqslant 0$ are not reached by any method at present available for calculating $\Delta f(s)$. So we proceed as follows: We insert the physical values of Re f(s) on the left of eq. (9) and of Im f(s) in the integral on the right of eq. (9). Thus we can deduce

$$\Delta(s) \equiv \frac{1}{2\pi i} \int_{\substack{(\text{unknown}\\ \text{cuts})}} \frac{\Delta f(s')ds'}{s' - s}$$

for low energy physical values of s.

This does not give us very much information and we improve matters considerably by using crossing symmetry. In eq. (8) T(s,t) appears. Now T(s,t) obeys a simple crossing relation of the form

$$T(s, t) = T(u, t) \quad (10)$$

where $s+t+u = 2M^2 + 2\mu^2$. For fixed t, large s and small u related, and by using eq. (10) we can find $T(s,t(x))$ for $0 \leq s \leq (M-\mu)^2$ in terms of the physical values of $T(s,t)$. In turn we can find the S-wave (or any other) amplitude $f(s)$ on $0 \leq s \leq (M-\mu)^2$, in terms of the physical values of $f_1(s)$. The catch is that in order to find the S-wave amplitude on $0 \leq s \leq (M-\mu)^2$ we must know other partial waves besides the S-wave amplitude on the physical region. However it can be done easily enough for values of s near the $(M-\mu)^2$ end of $0 \leq s \leq (M-\mu)^2$. This procedure gives us Re $f(s)$ (and Im $f(s)$) on $0 \leq s \leq (M-\mu)^2$ so we can now find $\Delta(s)$ on $0 \leq s \leq (M-\mu)^2$ (or at least on the part near $(M-\mu)^2$) in terms of experimental quantities.

Before describing the results of this procedure we must discuss charge properties. Let $T_+(s,t)$, $T_-(s,t)$ be the amplitudes for $\pi^+p \to \pi^+p$ and $\pi^-p \to \pi^-p$ respectively.

We define

$$T^{(+)} = \frac{1}{2}(T_- + T_+)$$

$$T^{(-)} = \frac{1}{2}(T_- - T_+) \tag{11}$$

The relation between the s-channel and the t-channel of Fig. 6 is shown in Fig. 7. We bend around one pion line and one nucleon line and look at the graph from left to right instead of from bottom to top. If we do this for $T^{(+)}$ we use both Figs. 7a and 7b, and the corresponding graph Fig. 7c is symmetric in the charges of the two pions π_1, π_2. Hence $T^{(+)}$ is related to $\pi + \pi \to N + \bar{N}$ in the $T = 0$ state, since only isospin $T = 0$ and $T = 1$ are allowed for $N + \bar{N}$ and two pion states with $T = 0$, $T = 1$ are respectively symmetric and antisymmetric under exchange of the pions. So $T^{(+)}$ is related to

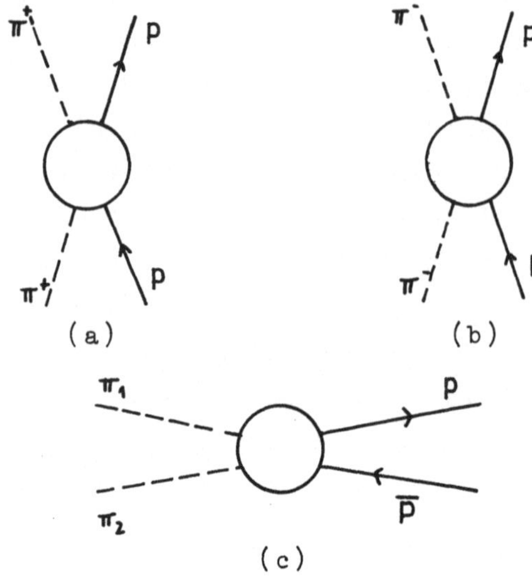

Fig. 7

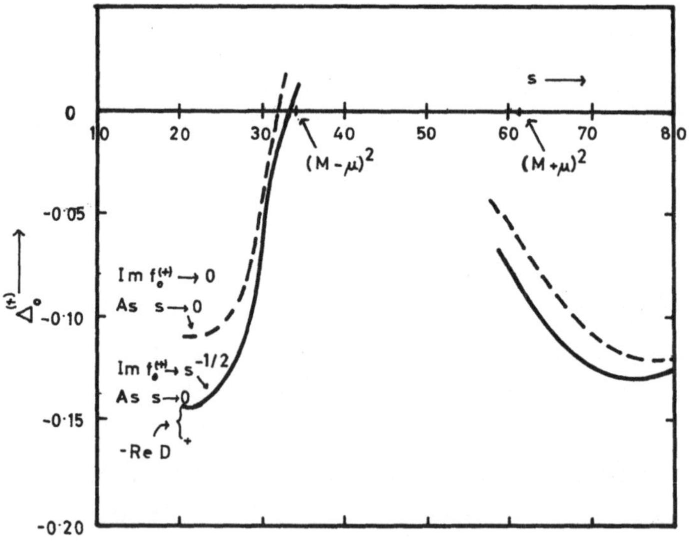

Fig. 8

The values of the discrepancy $\Delta_o^{(+)}(s)$. The two curves correspond to different behaviours of Im $f_o^{(+)}(s)$ as $s \to 0$.

the T = 0 pion pair, and $T^{(-)}$ to the T = 1 pion pair. Moreover the T = 0 pion pair has angular momentum J = 0,2,... while the T = 1 pair has J = 1,3,... So instead of eq. (9) we use the analogous relations for Re $f^{(+)}(s)$ and Re $f^{(-)}(s)$. Then we get the results for $\Delta^{(+)}(s)$ and $\Delta^{(-)}(s)$ shown in fig. 8 and 9.

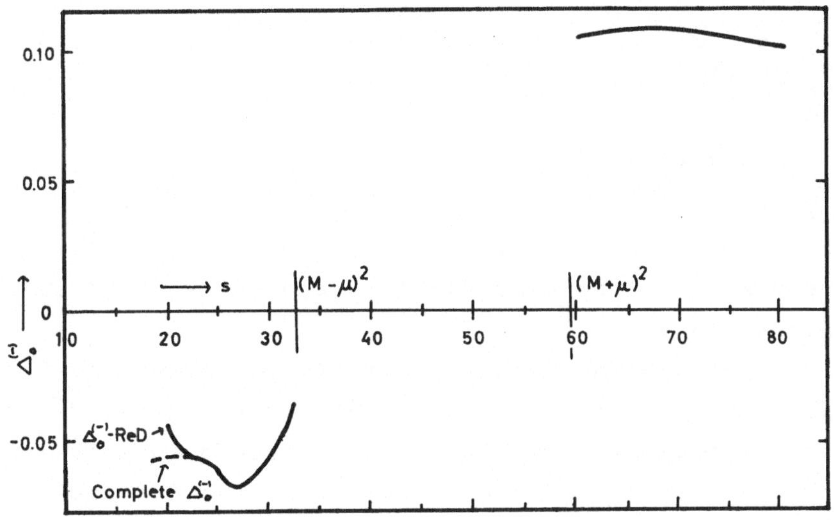

Fig. 9

The values of the discrepancy $\Delta_0^{(-)}(s)$. The effect of including π-N D-waves on the crossed physical cut is shown.

3. Interpretation of $\Delta^{(+)}(s)$ and $\Delta^{(-)}(s)$

Consider first $\Delta^{(+)}(s)$. On either side of $s = M^2 - \mu^2 = 44$ units ($\mu = 1$), $\Delta^{(+)}(s)$ rises steeply. This shows that $f^{(+)}(s)$ must have a strong singularity across the front of the circle. This can only be due to a low energy interaction in the T=0, J=0 π-π state. Now if we assume that there is such an interaction it is easy to show that its contribution to $\Delta^{(+)}(s)$ will be of the form

$$C(s) = \frac{(\text{const})}{(s-\text{Re } s_1)^2 + (\text{Im } s_1)^2}$$

where s_1 is some point on the circle and not far from $S=M^2-\mu^2$.

Clearly $\Delta^{(+)}(s)$ is of the form

$$\Delta^{(+)}(s) = C(s) + \frac{a}{s+b}$$

where $b > 0$, $a < 0$. The term $a/(s+b)$ represents a strong short range repulsion (it is a repulsion because $a < 0$ and is short range because $b > 0$).

Next we can use the magnitude of $C(s)$ to determine the phase δ_0^0 of T = 0, J = 0 π-π scattering at low energies. For this purpose we use a relativistic effective range formula for cot δ_0^0, insert it in an Omnes equation giving the amplitude for $\pi + \pi \rightarrow N + \overline{N}$ and fit to $C(s)$. The best fit gives a phase δ_0^0 which reaches a maximum of about 30° at a total 2π energy around 350 MeV, and then fall off. The scattering length is $a_0 \simeq 1.3$ natural units. This strongly attracting T=0, J=0 pion pair is what we called $(\pi\pi)_0$ above. A phase δ_0^0 going through 90° at a low energy would not give a decent fit to $C(s)$ so there is really no σ-meson.

Now $\Delta^{(-)}(s)$ is roughly of the form

$$D(s) = (\text{const}) \frac{(s-\text{Re } s_2)}{(s-\text{Re } s_2)^2 + (\text{Im } s_2)^2}$$

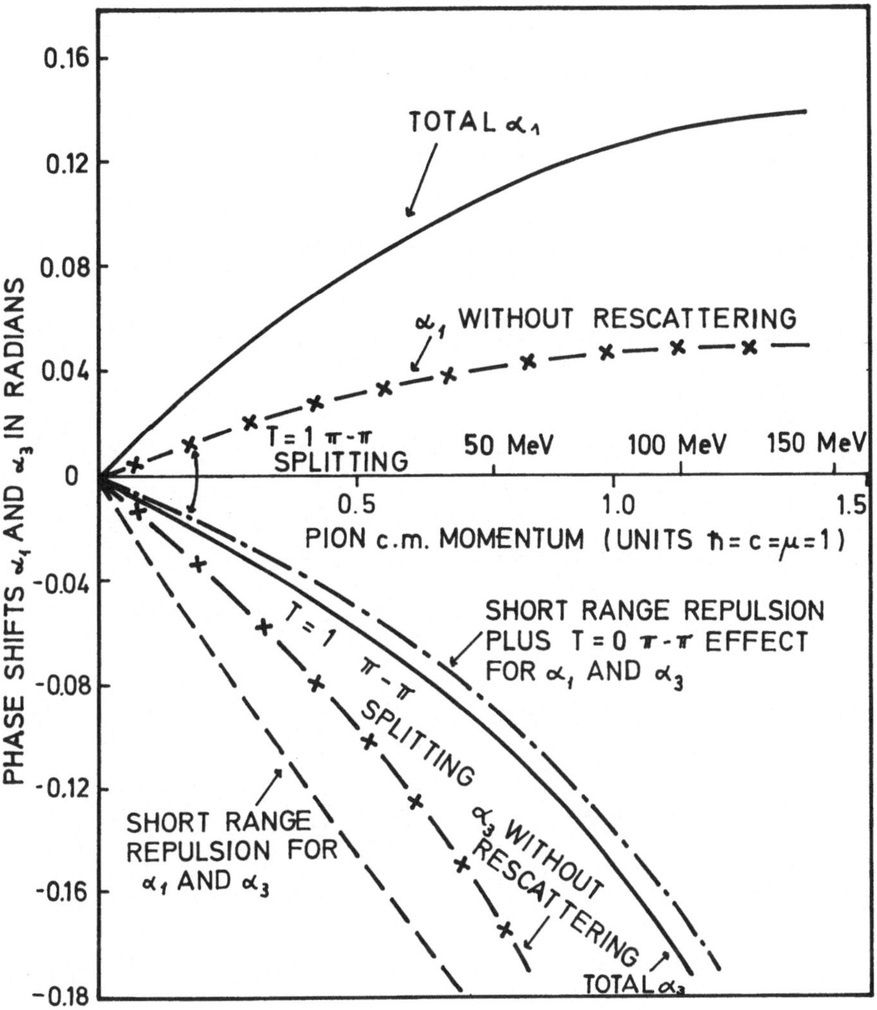

Fig. 10

where s_2 is on the front part of the circle. This form is just what we expect from the ρ-meson contribution to Fig. 6. The magnitude of D(s) enables us to determine the ρ coupling constants (assuming that the ratio of the tensor to vector coupling is given by the nucleon gyromagnetic ratio i.e. $G_{\rho NN}/f_{\rho NN} = -1.85$). Our result is

$$\frac{f_{\rho\pi\pi}\ f_{\rho NN}}{4\pi} = 2.85 \pm 0.3 \quad .$$

From $\rho \to \pi + \pi$ decay Sakurai gets $f^2_{\rho\pi\pi}/4\pi = 2.1$ so the value from π-N scattering is in fair agreement with Sakurais conserved vector current idea which requires $f_{\rho NN} = f_{\rho\pi\pi}$. We notice that in the $\Delta^{(-)}$ case there is very little short range interaction. Finally in fig. 10 we show the dynamical picture of π-N low energy S-wave scattering which comes from the above analyses.

III. The Peripheral Method

1. Long and short range Interactions

The longest range parts of the π-N interaction are shown in Fig. 11. These are (a) nucleon exchange, (b) N^* - exchange,

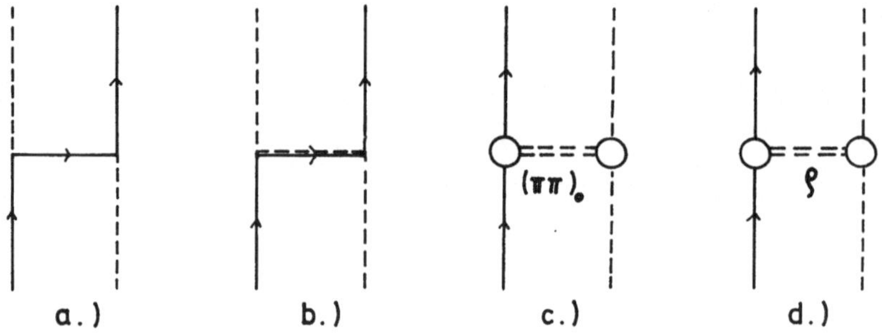

Fig. 11

(c) $(\pi\pi)_o$-exchange, (d) ρ-exchange. (N^* is the N_{33} resonance). Among these (a) and (c) are long range and (b) and (d) are medium range. In conventional potential theory terminology the long range parts have range $\simeq 10^{-13}$cm and the medium range parts have range $\simeq 0.2 - 0.3 \times 10^{-13}$cm. In addition to these four processes there are numerous short range processes due to multi-pion exchanges, nucleon anti-nucleon pair effects, exchange of higher π-N isobars etc. These are so numerous and so complicated that we cannot hope to calculate them. Moreever we have no method for the direct calculation of the discontinuity on the cut $-\infty \leqslant s \leqslant 0$ (Fig. 4) and no reliable method for direct calculation of the arc $|\arg s| > 66^\circ$ of the circle. The only thing we can do is to find some way of suppressing these shorter range parts of the π-N interaction.

Let us think about it physically. For a given energy the wave function corresponding to high orbital angular momentum will penetrate much less into the region of short range interactions than will a wave function corresponding to low orbital angular momentum. This is due to the centrifugal potential barrier which keeps out the higher orbital angular momentum partial waves. We can make use of this as follows.

In eq. (1) $\delta(s) \to q^{2l+1}$ for $q \to 0$ so $f(s) \sim q^{2l}$ for $q \to 0$. Thus we can work with the function

$$F_l(s) = f_l(s)/q^{2l} \tag{12}$$

instead of $f_l(s)$ itself. Now $|F_l(s)|$ is bounded as $q \to 0$ and the presence of q^{2l} in the denominator automatically ensures the correct behaviour of $F_l(s)$ as $q \to 0$. On the other hand $|f_l(s)| \sim q^{2l}$ as $q \to 0$ and if we used $f_l(s)$ it would be necessary to impose $l - 1$ conditions to get the correct behaviour as $q \to 0$. Using $F_l(s)$ - which we call the reduced amplitude - instead of $f_l(s)$ is thus equivalent to incorporating a large part of the centrifugal barrier effect.

2. The Function $q^2(s)$ and suppression of the short range interactions

The relation between s and the momentum in the c.m. system q is

$$s = [(M^2 + q^2)^{1/2} + (\mu^2 + q\ )^{1/2}]^2 \tag{13a}$$

On inverting this relation we get

$$q^2(s) = \frac{[s - (M + \mu)^2][s - (M - \mu)^2]}{4\ s} \ . \tag{13b}$$

Thus $q^{-2}(s)$ is a regular analytic function of s with simple poles at $s = (m + \mu)^2$ and at $s = (M - \mu)^2$. The former gives the required behaviour at the physical threshold, and the latter pole gives no real trouble.

The singularities of $F_l(s)$ thus lie in the same position as the singularities of $f_l(s)$ and except for $s = (M - \mu)^2$ there are no new singularities. On Fig. 4 we show the values of $q^2(s)$ on the cuts (in units $\hbar = c = \mu = 1$ as usual). The important point is that $|q^2(s)|$ is large on all those parts of cuts which give rise to short range interactions i.e. on these parts which are far from the physical threshold $s =$ $= (M + \mu)^2$. In fact $|q^2|$ is large except near $s = M^2 - \mu^2$ on the front of the circle, and on the right hand part of the cut $0 < s \leqslant (M - \mu)^2$.

Thus we set up the dispersion relation for $F_l(s)$

$$\mathrm{Re}\ F_l(s) = \frac{1}{\pi} P \int_{(M+\mu)^2} \frac{\mathrm{Im}\ F_l(s')}{s' - s} ds' + \frac{1}{2\pi i} \int_{\substack{\text{(unphysical}\\ \text{cuts)}}} \frac{\Delta F_l(s')}{s' - s} ds' \tag{14}$$

($\Delta F_l(s)$ is the discontinuity in $F_l(s)$ across the cut at s). The contribution of the short range interactions are suppressed in this dispersion relation. We can look at the situation in another way. Consider the impact parameter R.

The impact parameter is given by

$$R^2 = l(l + 1)/q^2 \ , \tag{15}$$

where l is the orbital angular momentum. We cannot use this picture unless we are careful not to violate the uncertainty principle, since the uncertainty in the pions transverse momentum is $\Delta q \geqslant 1/R$ we can only use this picture for $l \geqslant 2$. Now our unit of length is 1.4×10^{-13} cm and if we require

$R \geqslant 0.75$ (i.e. 1.05×10^{-13}cm) then we can use (15) to find the maximum values of q which are allowed for various l. We thus get upper energy limits of 610 MeV for D-waves and 1.1 GeV for F-waves. With R in this range we do not expect that interactions of range $< 0{,}2 \times 10^{-13}$cm will have much effect on the scattering. The range of validity of our method is somewhat greater than what has just been indicated. This technique of using $F_l(s)$ and eq. (14) we call the peripheral method. Clearly it does not apply to S-waves, and in the case of P waves it may not be valid over a large energy range.

3. Peripheral Method Calculations

We use eq. (14) and for the unphysical cut contributions we put in the four basic exchange processes shown in Fig. 11. Precisely, we calculate (a) for the cut $(M - \mu^2/M)^2 \leqslant s \leqslant M + 2\mu^2$. (b) for the cut $0 < s < (M - \mu)^2$ and we use no other term on that cut except that for the higher l we need an S-wave π-N contribution near $s = (M - \mu)^2$; (c) and (d) are used for the arc $|\arg s| < 66^\circ$ of the circle. We call this procedure the DHL approximation (notice that there is no short range interaction). With this we determine

$$F_l'(s) \equiv \frac{1}{2\pi i} \int_{\substack{\text{unphysical}\\ \text{cuts}}} \frac{\Delta F(s')}{s' - s}\, ds' \qquad (16)$$

for l = 1,2,3, i.e. P, D and F-waves. There are various ways of testing the accuracy of this procedure. For P-waves $F'(s)$ should be accurate up to 300 MeV, and qualitatively correct up to 800 MeV or so. For D-waves it should be accurate up to 600 MeV and qualitatively correct up to above 1 GeV, and for F-waves it should be reasonably accurate up to 1.5 GeV.

4. Systematics of the Interactions

The behaviour of $F_l'(s)$ as we change the total angular momentum J and the isospin T, keeping l fixed, is very important. Consider any one of the four basic exchange processes (a) - (d) in Fig. 11. Work out the contribution of this particular exchange interaction α to $F_l'(s)$, and call it $(F_l'(s))_\alpha$. Then we find that for fixed l the ratios

$$\frac{(F_l'(s))_\alpha|T = 3/2, J}{(F_l'(s))_\alpha|T = 1/2, J} \quad \text{and} \quad \frac{(F_l'(s))_\alpha|T, J = l + 1/2}{(F_l'(s))_\alpha|T, J = l - 1/2}$$

are roughly independent of s over the range of validity of our method; this holds for α = (a), (b), (c), (d) and l = 1,2,3.

We call these the isospin and spin ratios, and we shall soon see that their values determine many of the main features of low and medium energy π-N scattering.

The ratios are as follows (with an obvious notation)

a. N-exchange

$$\frac{T = 1/2}{T = 3/2} = -1/2 \qquad \frac{J = l - 1/2}{J = l + 1/2} \simeq -\frac{1}{21} \qquad (l = 1,2,3)$$

Thus the N-exchange interaction is therefore strongest in the states T = 3/2, J = l + 1/2. Also, a positive $F_l'(s)$ is an attraction and a negative $F_l'(s)$ is a repulsion. Now the N-exchange interaction reserves sign each time we increase l by unity, i.e. the contribution to $F_l'(s)$ contains a factor $(-1)^l$. This is easily understood from the spatial exchange nature of the graph in Fig. 11a. We find that N-exchange is attractive in P_{33} and F_{37} and repulsive in D_{35}.

b. N^*-exchange

$$\frac{T = 3/2}{T = 1/2} \simeq 1/4 \qquad \frac{J = l + 1/2}{J = l - 1/2} \simeq 1/4$$

Again there is a spatial exchange factor $(-1)^l$ and the interaction is strongest and attractive in P_{11} and F_{15}, while it is repulsive in D_{13}.

However N^*-exchange is never the dominant interaction.

c. $(\pi\pi)_o$-exchange

The relevant parameters are found from the S-wave π-N analysis above. Also the ratios are

$$\frac{T = 3/2}{T = 1/2} = 1 \quad , \quad \frac{J = l + 1/2}{J = l - 1/2} \simeq 1 \quad .$$

There is no exchange factor and the interaction is always attractive, as we would expect for a scalar exchange.

Clearly it is independent of isospin and almost independent of spin. The interaction is only really important at low energies.

d. ρ-exchange

Again the parameters are obtained from the S-wave π-N analysis. We find

$$\frac{T = 3/2}{T = 1/2} = -1/2 \; , \; \frac{J = l + 1/2}{J = l - 1/2} \simeq \underset{(l=1)}{-1/5} \; , \; \underset{(l=2)}{-1/3} \; , \; \underset{(l=3)}{-1/3}$$

There is no exchange factor and the interaction is strongest and attractive in P_{11}, D_{13}, F_{15}.

Clearly the states with strong interactions are T = 3/2, J = l + 1/2 (l odd)(N-exchange) and T = 1/2, J = l - 1/2 (ρ-exchange).

5. The Nucleon Isobars

How large does $F_l'(s)$ have to be to produce a resonance. We can answer this by looking at eq. (14) and noting that

$$\mathrm{Re}\, F_l(s) = \eta\, \frac{\sin 2\alpha(s)}{2q^{2l+1}} \quad .$$

Thus if $Q_l(s) = q^{2l+1} F_l'(s)$ rises (as s increases in the physical region) and exceeds 1/2, the physical or rescattering integral must become negative. The rescattering integral will be positive near threshold and later pass through zero in the case of a resonance.

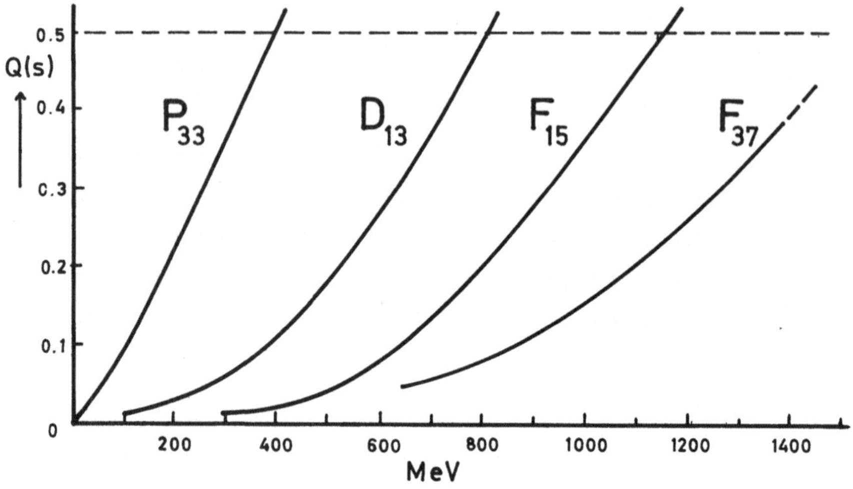

Fig. 12

In Fig. 12 we plot $Q_l(s)$ for P_{33}, D_{13}, F_{15}, F_{37}. (We omit P_{11} because it is really a two-channel problem.) We expect these amplitudes to have resonances at energies somewhat below (200 MeV below) the energies for which $Q_l(s) = 1/2$. The latter are 380 MeV, 810 MeV, 1090 MeV, $\sim$ 1.5 GeV for P_{33}, D_{13}, F_{15} and F_{37}, respectively. The resonances occur experimentally at 200 MeV, 600 MeV, 900 MeV and 1.35 GeV, respectively. Thus we have good agreement with the picture that these resonances are due to strong attractive interactions in the elastic channel. The analysis of the experimental data shows that these resonances are elastic (P_{33}) or else pseudo-elastic, as we would expect from the above theory.

If we examine $Q_l(s)$ for the other P, D and F-waves (except P_{11}) we find that $Q_l(s)$ is in many cases negative, because the net interaction is repulsive. When $Q_l(s)$ is positive, it never approaches near to 1/2 in the region of validity of the calculations. Analysis of the experimental data up to 1 GeV shows that there are no other P, D or F-wave resonances, except for a possible resonance in P_{11} and for a resonance in D_{15} at 900 MeV. However this last has $\eta \approx 0$ at the resonance and it seems clear that it is an inelastic resonance (i.e it is of the type shown in Fig. 3b).

Thus it is no way in conflict with our theory. We interpret this D_{15} resonance as being primarily due to interactions in some other channel.

6. Prediction of the small Phase Shifts

We consider the cases where $F_l(s)$ is either negative (repulsive) or small and positive (weak attractive). These partial waves are P_{13}, P_{31}; D_{33}, D_{35}, D_{15}; F_{17}, F_{35}. We can use the DHL method to predict the phase shifts in these cases. We use an iteration method for solving eq. (14). First use

$$\text{Re } F_l(s) = F_l'(s) \tag{17}$$

i.e. we ignore rescattering because unless the inelasticity is large rescattering will be small in these cases. Eq. (17) then gives an approximation which is good at low energies.

However to do better we have to know $\eta(s)$ (from analysis of experiments). Then eq. (17) gives a first approximation for $\eta(s)$. Substituting that,and again using $\eta(s)$ we can find Im F(s) and so at the second state we can estimate the rescattering correction.

This is still not quite good enough, because in the case of these repulsive or weak interactions, the short range interaction, though it is small, may give noticable effects. We can estimate the short range effects by using a unitary sum rule. Letting $s \to \infty$ in eq. (14) and noting that $|\text{Re } F_l(s)| < 1/2\, q^{2l+1}$ by unitarity, we see that for $l \geqslant 1$ we must have

$$\frac{1}{\pi} \int_{(M+\mu)^2} \text{Im } F_1(s')ds' + \frac{1}{2\pi i} \int_{\substack{\text{unphysical}\\ \text{cuts}}} \Delta F_1(s')ds' = 0. \tag{18}$$

The second integral in eq. (18) is, as it were, the sum of the residues of the unphysical poles. Of these we do not know the short range poles. Using eq. (17) (or some similar approximation) we can estimate

$$\frac{1}{\pi} \int_{(M+\mu)^2}^{\infty} \text{Im } F_1(s') \, ds'$$

(for this we must know the inelasticely parameter $\eta(s)$if inelasticity becomes important at moderate energies). Thus eq. (18) gives the sum of the residues, and hence the sum of the short range residues.

Next we lump all the short range poles (or cuts) into one short range pole. By eq. (18) we know its residue. Where do we put this pole? Clearly it should be on the real axis and should be on the negative side of s = o. Looking at $|q^2|$ on $-\infty \leqslant s \leqslant 0$ (Fig. 4) we see that $|q^2|$ is least as $s = M^2-\mu^2$, so this is the natural place to put our short range pole. (Actually it is easy to see that the exact position is not important since the short range effect is only important at fairly high energies).

Thus if $\alpha_{DHL}(s)$ is the result of approximation (17) we now have the correction $\Delta_S\alpha(s)$ due to short range effects. Next we use

$$\alpha_{DHL} + \Delta_S\alpha$$

to estimate the rescattering correction as indicated above. This gives a correction $\Delta_R\alpha$ and finally

$$\alpha_F(s) = \alpha_{DHL}(s) + \Delta_S\alpha(s) + \Delta_R\alpha(s) \quad . \tag{19}$$

Our analysis of the inelasticities shows that we should be

able to carry out this procedure up to 500 - 600 MeV and possibly higher. The results are shown in figs. 13 - 19 together with the results of the experimental analysis of Auvil, Donnachie, Lea and Lovelace.

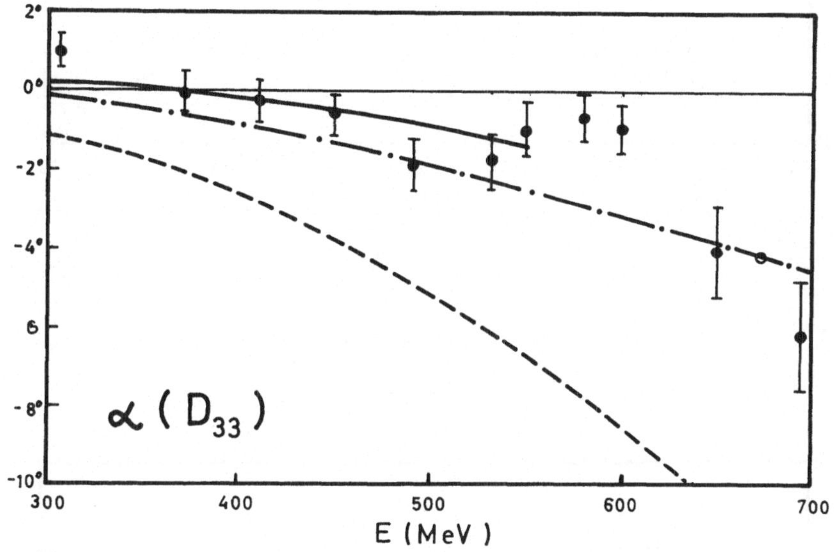

Fig. 13

- - - - - - - α_{DHL}

—·—·— $\alpha_{DHL} + \Delta_S \alpha$

——— α_F

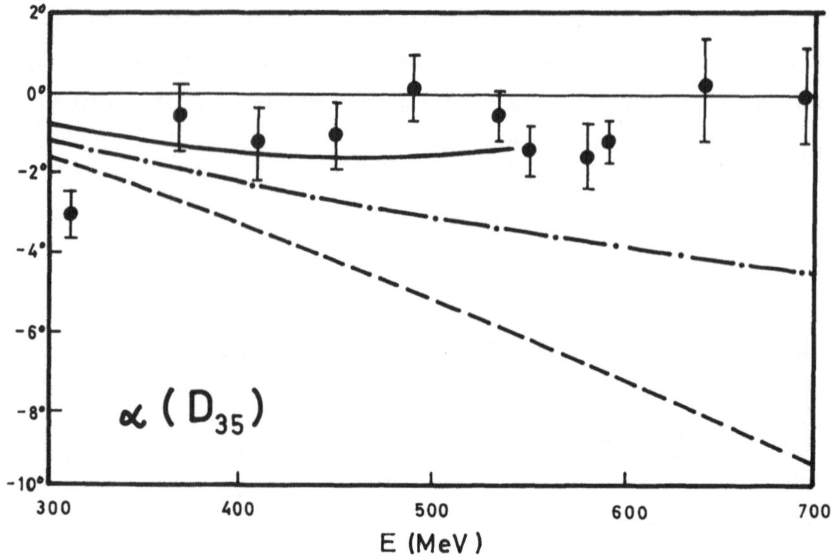

Fig. 14

------- α_{DHL}

—·—·— $\alpha_{DHL}+\Delta_S\alpha$

——— α_F

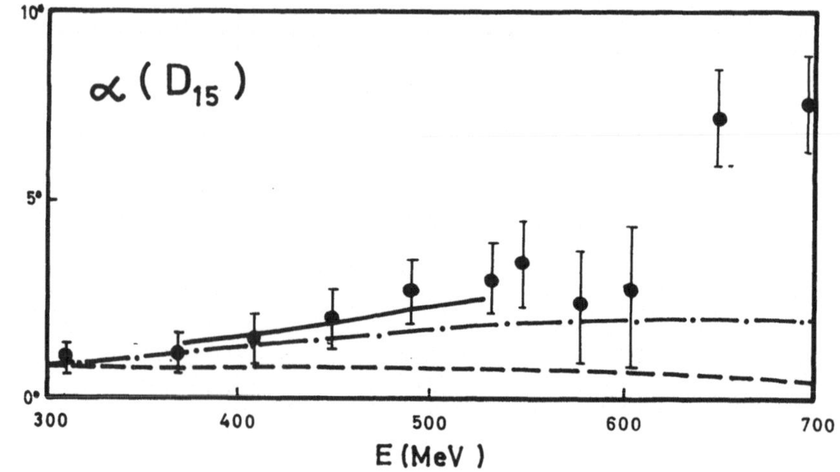

Fig. 15

------- α_{DHL}

—·—·— $\alpha_{DHL}+\Delta_S\alpha$

——— α_F

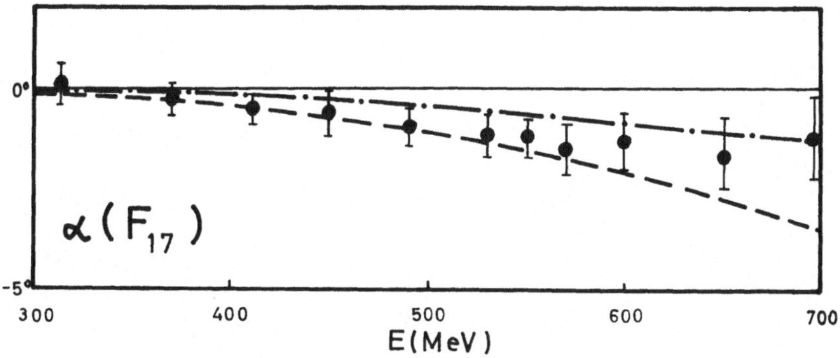

Fig.16

------- α_{DHL}

—·—·— $\alpha_F = \alpha_{DHL} + \Delta_S\alpha$

($\Delta_R\alpha$ is negligible)

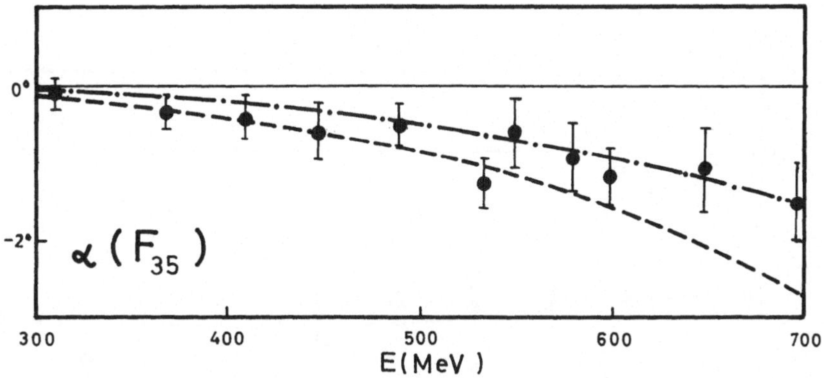

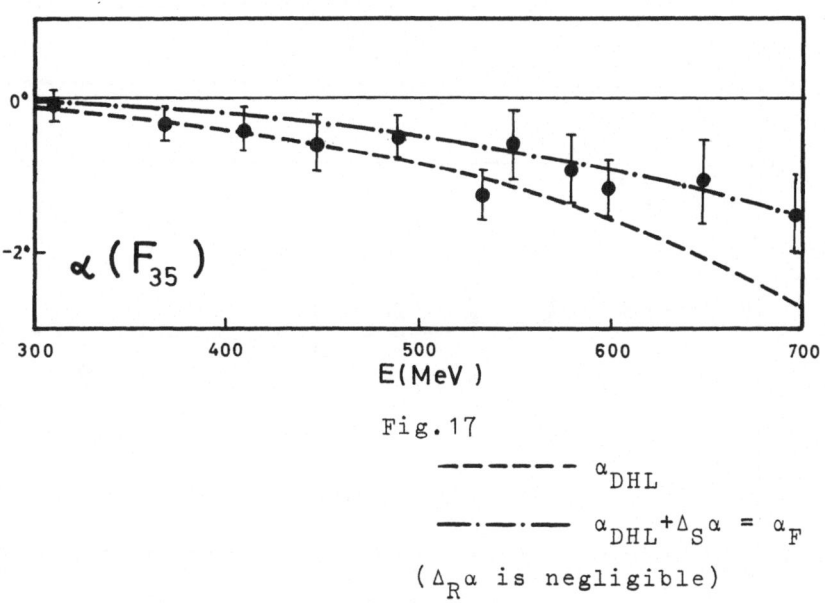

Fig.17

------- α_{DHL}

—·—·— $\alpha_{DHL} + \Delta_S\alpha = \alpha_F$

($\Delta_R\alpha$ is negligible)

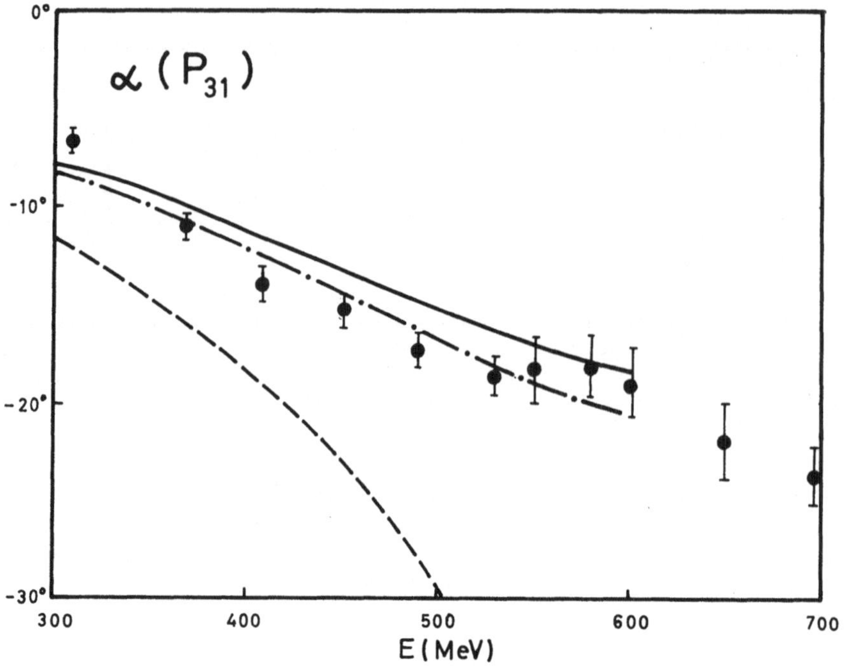

Fig. 18

- - - - - - α_{DHL}

—·—·— $\alpha_{DHL}+\Delta_S\alpha$

——— α_F

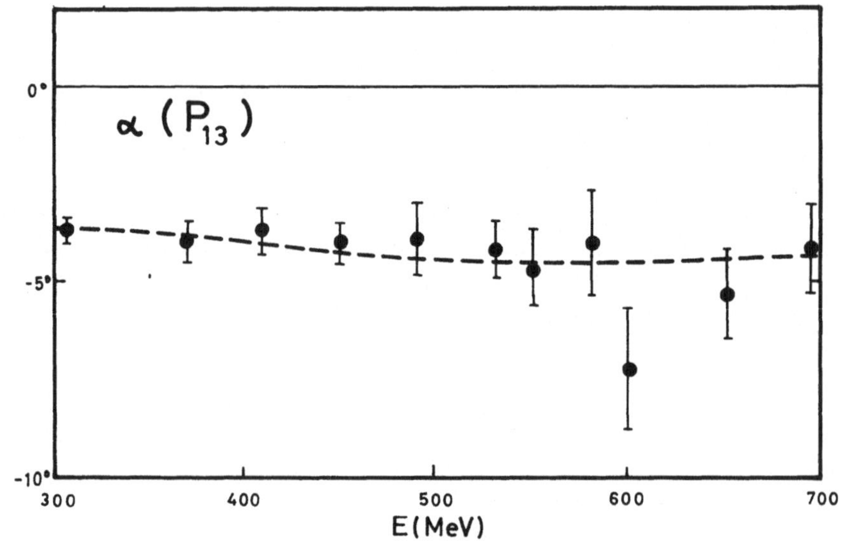

Fig. 19

- - - - - $\alpha_F = \alpha_{DHL}$ ($\Delta_S\alpha$ and $\Delta_R\alpha$ are negligible in this case)

IV. Outlook

What remains to be done.

1. We have analysed the S-waves at low-energy, but it would be desirable to have some understanding of the origin of the very strong short range repulsion. This might turn out to be a profound question.

2. We have not treated P_{11}. Phenomenological analysis (like that used for the S-waves) shows that this is the only P-wave which has an appreciable short range interaction. Moreover we saw early on that the presence of the channel $N + (\pi\pi)_0$ is important, and P_{11} is really a two-or many channel problem.

3. Exact solutions for D_{13}, F_{15}, F_{37} in the resonance region are wanted (the variational method gives a very good solution for P_{33}).

4. Phase shift predictions should be extended above 500 - 600 MeV, and inelasticity predictions should be made. Again we have two - or many-channel problems and since we believe N/D is not adequate we require new techniques.

References

J. Hamilton, Review Article in "Strong Interactions and High Energy Physics", Scottish Summer School (1963).

Donnachie, Hamilton and Lea, Phys. Rev., 135, B 515 (1964)
Donnachie, Hamilton, Ann. of Phys. 31, 410 (1965)
Donnachie, Hamilton, Phys. Rev. 138, B 678 (1965).

Experimental Analysis: Anvil, Donnachie, Lea and Lovelace, Phys. Lett. 12, 76 (1964).

DEGENERATE REPRESENTATIONS OF NON-COMPACT LIE GROUPS†

By

R. RACZKA

International Centre for Theoretical Physics, Trieste
and
Institute of Nuclear Research, Warsaw

Contents

1. Problems in the Theory of Representations of Non-Compact Groups

Suppose, that we have selected a certain non-compact group G as the higher symmetry group of some physical system. Then to regard all consequences and restrictions which follow from

† Lecture given at the V. Internationalen Universitätswochen für Kernphysik, Schladming, 24 February - 9 March 1966.

the introduced symmetry we should examine the following questions:

1. What is the system of invariant operators which generate the ring of invariant operators? Have we the possibility of a reduction of this system if we have no convenient interpretation for these operators as physical observables?

2. What is the maximal system of commuting operators in the fixed representation space which we shall interpret as physical observables? What is the shape and range of their spectra?

3. What are the properties of the basic functions in the representation space, which we would like to identify with physical states?

4. What are the properties of the direct product of two representations (Clebsch-Gordan series) and two basic vectors (Clebsch-Gordan coefficients)?

5. What are the properties of the decomposition of a considered representation with respect to a maximal compact subgroup (which is often an initial symmetry group)?

We cannot answer most of these questions if we use the method of global construction of representations which is mainly employed by mathematicians [1] [2]. In this series of lectures we shall present a new approach, which enables the solution of all the mentioned problems. The main idea is based on regarding the geometries of the considered groups and related spaces. The main advantage is rooted in the fact that the analysis of the properties of 1. - 5. of the representations is reduced to the investigations of the properties of second order ordinary differential equations.

2. Group Representations and Geometry of Symmetric Spaces

Unitary representations of an arbitrary semi-simple Lie group can be realized in a Hilbert space $H(X)$ of functions with a domain X being some homogeneous space of the type

$$X_i = G/G_i$$

where G_i is a closed subgroup of G. Properties of representations realized in the Hilbert space H(X) are closely related with the geometric properties of the group G and the symmetric space X . Namely we have:

1. The number of different principal series of irreducible unitary representations is equal to the number of non-isomorphic Cartan subgroups of a considered semi-simple Lie group G. [2].

2. If a given Cartan subgroup is isomorphic to a direct product of a k-dimensional linear space and r-dimensional torus then there exist a corresponding series of irreducible unitary representations determined by k-real numbers and r-integers [2]. Hence if a group G has a compact Cartan subgroup, there exists a discrete principal series of irreducible unitary representations characterized only be integers.

3. If on a manifold X on which a principal series of representations is realized there exists a double-point invariant measure then we can construct also on this manifold a supplementary series of representations [1].

4. Besides principal and supplementary series of representations there exists still a large number of so-called degenerate principal and degenerate supplementary series of representations. To describe them let us introduce the notion of the rank of a symmetric space X as the number of independent invariants of any pair of points of X with respect to action of the group G on the space X. The key theorem which relates the geometry of a homogeneous space X and the properties of representations realized on the Hilbert space H(X) is the fundamental Gelfand theorem [3]. This theorem states that the number of independent invariant differential operators, whose eigenvalues determine irreducible unitary representations, realized in the space H(X) is equal to the rank of the space X [4]. Thus we have as many different principal degenerate series of representations as symmetric spaces X with different rank $k(1 \leq k < n = \text{rank of } G)$. To any principal degenerate series there corresponds a supplementary degenerate series

realized in a Hilbert space H(X) with a double-point invariant measure on the space X.

The classification of symmetric spaces with a compact stability group was given by Cartan. In table 1 we give as an example the Cartan symmetric spaces related to SU(p,q) group.

Table 1

X_i	Rank	Dimension of X_i
SU(n)/SO(n)	n - 1	$\frac{1}{2}(n-1)(n+2)$
SU(2n)/Sp(n)	n - 1	(n - 1)(2n + 1)
SU(p+q)/S[U(p)×U(q)]	min(p,q)	2pq
SU(p,q)/S[U(p)×U(q)]	min(p,q)	2pq

The classification of symmetric spaces with the non-compact stability group was performed by Rosenfeld and Fedenko [6]. For U(p,q) group one class of such spaces is given by

$$X_{r,s} = SU(p,q)/S[U(r,s) \times U(p-r,q-s)] \; . \tag{2.1}$$

Consider for example some higher symmetry group of elementary particle physics e.g. SU(6,6) group. This group of rank eleven has a principal series of representations determined by eleven invariant numbers and the maximal set of commuting operators contains 77 operators. Of course we have not so many physical observables which we may relate with these 77 operators so we are interested in the maximal reduction of the invariant numbers which determine a representation and of the number of commuting operators. This problem can be easily solved using the Gelfand theorem. Namely if we would like to get the representation of SU(6,6) which is determined only by six invariant numbers we should realize the representation of SU(6,6) in the Hilbert space H(X) with the domain X of the type

$$X_{6,0} = SU(6,6)/S[U(6) \times U(6)] \quad . \tag{2.2}$$

If we would like to reduce maximally the number of invariant numbers we can consider the series of representations realized in H(X) with the domain X of the rank one e.g.

$$X_{1,0} = SU(6,6)/S[U(1) \times U(5,6)] \; ;$$

$$X_{0,1} = SU(6,6)/S[U(1) \times U(6,5)] \; . \tag{2.3}$$

The corresponding system of commuting operators is reduced maximally (see Section 8).

If we consider properties of a higher symmetry group of known physical system, e.g. H-atom, harmonic oscillator, rigid rotator (in n-dimension) and so on, we remark that the representations of the considered group are determined only by one or two invariant numbers although the rank of the considered groups can be arbitrarily large. These examples strongly suggest that the degenerate representations of the non-compact higher symmetry groups are most important in quantum mechanics and elementary particle physics.

In this series of lectures we shall investigate in detail the properties of degenerate representations of two classes of non-compact groups namely SO(p,q) and U(p,q) groups, which seem most important for physical applications. If we investigate the properties of most degenerate representations, which are determined by one invariant number then we have important information about the structure of the ring of invariant differential operators [5]. Namely, this ring is generated by the second order Laplace-Beltrami operator of the form

$$\Delta(X) = \frac{1}{\sqrt{|\bar{g}|}} \partial_\alpha \, g^{\alpha\beta}(X) \sqrt{|\bar{g}|} \, \partial_\beta \tag{2.4}$$

where $g_{\alpha\beta}(X)$ is a Riemannian metric tensor on the space X and $\bar{g} = \det\{g_{\alpha\beta}\}$. Thus the construction of the most degenerate irreducible unitary representations may be reduced to the following [7].

i) Construction of a convenient coordinate system on the space X, in which the metric tensor $g_{\alpha\beta}(X)$ is diagonal.

ii) Solution of the eigenvalue problem for the Laplace-Beltrami operator

$$\Delta(X)\ \psi^{\lambda} = \lambda\psi^{\lambda} \tag{2.5}$$

iii) Proof of the irreducibility and unitarity of the representation related with the set of harmonic functions. In the case of U(p,q) we consider the degenerate representations determined by eigenvalues of two invariant operators

$$Q_1 = \hat{M} = \sum_{i=1}^{p+q} L_i \quad \text{and} \quad Q_2 = \Delta(X) \quad [8] .$$

In this case the representation space is spanned by harmonic functions which are simultaneous eigenfunctions of

$$\Delta(X)\ \psi^{\lambda,M} = \lambda\psi^{\lambda,M} \tag{2.6}$$

and

$$\hat{M}\psi^{\lambda,M} = M\psi^{\lambda,M} \quad .$$

3. Discrete Most Degenerate Representations of SO(p,q) Groups (p ≥ q > 2)

As homogeneous spaces X we may take the quotient spaces G/G_o with a compact or non-compact stability group G_o. The homogeneous spaces of rank k with the compact stability group related with SO(p,q) groups are ([5] Chap. IX), [9]:

$$X_k = SO_o(p,q)/[SO_o(p) \times SO_o(q)] \quad .$$

Since the rank k of these Cartan symmetric spaces is equal to min (p,q) we may construct in these spaces the most degenerate representation only of the Lorentz type groups SO(p,1). For an arbitrary SO(p,q) group we have to consider more general spaces of rank one. We may take these spaces as homogeneous spaces of rank one of the following form [6] :

$$X_{+}^{p+q-1} = SO_{o}(p,q)/SO_{o}(p-1,q) \quad \text{and}$$

$$X_{-}^{p+q-1} = SO_{o}(p,q)/SO_{o}(p,q-1) \quad , \tag{2.2}$$

where the superscript p+q-1 denotes the dimension of the space $X_{\pm}^{p+q-1}$.

To choose a suitable coordinate system we have to introduce some convenient model of the space $X_{\pm}^{p+q-1}$ (2.2). This means we have to introduce a manifold, which has the same dimension and the same stability group as $X_{\pm}^{p+q-1}$ itself and on which the group SO(p,q) acts transitively.

For the space X_{+}^{p+q-1} such a model can be realized by the hyperboloid H_{q}^{p} determined by the equation

$$(x^1)^2 + \ldots + (x^p)^2 - (x^{p+1})^2 - \ldots - (x^{p+q})^2 = 1 \tag{3.1}$$

As an appropriate model for the space X_{-}^{p+q-1} we take the hyperboloid H_{p}^{q} defined by the equation

$$(x^1)^2 + \ldots + (x^q)^2 - (x^{q+1})^2 - \ldots - (x^{q+p})^2 = 1 \tag{3.2}$$

If we introduce internal coordinates $\Omega = \{\theta^1, \ldots, \theta^{p+q-1}\}$ on the space H_{q}^{p}(or H_{p}^{q}), which is imbedded in the flat Minkowski space M^{p+q}, then the metric tensor $g_{\alpha\beta}(H_{q}^{p})$ on the hyperboloid H_{q}^{p} is induced by the metric tensor $g_{ab}(M^{p+q})$ on the Minkowski space M^{p+q} and is defined as

$$g_{\alpha\beta}(H_{q}^{p}) = g_{ab}(M^{p+q}) \,.\, \partial_{\alpha}x^{a}(\Omega) \,.\, \partial_{\beta}x^{b}(\Omega) \quad , \tag{3.3}$$

where $a,b = 1,2,\ldots, p+q$ and $\alpha,\beta = 1,2,\ldots, p+q-1$.

Generally, we may choose a large number of different coordinate systems on the hyperboloid H_{q}^{p}(or H_{p}^{q}), in which the

Laplace-Beltrami operator can be separated. However, as follows from our previous work [7], the most convenient coordinate system is the biharmonic one because in this system the maximal Abelian compact subalgebra of the considered SO(p,q) group is automatically contained in the maximal set of commuting operators.

The biharmonic coordinate system on the hyperboloid $H^p_q(3,1)$ is constructed as follows:

$$x^k = x'^k \operatorname{ch}\theta, \quad k = 1,2,\dots,p,$$
$$x^{p+1} = x'^l \operatorname{sh}\theta, \quad l = 1,2,\dots,q, \quad \theta \in [0,\infty), \quad (3.4)$$

where the form of the x'^k and x'^l depends on whether p and q are even or odd. We must distinguish four cases:

$$\begin{array}{llll} \text{(i)} & p = 2r; & q = 2s, & \\ \text{(ii)} & p = 2r; & q = 2s+1, & \\ \text{(iii)} & p = 2r+1; & q = 2s, & r,s = 1,2,\dots \\ \text{(iv)} & p = 2r+1; & q = 2s+1, & \end{array} \quad (3.5)$$

Then, if p is even ($p = 2r$), the corresponding x'^k ($k = 1,2,\dots 2r$) are given by recursion formulae

for $r = 1$
$$x'^1 = \cos\varphi^1$$
$$x'^2 = \sin\varphi^1, \qquad \varphi^1 \in [0,2\pi)$$

for $r > 1$
$$x'^i = x''^i \sin\vartheta^r \qquad i = 1,2,\dots,2r-2,$$
$$x'^{2r-1} = \cos\varphi^r \cos\vartheta^r, \qquad \varphi^j \in [0,2\pi),\ j=1,2,\dots r,$$
$$x'^{2r} = \sin\varphi^r \cos\vartheta^r, \qquad \vartheta^k \in [0,\tfrac{\pi}{2}],\ k=2,3,\dots,r \quad (3.6)$$

and if p is odd ($p = 2r+1$) we first construct the x^{*i}, $i=1,2,\dots,2r$, by using the above-mentioned method for $p = 2r$; we then obtain the corresponding x'^k, $k=1,2,\dots,2r+1$, as

$$x'^i = x^{*i} \sin\vartheta^{r+1} \qquad i = 1,2,\dots,2r$$
$$x'^{2r+1} = \cos\vartheta^{r+1}, \qquad \vartheta^{r+1} \in [0,\pi]. \quad (3,7)$$

The recursion formulae for x'^{l}, q even or odd, are the same as those for x'^{k}, p even or odd respectively, except angles φ^{i}, ϑ^{j} in x'^{k} are replaced by $\tilde{\varphi}^{i}$, $\tilde{\vartheta}^{j}$.

Choosing the parametrization $\Omega \equiv \{\omega,\tilde{\omega},\theta\}$ on the hyperboloid H^{p}_{q} as [10]

$$\omega \equiv \{\varphi^{1},\dots,\varphi^{[\frac{p}{2}]},\vartheta^{2},\dots,\vartheta^{\{\frac{p}{2}\}}\},\quad \tilde{\omega} \equiv \{\tilde{\varphi}^{1},\dots,\tilde{\varphi}^{[\frac{q}{2}]},\tilde{\vartheta}^{2},\dots,\tilde{\vartheta}^{\{\frac{q}{2}\}}\} \tag{3.8}$$

and denoting

$$\{\partial_{\gamma}\} \equiv \left\{\frac{\partial}{\partial\varphi^{1}},\frac{\partial}{\partial\vartheta^{2}},\dots\frac{\partial}{\partial\varphi^{[\frac{p}{2}]}},\frac{\partial}{\partial\vartheta^{\{\frac{p}{2}\}}},\frac{\partial}{\partial\tilde{\varphi}^{1}},\frac{\partial}{\partial\tilde{\vartheta}^{2}},\dots\frac{\partial}{\partial\tilde{\varphi}^{[\frac{q}{2}]}},\frac{\partial}{\partial\tilde{\vartheta}^{\{\frac{q}{2}\}}},\frac{\partial}{\partial\theta}\right\}, \tag{3.9}$$

$$\gamma = 1,2,\dots,p+q-1$$

we can calculate the metric tensor $g_{\alpha\beta}(H^{p}_{q})$ as well as the Laplace-Beltrami operator $\Delta(H^{p}_{q})$ on the Hilbert space $\mathcal{H}(H^{p}_{q})$.

Since in all four cases (3.5) the variables in the Laplace-Beltrami operator (2,1) are separated in the same way due to the properties of the metric tensor (3,3), we can write the operator $\Delta(H^{p}_{q})$ in the form

$$\Delta(H^{p}_{q}) = \frac{-1}{ch^{p-1}\theta\, sh^{q-1}\theta}\,\frac{\partial}{\partial\theta}\, ch^{p-1}\theta\, sh^{q-1}\theta\,\frac{\partial}{\partial\theta}\;\frac{\Delta(X^{p-1})}{ch^{2}\theta} - \frac{\Delta(X^{q-1})}{sh^{2}\theta} \tag{3.10}$$

where $\Delta(X^{p-1})$ $[\Delta(X^{q-1})]$ is the Laplace-Beltrami operator of the rotation group SO(p) [SO(q)]. If we represent the eigenfunctions of $\Delta(H^{p}_{q})$ as a product of the eigenfunctions of $\Delta(X^{p-q})$, $\Delta(X^{q-1})$ and a function $\psi^{\lambda}_{l_{\{\frac{p}{2}\}},\tilde{l}_{\{\frac{q}{2}\}}}(\theta)$, we obtain the following equation:

$$\left[\frac{-1}{ch^{p-1}\theta\, sh^{q-1}\theta}\cdot\frac{d}{d\theta}\, ch^{p-1}\theta\, sh^{q-1}\theta\frac{d}{d\theta} - \frac{l_{\{\frac{p}{2}\}}(l_{\{\frac{p}{2}\}}+p-2)}{ch^{2}\theta} + \frac{\tilde{l}_{\{\frac{q}{2}\}}(\tilde{l}_{\{\frac{q}{2}\}}+q-2)}{sh^{2}\theta} - \lambda\right]\psi^{\lambda}_{l_{\{\frac{p}{2}\}},\tilde{l}_{\{\frac{q}{2}\}}}(\theta) = 0 \tag{3.11}$$

where $l_{\{p/2\}}(l_{\{p/2\}} + p - 2)$ $[l_{\{q/2\}}(l_{\{q/2\}} + q - 2)]$ are eigenvalues of the operator $\Delta(X^{p-1})$ $[\Delta(X^{q-1})]$ with $l_{\{p/2\}}$ $[l_{\{q/2\}}]$ being certain non-negative integers for $p > 2[q > 2]$.

A discrete series of representations exist if there exist solutions of (3.11) which are square integrable functions

$\Psi^{\lambda}_{l_{\{\frac{p}{2}\}}\tilde{l}_{\{\frac{q}{2}\}}}(\theta)$ $\theta \in (0, \infty)$, with respect to the measure

$$d\mu(\theta) = ch^{p-1}\theta \; sh^{q-1}\theta d\theta \tag{3.12}$$

which is induced by the measure $d\mu(\Omega)$ on the hyperboloid H^p_q [11]:

$$d\mu(\Omega) = (\overline{g}(H^p_q))^{1/2} d\Omega = d\mu(\omega)d\mu(\tilde{\omega}) \; ch^{p-1}\theta \; sh^{q-1}\theta d\theta \tag{3.13}$$

As the differential equation (3.11) has meromorphic coefficients which are regular in the interval $(0,\infty)$, any two linearly independent solutions are also regular analytic in this interval [12]. Since at the origin and at infinity the coefficients are singular, the solutions are not generally regular there and we easily find two essentially distinct behaviours of the solutions at the origin:

$$\Psi^{o}_{1} \sim \theta^{\tilde{l}_{\{\frac{q}{2}\}}}, \qquad \Psi^{o}_{2} \sim \theta^{-\tilde{l}_{\{\frac{q}{2}\}} - q + 2}$$

and at infinity

$$\Psi^{\infty}_{1,2} \sim \exp\left\{-\left(\frac{p+q-2}{2}\right) \pm \sqrt{\left(\frac{p+q-2}{2}\right)^2 - \lambda}\right\}\theta$$

The only satisfactory solution, i.e. the solution which is square integrable with respect to our measure $d\mu(\theta)$ (3.12), is that which behaves like $\psi^{o}_{1}(\theta)$ at the origin and like $\psi^{\infty}_{2}(\theta)$ at infinity. It turns out that the solution of (3.11) with these properties is

$$\Psi^{\lambda}_{l_{\{\frac{p}{2}\}},\tilde{l}_{\{\frac{q}{2}\}}}(\theta) = th^{\tilde{l}_{\{\frac{q}{2}\}}}\theta \cdot ch^{-\left(\frac{p+q-2}{2} + \sqrt{\left(\frac{p+q-2}{2}\right)^2 - \lambda}\right)}\theta \; {}_2F_1\left(-n + l_{\{\frac{p}{2}\}} + \frac{p-2}{2}, -n; \tilde{l}_{\{\frac{q}{2}\}} + \frac{q}{2}; th^2\theta\right),$$

where a non-negative integer n is connected with our $l_{\{p/2\}}$, $\tilde{l}_{\{q/2\}}$ and λ by the condition that ${}_2F_1$ be a polynomial, i.e.

$$l_{\{\frac{p}{2}\}} - \tilde{l}_{\{\frac{q}{2}\}} - 2n = \frac{p+q-2}{2} + \sqrt{\left(\frac{p+q-2}{2}\right)^2 - \lambda} - p + 2, \quad n = 0,1,2,\ldots \tag{3.14}$$

From this restrictive condition we can find that the discrete spectrum of the operator $\Delta(H^p_q)$ is of the form

$$\lambda = - L(L + p + q - 2) \qquad L = 1,2,\ldots \tag{3.15}$$

where

$$L = l_{\{\frac{p}{2}\}} - l_{\{\frac{q}{2}\}} - q - 2n \qquad \begin{aligned} l_{\{\frac{p}{2}\}} &= q+1,\ q+2,\ldots \\ \tilde{l}_{\{\frac{q}{2}\}} &= 0,1,2,\ldots,\ l_{\{\frac{p}{2}\}} - q - 1 \\ n &= 0,1,\ldots,\left\{\frac{l_{\{\frac{p}{2}\}} - \tilde{l}_{\{\frac{q}{2}\}} - q}{2}\right\} - 1 \end{aligned} \tag{3.16}$$

Of course it does not mean that n is limited as $l_{\{p/2\}}$ is an arbitrary integer larger than 2.

Summarizing, we have proved that there exist discrete most degenerate series of representations of an arbitrary SO(p,q) group ($p \geq q > 2$) on the Hilbert space $\mathcal{H}^L(H^p_q)$, i.e. on the space of square integrable functions

$$\Psi^{\lambda, l_2,\ldots,l_{\{\frac{p}{2}\}}, \tilde{l}_2,\ldots,\tilde{l}_{\{\frac{q}{2}\}}}_{m_1,\ldots,m_{[\frac{p}{2}]}, \tilde{m}_1,\ldots,\tilde{m}_{[\frac{q}{2}]}}(\omega, \tilde{\omega}, \theta)$$

with respect to the measure (3.13) and with λ given in (3.15). We shall denote such representations by $D^L(H^p_q)$.

The basis of the Hilbert space $\mathcal{H}^L(H^p_q)$ is formed by the orthonormal functions:

$$Y^{L, l_2,\ldots,l_{\{\frac{p}{2}\}}, \tilde{l}_2,\ldots\tilde{l}_{\{\frac{q}{2}\}}}_{m_1,\ldots,m_{[\frac{p}{2}]}, \tilde{m}_1,\ldots,\tilde{m}_{[\frac{q}{2}]}}(\omega, \tilde{\omega}, \theta) = Y^{l_2,\ldots,l_{\{\frac{p}{2}\}}}_{m_1,\ldots,m_{[\frac{p}{2}]}}(\omega) \cdot Y^{\tilde{l}_2,\ldots,\tilde{l}_{\{\frac{q}{2}\}}}_{\tilde{m}_1,\ldots,\tilde{m}_{[\frac{q}{2}]}}(\tilde{\omega}) \cdot V^L_{l_{\{\frac{p}{2}\}}, \tilde{l}_{\{\frac{q}{2}\}}}(\theta) \tag{3.17}$$

where

$$Y^{l_2,\dots,l_{\{\frac{p}{2}\}}}_{(\omega)\,m_1,\dots,m_{[\frac{p}{2}]}} = \begin{cases} Y^{l_2,\dots,l_r}_{(\omega)\,m_1,\dots,m_r} = \frac{1}{\sqrt{N_r}} \prod_{k=2}^{r} \sin^{2-k}\vartheta^k \cdot d^{J_k}_{M_k;M_k'}(2\vartheta^k) \cdot \prod_{k=1}^{r} e^{im_k\varphi^k} & \text{if } p=2r \\ Y^{l_2,\dots,l_{r+1}}_{(\omega)\,m_1,\dots,m_r} = \frac{1}{\sqrt{N_{r+1}}} \sin^{1-r}(\vartheta^{r+1}) \cdot d^{J_{r+1}}_{M_{r+1};0}(\vartheta^{r+1}) \prod_{k=2}^{r} \sin^{2-k}(\vartheta^k) \times & \\ \quad \times\, d^{J_k}_{M_k;M_k'}(2\vartheta^k) \cdot \prod_{k=1}^{r} e^{im_k\varphi^k} & \text{if } p=2r+1 \end{cases}$$

(3.18)

are eigenfunctions of $\Delta(X^{p-1})$ derived in Appx.of [7]. $Y^{\tilde{l}_2,\dots,\tilde{l}_{\{\frac{q}{2}\}}}_{(\tilde{\omega})\,\tilde{m}_1,\dots,\tilde{m}_{[\frac{q}{2}]}}$ are eigenfunctions of $\Delta(X^{q-1})$ expressed as the product of the usual d-functions of angular momenta and exponential functions exactly as (3.18), but with variables $\tilde{\varphi}^i$, $\tilde{\vartheta}^j$ and $\tilde{l}_k$, $\tilde{m}_l$ instead of φ^i, ϑ^j and l_k, m_l respectively, and

$V^L_{l_{\{\frac{p}{2}\}},\tilde{l}_{\{\frac{q}{2}\}}}(\theta)$ is the solution of (3.11) given by

$$V^L_{l_{\{\frac{p}{2}\}},\tilde{l}_{\{\frac{q}{2}\}}}(\theta) = \frac{1}{\sqrt{N}}\, th^{\tilde{l}_{\{\frac{q}{2}\}}}(\theta)\, ch^{-(L+p+q-2)}(\theta)\; {}_2F_1\left(\frac{p+q-2+l_{\{\frac{p}{2}\}}+\tilde{l}_{\{\frac{q}{2}\}}+L}{2}, \frac{L+q+\tilde{l}_{\{\frac{q}{2}\}}-l_{\{\frac{p}{2}\}}}{2}; \tilde{l}_{\{\frac{q}{2}\}}+\frac{q}{2}, th^2\theta\right), \tag{3.19}$$

where for a definite representation L is fixed and $l_{\{p/2\}}$, $\tilde{l}_{\{q/2\}}$ are restricted by the condition that ${}_2F_1$ be a polynomial, i.e.

$$l_{\{p/2\}} - \tilde{l}_{\{q/2\}} = L + q + 2n\,, \qquad n = 0,1,2,\dots \tag{3.20}$$

N_r, N_{r+1}, N are normalization factors given by

$$N_r = 2\pi^r \prod_{k=2}^{r} \frac{1}{l_k+k-1}\,, \qquad N_{r+1} = 4\pi^r \frac{1}{2(l_{r+1}+r)-1} \prod_{k=2}^{r} \frac{1}{l_k+k-1}\,, \tag{3.21}$$

$$N = \frac{\Gamma\left(\frac{1}{2}(l_{\{\frac{p}{2}\}} - \tilde{l}_{\{\frac{q}{2}\}} - L - q + 2)\right)\Gamma^2\left(\tilde{l}_{\{\frac{q}{2}\}} + \frac{q}{2}\right)\Gamma\left(\frac{1}{2}(L - \tilde{l}_{\{\frac{q}{2}\}} + l_{\{\frac{p}{2}\}} + p)\right)}{2\left(L + \frac{p+q-2}{2}\right)\Gamma\left(\frac{1}{2}(l_{\{\frac{p}{2}\}} + \tilde{l}_{\{\frac{q}{2}\}} + L + p + q - 2)\right)\Gamma\left(\frac{1}{2}(l_{\{\frac{p}{2}\}} + \tilde{l}_{\{\frac{q}{2}\}} - L)\right)} \qquad (3.22)$$

and the indices J_k, M_k. M'_k are defined in the Appx. as:

$$J_k = \frac{1}{2}(l_k + k - 2) \ , \ M_k = \frac{1}{2}(m_k + l_{k-1} + k - 2),$$

$$M'_k = \frac{1}{2}(m_k - l_{k-1} - k + 2) \qquad \text{for } k = 2,3,\ldots r \qquad (3,23)$$

$$J_{r+1} = l_{r+1} + r - 1 \qquad , M_{r+1} = l_r + r - 1 \ .$$

l_k, $k = 2,\ldots, r+1$, are non-negative integers, m_k, $k = 1,\ldots r$, are integers restricted as follows (See Appx. of [7])

$$|m_2| + |m_1| = l_2 - 2n_2, |m_3| + l_2 = l_3 - 2n_3 \ , \ \ldots \ ,$$

$$|m_r| + l_{r-1} = l_r - 2n_r \ , \ l_r = l_{r+1} - n_{r+1} \ ,$$

$$n_k = 0,1,\ldots, \{l_k/2\} \ ,$$

$$k = 2,3,\ldots, r+1 \ . \qquad (3,24)$$

There exists also a discrete series of representations $D^L(H^q_p)$ on the Hilbert space $\mathcal{H}^L(H^q_p)$ with H^q_p given in (3,2). This series is obtained from the previous one $D^L(H^p_q)$ formally by exchanging p, $l_{\{p/2\}}$ for q, $\tilde{l}_{\{q/2\}}$ respectively and vice versa. The most degenerate representations $D^L(H^q_p)$ created on $\mathcal{H}^L(H^q_p)$ are not equivalent except in the case $p = q$, when both Hilbert spaces coincide.

Finally, we would like to mention that the representations $D^L(H^q_p)$ and $D^L(H^p_q)$ are irreducible and unitary as it will be proved in Section 6.

4. Discrete most Degenerate Representations of SO(p,2) Groups (p ≧ 2)

For the "de Sitter type groups" SO(p,2) the homogeneous spaces are [6]

$$X_{+}^{p+1} = SO_o(p,2)/SO_o(p-1,2) \qquad \text{and}$$

$$X_{-}^{p+1} = SO_o(p,2)/SO_o(p,1) \tag{4.1}$$

and can be represented by hyperboloids H_2^p and H_p^2 respectively.

The biharmonic coordinate system is introduced again in the same way as in Section 3. Hence for the Laplace-Beltrami operator on the Hilbert space $\mathcal{H}(H_p^2)$ we obtain

$$\Delta(H_p^2) = \frac{-1}{ch\theta\, sh^{p-1}\theta}\frac{\partial}{\partial\theta}\, ch\theta\, sh^{p-1}\theta\, \frac{\partial}{\partial\theta} + \frac{1}{ch^2\theta}\frac{\partial^2}{(\partial\tilde{\varphi}^1)^2} - \frac{\Delta(X^{p-1})}{sh^2\theta} \tag{4.2}$$

The main difference with respect to the equation for $\Delta(H_p^q)$ is rooted in the fact that instead of the operator $\Delta(X^{q-1})$, $q > 2$, appearing there, in equation (4.2) the operator $\Delta(X^1) = \partial^2/(\partial\tilde{\varphi}^1)^2$ appears, which has eigenvalues $-(\tilde{m}_1)^2$ with $\tilde{m}_1$ an arbitrary integer. By using the same procedure as in Section 3 we obtain finally the following equation for the function of θ

$$\left[\frac{-1}{ch\theta\, sh^{p-1}\theta}\frac{d}{d\theta}\, ch\theta\, sh^{p-1}\theta\, \frac{d}{d\theta} - \frac{(\tilde{m}_1)^2}{ch^2\theta} + \frac{l_{\{\frac{p}{2}\}}(l_{\{\frac{p}{2}\}}+p-2)}{sh^2\theta} - \lambda\right]\psi^{\lambda}_{\tilde{m}_1, l_{\{\frac{p}{2}\}}}(\theta) = 0 \tag{4.3}$$

The discrete series of representations exist again due to the fact that there exist solutions of (4.3) which are square integrable in $\theta \in (o,\infty)$ with respect to the measure $d\mu(\theta) = sh^{p-1}\theta . ch\theta . d\theta$.

The discrete spectrum of the operator $\Delta(H_p^2)$ looks like

$$\lambda = - L(L + p) , \qquad L = 1,2,\ldots , \tag{4.4}$$

where

$$L = |\tilde{m}_1| - |l_{\{\frac{p}{2}\}}| - 2p - p , \qquad \begin{aligned} |\tilde{m}_1| &= p+1, p+2, \ldots , \\ |l_{\{\frac{p}{2}\}}| &= 0,1,\ldots, |\tilde{m}_1| - p - 1 \\ n &= 0,1,\ldots, \left\{\frac{|\tilde{m}_1| - |l_{\{\frac{p}{2}\}}| - p}{2}\right\} - 1 \end{aligned} \tag{4.5}$$

In a definite representation the value L is fixed and equation (4.5) imposes the following restriction on $|\tilde{m}_1|$:

$$|\tilde{m}_1| \geqq L + p . \tag{4.6}$$

Since generators of an SO(p,q) group can change the quantum number $\tilde{m}_1$ only by one (see section 6), we create on $\mathcal{H}^L(H_p^2)$ two discrete non-equivalent series of representations. One of them, corresponding to $\tilde{m}_1 \geq L + p$, we denote by $D_+^L(H_p^2)$ and the other with $\tilde{m}_1 \leq -(L+p)$ by $D_-^L(H_p^2)$.

The representations $D_\pm^L(H_p^2)$ are representations on different invariant subspaces of the Hilbert space $\mathcal{H}^L(H_q^p)$ with basis formed by the following orthogonal functions

$$Y^{L,l_2,\ldots,l_{\{\frac{p}{2}\}}}_{m_1,\ldots,m_{[\frac{p}{2}]},\tilde{m}_1}(\omega,\tilde{\varphi}^1,\theta) = Y^{l_2,\ldots,l_{\{\frac{p}{2}\}}}_{m_1,\ldots,m_{[\frac{p}{2}]}}(\omega) \frac{e^{i\tilde{m}_1\tilde{\varphi}^1}}{\sqrt{2\pi}} \cdot V^{L}_{\tilde{m}_1,l_{\{\frac{p}{2}\}}}(\theta) \tag{4.7}$$

where $Y^{l_2,\ldots,l_{\{\frac{p}{2}\}}}_{m_1,\ldots,m_{[\frac{p}{2}]}}(\omega)$ is given in (3.18) and

$$V^{L}_{\tilde{m}_1,l_{\{\frac{p}{2}\}}}(\theta) = \frac{1}{\sqrt{N}}\, th\,\theta^{|l_{\{\frac{p}{2}\}}|} \cdot ch\,\theta^{-(L+p)} \cdot {}_2F_1\left(\frac{-|\tilde{m}_1| + |l_{\{\frac{p}{2}\}}| + L + p}{2}, \frac{|\tilde{m}_1| + |l_{\{\frac{p}{2}\}}| + L + p}{2}, |l_{\{\frac{p}{2}\}}| + \frac{p}{2}; th^2\theta\right)$$

$$N = \frac{\Gamma\left(\frac{|\tilde{m}_1| - |l_{\{\frac{p}{2}\}}| - L - p + 2}{2}\right)\Gamma\left(|l_{\{\frac{p}{2}\}}| + \frac{p}{2}\right)\Gamma\left(\frac{L - |l_{\{\frac{p}{2}\}}| + |\tilde{m}_1| + 2}{2}\right)}{\left(L + \frac{p}{2}\right)\Gamma\left(\frac{|\tilde{m}_1| + |l_{\{\frac{p}{2}\}}| + L + p}{2}\right)\Gamma\left(\frac{|\tilde{m}_1| + |l_{\{\frac{p}{2}\}}| - L}{2}\right)} \tag{4.8}$$

where for definite representation L is fixed and $|\tilde{m}_1|$, $l_{\{p/2\}}$ are restricted by condition that ${}_2F_1$ be a polynomial, i.e.

$$|\tilde{m}_1| - |l_{\{p/2\}}| = L + p + 2n \quad , \qquad n = 0,1,\ldots \qquad (4,9)$$

the discrete series of representations on the Hilbert space $\mathcal{H}^L(H_2^p)$ are constructed by the same method, but (except p=2) we obtain only one series because now $l_{\{p/2\}}$ plays the role of $\tilde{m}_1$ and for $p > 2$ $l_{\{p/2\}}$ is a non-negative integer. For $p=2(l_1 \equiv m_1)$ we find again two discrete non-equivalent series as both Hilbert spaces $\mathcal{H}^L(H_p^2)$ and $\mathcal{H}^L(H_2^p)$ coincide.

5. Discrete most Degenerate Representations of SO(p,1) Groups

The homogeneous spaces of rank one for the Lorentz type groups are [6]

$$X_+^p = SO_o(p,1)/SO_o(p-1,1) \quad \text{and} \quad X_-^p = SO_o(p,1)/SO_o(p) \ , \qquad (5,1)$$

where the X_-^p space is the Cartan symmetric one. As their models we take the hyperboloids H_1^p and H_p^1 respectively.

The biharmonic coordinates on H_1^p and H_p^1 are introduced again by the method explained in Section 3, but on the hyperboloid H_1^p the range of θ is $(-\infty,\infty)$. On the hyperboloid H_p^1 the range of θ is from zero to infinity since we restrict ourselves to the upper sheet of the hyperboloid H_p^1 of course, the upper sheet of H_p^1 is a transitive manifold only under the proper $SO_o(p,1)$ group, i.e. under the group of transformations $g = (g_{ik})$, for which g_{11} is positive.

The Laplace-Beltrami operator on the Hilbert space $\mathcal{H}(H_1^p)$ has the form

$$\Delta(H_1^p) = \frac{-1}{ch^{p-1}\theta}\frac{\partial}{\partial\theta} ch^{p-1}\theta \frac{\partial}{\partial\theta} + \frac{\Delta(X^{p-1})}{ch^2\theta} \ , \qquad \theta \in (-\infty,\infty), \quad (5.2)$$

where $\Delta(X^{p-1})$ is the Laplace-Beltrami operator for the SO(p) group given in the Appx. of [7].

The eigenvalue problem of $\Delta(H_1^p)$ is reduced to

$$\left[\frac{-1}{ch^{p-1}\theta}\frac{d}{d\theta}ch^{p-1}\theta\frac{d}{d\theta} - \frac{l_{\{\frac{p}{2}\}}(l_{\{\frac{p}{2}\}}+p-2)}{ch^2\theta} - \lambda\right]\overset{\lambda}{\underset{l_{\{\frac{p}{2}\}}}{\psi}}(\theta) = 0 \qquad (5.3)$$

Analogously to the previous cases, we find the discrete spectrum of $\Delta(H_1^p)$ to be of the form

$$\lambda = -L(L+p-1)\ , \qquad L = 0,1,2,\ldots\ , \qquad (5.4)$$

where

$$L = |l_{\{\frac{p}{2}\}}| - 1 - n \qquad |l_{\{\frac{p}{2}\}}| = 1,2,\ldots \qquad n = 0,1,\ldots,\left\{\frac{|l_{\{\frac{p}{2}\}}|-1}{2}\right\} \qquad (5.5)$$

$l_{\{p/2\}}$ a positive integer for $p > 2$, and for $p = 2$ an arbitrary non-zero integer, m_1. Hence, there is an exceptional case for $p = 2$ and we obtain again two types of discrete non-equivalent series of representations $D_+^L(H_1^2)$ and $D_-^L(H_1^2)$ on different invariant subspaces of the Hilbert space $\mathcal{H}^L(H_1^2)$. For SO(2,1) these results were obtained by Bargmann [13], for SO(3,1) by Gel'fand and Graev [14] and for SO(4,1) by Dixmier [15].

The basis of the Hilbert space $\mathcal{H}^L(H_1^p)$ is formed by the orthonormal functions

$$\overset{L,l_2,\ldots,l_{\{\frac{p}{2}\}}}{\underset{m_1,\ldots,m_{[\frac{p}{2}]}}{Y}}(\omega,\theta) = \begin{cases} {}_1\overset{L,l_2,\ldots,l_{\{\frac{p}{2}\}}}{\underset{m_1,\ldots,m_{[\frac{p}{2}]}}{Y}}(\omega,\theta) = \overset{l_2,\ldots,l_{\{\frac{p}{2}\}}}{\underset{m_1,\ldots,m_{[\frac{p}{2}]}}{Y}}(\omega)\cdot {}_1\overset{L}{\underset{l_{\{\frac{p}{2}\}}}{V}}(\theta) & \text{if } L-l_{\{\frac{p}{2}\}} = -(2n+1), \\ {}_2\overset{L,l_2,\ldots,l_{\{\frac{p}{2}\}}}{\underset{m_1,\ldots,m_{[\frac{p}{2}]}}{Y}}(\omega,\theta) = \overset{l_2,\ldots,l_{\{\frac{p}{2}\}}}{\underset{m_1,\ldots,m_{[\frac{p}{2}]}}{Y}}(\omega)\cdot {}_2\overset{L}{\underset{l_{\{\frac{p}{2}\}}}{V}}(\theta) & \text{if } L-l_{\{\frac{p}{2}\}} = -(2n+2), \\ & n = 0,1,2,\ldots \end{cases} \qquad (5.6)$$

where $Y^{l_2,\dots,l_{\{\frac{p}{2}\}}}_{m_1,\dots,m_{[\frac{p}{2}]}}(\omega)$ is explicitly given in equation (3,18) and

$$ {}_1V^L_{l_{\{\frac{p}{2}\}}}(\theta) = \frac{ch\theta^{-(L+p-1)}}{\sqrt{{}_1N}} \cdot {}_2F_1\left(\frac{L+l_{\{\frac{p}{2}\}}+p-1}{2}, \frac{L-l_{\{\frac{p}{2}\}}+1}{2}; \frac{1}{2}; th^2\theta\right) \tag{5.7} $$

$$ {}_2V^L_{l_{\{\frac{p}{2}\}}}(\theta) = \frac{-2}{\sqrt{{}_2N}} th\theta \cdot ch\theta^{-(L+p-1)} \cdot {}_2F_1\left(\frac{L+l_{\{\frac{p}{2}\}}+p}{2}, \frac{L-l_{\{\frac{p}{2}\}}+2}{2}; \frac{3}{2}; th^2\theta\right) $$

Here, normalization factors ${}_1N$, ${}_2N$ are of the form

$$ {}_1N = \frac{2\pi\Gamma\left(\frac{l_{\{\frac{p}{2}\}}-L+1}{2}\right)\Gamma\left(\frac{l_{\{\frac{p}{2}\}}+L+p}{2}\right)}{\left(L+\frac{p-1}{2}\right)\Gamma\left(\frac{l_{\{\frac{p}{2}\}}+L+p-1}{2}\right)\Gamma\left(\frac{l_{\{\frac{p}{2}\}}-L}{2}\right)}, \quad {}_2N = \frac{4(2L+p-1)\,{}_1N}{(l_{\{\frac{p}{2}\}}-L)(l_{\{\frac{p}{2}\}}+L+p-2)(2L+p+1)} \tag{5.8} $$

and for definite representation L is fixed and $l_{\{p/2\}}$ must satisfy the restrictive condition that ${}_2F_1$ be a polynomial, i.e.

$$ l_{\{p/2\}} = L + 1 + n \quad , n = 0,1,2,\dots \tag{5.9} $$

The discrete series of representations on the Hilbert space $\mathcal{H}^L(H^1_p)$ does not exist, because the Laplace-Beltrami operator

$$ \Delta(H^1_p) = \frac{-1}{sh^{p-1}\theta}\frac{d}{d\theta}\, sh^{p-1}\theta\, \frac{d}{d\theta} - \frac{l_{\{\frac{p}{2}\}}(l_{\{\frac{p}{2}\}}+p-2)}{sh^2\theta} $$

$\theta \in [0,\infty)$, has no discrete spectrum.

6. Irreducibility and Unitarity

The irreducibility of the representations related with the set of the harmonic functions (3.17), (4.7) or (5.6) was proved in [7].

Unitarity:

The representation T_g of a group element $g \varepsilon SO(p,q)$ on the Hilbert space $^L(H^p_q)$ is determined by the left-translation:

$$T_g Y^{L, l_2, \ldots, l_{\{\frac{p}{2}\}}, \tilde{l}_2, \ldots, \tilde{l}_{\{\frac{q}{2}\}}}_{m_1, \ldots, m_{[\frac{p}{2}]}, \tilde{m}_1, \ldots, \tilde{m}_{[\frac{q}{2}]}}(\Omega) = Y^{L, l_2, \ldots, l_{\{\frac{p}{2}\}}, \tilde{l}_2, \ldots, \tilde{l}_{\{\frac{q}{2}\}}}_{m_1, \ldots, m_{[\frac{p}{2}]}, \tilde{m}_1, \ldots, \tilde{m}_{[\frac{q}{2}]}}(g\Omega) \qquad (6.13)$$

as follows from the representation of the Lie algebra given by (6,2) and (6,3). Here the symbol $g\Omega$ represents the set of

parameters $\varphi'_1, \ldots \vartheta'^{\{\frac{q}{2}\}}, \theta'$ of the point $\Omega' = g\Omega$ on H^p_q and $Y^{L, l_2, \ldots, l_{\{\frac{p}{2}\}}, \tilde{l}_2, \ldots, \tilde{l}_{\{\frac{q}{2}\}}}_{m_1, \ldots, m_{[\frac{p}{2}]}, \tilde{m}_1, \ldots, \tilde{m}_{[\frac{q}{2}]}}(\Omega)$ is a harmonic function defined in the expressions (3.17), (4.7) or (5.6). Therefore, the unitarity follows from the left-invariance of the measure $d\mu(\Omega)$ on the corresponding hyperboloid H^p_q.

7. Conclusions

We devote this section to a brief review and discussion of the derived representations $D^L(H^p_q)$ of the group SO(p,q).

A) The case $p \geq q > 2$ (Section 3).:

There exist two series of representations: $D^L(H^p_q)$ and $D^L(H^q_p)$ $L = 1,2,\ldots$ related to hyperboloids H^p_q(3,1) and H^q_p (3.2) respectively. The non-negative integers $l_{\{p/2\}}$, $\tilde{l}_{\{q/2\}}$ which determine the irreducible representations of the subgroup SO(p) and SO(q) respectively, are not independent as in the case of continuous most degenerate representations [16], but are restricted by

$$l_{\{\frac{p}{2}\}} - \tilde{l}_{\{\frac{q}{2}\}} - 2n = L + q \qquad \text{for } H^p_q, \qquad (7,1)$$

$$\tilde{l}_{\{\frac{q}{2}\}} - l_{\{\frac{p}{2}\}} - 2n = L + p \qquad \text{for } H^q_p, \qquad (7,2)$$

where $l_{\{p/2\}}$, $\tilde{l}_{\{q/2\}}$ and n range through every such triplet of non-negative integers which satisfy (7.1) and (7.2) respectively. These two conditions are illustrated graphically in Fig. 1 and Fig. 2 respectively. Every knot of the net in the figures represents a subspace $\mathcal{H}^L_{l_{\{p/2\}},\tilde{l}_{\{q/2\}}}$ of an irreducible representation of the maximal compact subgroup SO(p) × SO(q) determined by a pair of integers $l_{\{p/2\}}$ and $\tilde{l}_{\{q/2\}}$. Generators L_{ij} of the compact type act inside the subspace $\mathcal{H}^L_{l_{\{p/2\}},\tilde{l}_{\{q/2\}}}$. On the other hand, the generator B_{rs} of the noncompact type maps the subspace $\mathcal{H}^L_{l_{\{\frac{p}{2}\}},\tilde{l}_{\{\frac{q}{2}\}}}$ into four neighbouring subspaces $\mathcal{H}^L_{l+1,\tilde{l}+1}$, $\mathcal{H}^L_{l-1,\tilde{l}+1}$, $\mathcal{H}^L_{l+1,\tilde{l}-1}$, $\mathcal{H}^L_{l-1,\tilde{l}-1}$ graphically represented in Fig. 1 .

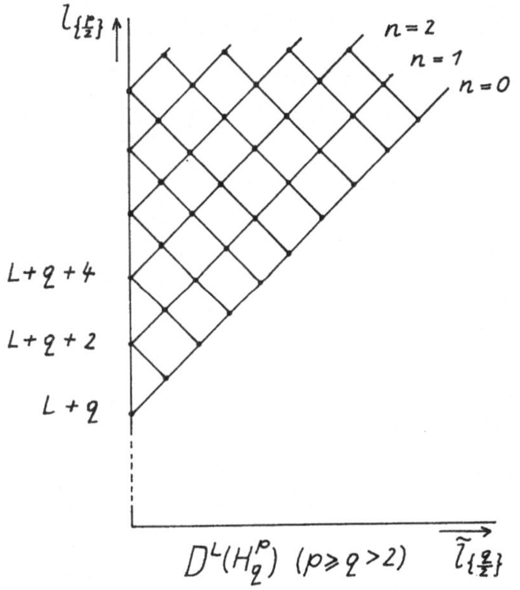

Fig. 1

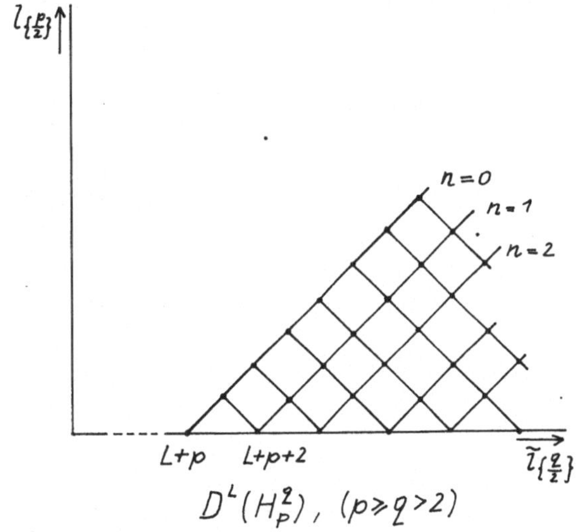

Fig. 2

All the representations $D^L(H^p_q)$ and $D^L(H^q_p)$ are inequivalent except for p = q when we have only one series of representations $D^L(H^p_p)$.

B) The case $p > q = 2$ (Section 4):

Generally, there exist three series of representations. Before proceeding with their description we wish to stress that the irreducible representations of the subgroup SO(2) are characterized by an integer m which also takes on negative values. Instead of the conditions (7.1) and (7.2) we have now

$$|l_{\{\frac{p}{2}\}}| - |\tilde{m}_1| - 2n = L + 2 \qquad \text{for } H_2^p \tag{7.3}$$

$$|\tilde{m}_1| - |l_{\{\frac{p}{2}\}}| - 2n = L + p \qquad \text{for } H_p^2 \tag{7.4}$$

where $|l_{\{p/2\}}|$, $|\tilde{m}_1|$ and n range through all such non-negative integers that (7.3) and (7.4) are satisfied respectively, It follows from these conditions and conclusions of Section 4, that there exists only one series $D^L(H_2^p)$ of representations related with the hyperboloid H_2^p , while there exist two series of representations $D_+^L(H_p^2)$ and $D_-^L(H_p^2)$ related with the hyperboloid H_p^2 . Their graphical representations are given in Fig. 3, Fig. 4 and Fig. 5 respectively.

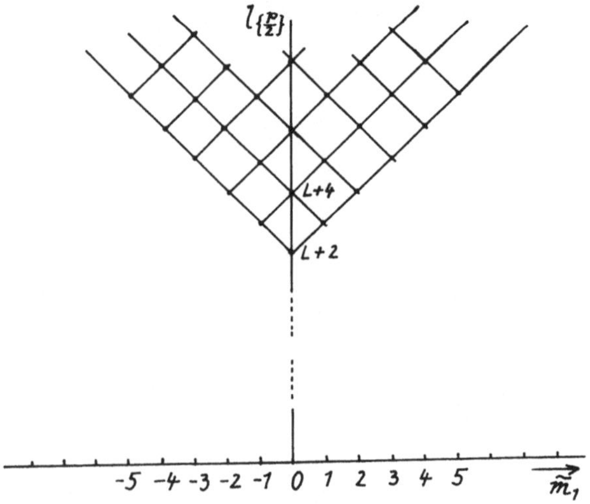

$D^L(H_2^p)$

Fig. 3

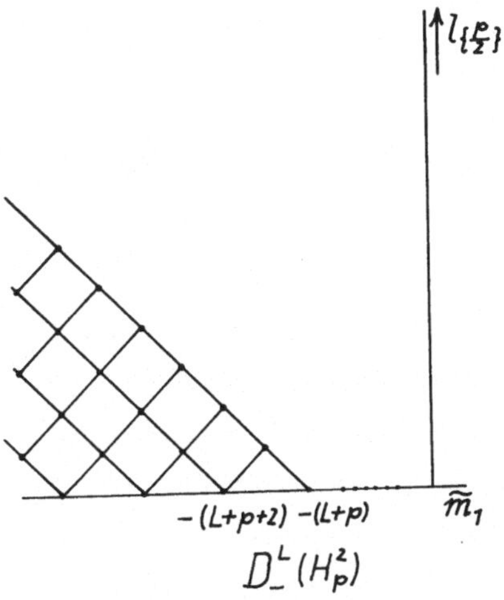

Fig. 4

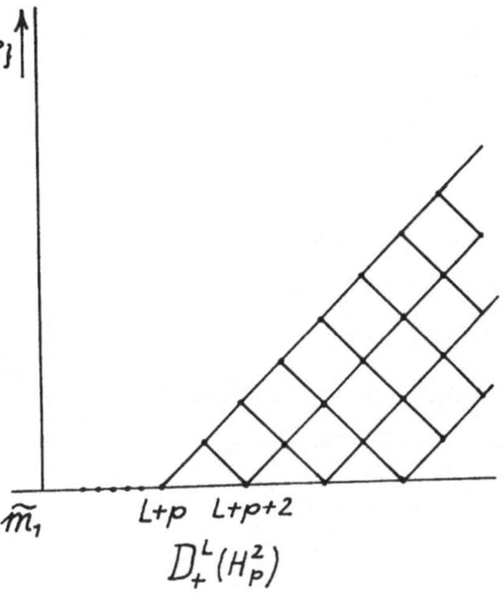

Fig. 5

The representations are inequivalent except for the case p = q = 2. In the latter case two subgroups SO(2) of the group

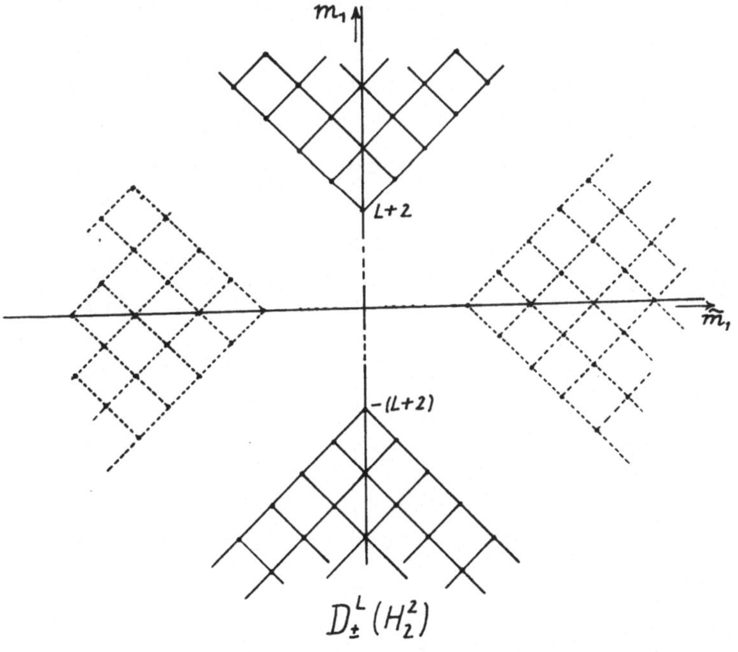

Fig. 6

SO(2,2) are indistinguishable and we have only two inequivalent representations drawn in solid lines in Fig. 6. The representations which appear after changing m_1 and $\tilde{m}_1$ are equivalent to a pair of previous representations. We represent them by dotted lines in Fig. 6.

C) The case q = 1 (Section 5):

Generally, there exists only one series of discrete most degenerate representations $D^L(H_1^p)$. However, in the case of SO(2,1) we have obtained two series of irreducible representations, i.e., $D_+^L(H_1^2)$ and $D_-^L(H_1^2)$. The condition on the number $l_{\{p/2\}}$, which determines the irreducible representations of the maximal compact subgroup SO(p), has the form

$$|l_{\{p/2\}}| = L + 1 + 2n \qquad n = 0,1,2,\ldots \tag{7.5}$$

It is interesting that we have found the discrete most degenerate representations even for the groups SO(p,q) with p and q odd, which have no discrete principal series of representations (see Table I of [7]) Let us explain this unexpected fact, for example, for the Lorentz group SO(3,1). The action of two Casimir operators $\Delta_1 = \vec{M}^2 - \vec{N}^2$ and $\Delta_2 = \vec{M} \cdot \vec{N}$ on the basis f_r^k of the Hilbert space, which realizes the irreducible representation, can be written in the form [17]:

$$\Delta_1 f_r^k = -2(k_0^2 + c^2 - 1)f_r^k \quad , \qquad k_0 = 0, \tfrac{1}{2}, 1, \tfrac{3}{2}, \ldots , \tag{7.6}$$

$$\Delta_2 f_r^k = -4ik_0 c f_r^k \quad , \; c = i\rho, \; \rho \in [0,\infty) . \tag{7.7}$$

If we take the hyperboloid H_1^3 as the domain of the functions f_r^k the second Casimir operator Δ_2 vanishes identically and the first one admits the discrete spectrum for c = 0

$$\Delta_1 f_r^k = -2(k_0^2 - 1) f_r^k \quad .$$

The result agrees with our result derived in Section 5, if we put $L = k_o - 1$.

For applications to physical problems with the SO(p,q) symmetry the derived discrete most degenerate representations $D^L(H^p_q)$ or $D^L(H^q_p)$ are especially convenient due to the following facts:

i) The maximal set of the commuting operators is maximally reduced in these representations of SO(p,q) groups. That is, for the discrete most degenerate representations of the SO(p,q) group the maximal set of commuting operators consists of

$$\Delta\,[SO(p,q)]\quad ,$$

$$C_p \equiv \begin{Bmatrix} \Delta[SO(p)],\Delta[SO(p-2)],\ldots,\Delta[SO(4)] & \text{for } p \text{ even} \\ \Delta[SO(p)],\Delta[SO(p-1)],\Delta[SO(p-3)],\ldots,\Delta[SO(4)] & \text{for } p \text{ odd} \end{Bmatrix}$$

$$\tilde{C}_q \equiv \begin{Bmatrix} \Delta[SO(q)],\Delta[SO(q-2)],\ldots,\Delta[SO(4)] & \text{for } q \text{ even} \\ \Delta[SO(q)],\Delta[SO(q-1)],\Delta[SO(q-3)],\ldots,\Delta[SO(4)] & \text{for } q \text{ odd} \end{Bmatrix}$$

$$H = \left\{ L_{2k,2k-1} = -\frac{\partial}{\partial\varphi^k},\ L_{2l,2l-1} = -\frac{\partial}{\partial\tilde{\varphi}^l},\quad \begin{matrix} k = 1,2,\ldots,[\frac{p}{2}] \\ l = [\frac{p}{2}]+1,\ldots,[\frac{p+q}{2}] \end{matrix} \right\}$$

where $\Delta[SO(p,q)]$ represents the Casimir operator of SO(p,q) and C_p and $\tilde{C}_q$ the sequence of corresponding Casimir operators of the maximal compact subgroup SO(p) × SO(q). The set H contains operators of the Cartan subalgebra except in the case p and q odd where H represents the maximal abelian compact subalgebra of SO(p,q) (See Table I of [7]).

The number of operators contained in the maximal set of commuting operators for the discrete most degenerate representations of SO(p,q) is equal to

$$N = p + q - 1$$

while the corresponding number for principal non-degenerate representations is

$$N' = \frac{r+l}{2} = \frac{1}{4}[N(N + 1) + 2l]$$

where r and l are the dimension and rank of SO(p,q) respectively.

ii) The additive quantum numbers may be related to the eigenvalues of the set H. It turns out that the set H is largest in the biharmonic coordinate system, which we have used.

iii) The eigenfunctions of the maximal commuting set of operators are given in explicit form by formulae (3.17), (4.7), (5.6) and the range of the numbers

$$L,\ l_2,\dots,l_{\{p/2\}},\ \tilde{l}_2,\dots,\tilde{l}_{\{q/2\}},m_1,\dots,m_{[p/2]}\tilde{m}_1,\dots,\tilde{m}_{[q/2]}$$

which may play the role of quantum numbers, is determined by (3.20), (3.24), (4.9) and (5.9).

The series of the continuous degenerate representations of SO(p,q) groups determined by a real number $\Lambda (0 \le \Lambda \le \infty)$ can be also constructed using this method [16]. The corresponding set of harmonic functions is given now by (e.g. for $p \geqslant q > 2$, $p = 2r$, $q = 2s+1$)

$$Y^{\Lambda, l_2,\dots,l_r,\tilde{l}_2,\dots,\tilde{l}_{s+1}}_{m_1,\dots,m_r,\dots,\tilde{m}_1,\dots,\tilde{m}_s}(\omega,\tilde{\omega},\theta) = V^{\Lambda}_{l_r+r-1,\tilde{l}_{s+1}+s-\frac{1}{2}}(\theta)\ Y^{l_2,\dots,l_r}_{m_1,\dots,m_r}(\omega)\ Y^{\tilde{l}_2,\dots,\tilde{l}_{s+1}}_{m_1,\dots,\tilde{m}_s}(\tilde{\omega})$$

where

$$V^{\Lambda}_{a,b}(\theta) = \frac{1}{\sqrt{N}}\, th^{b}(\theta)\, ch^{-1+i\Lambda}(\theta) \cdot {}_2F_1\left(\frac{b+1-i\Lambda-a}{2}, \frac{b+1-i\Lambda+a}{2}; b+1; th^2\theta\right)$$

and $Y^{l}_{m}(\omega)$ and $Y^{\tilde{l}}_{\tilde{m}}(\tilde{\omega})$ are given by formula (3.18). The functions

$V^{\Lambda}_{a,b}(\theta)$ fulfill the following orthogonality relation [16]

$$\int \bar{V}^{\Lambda}_{a,b}(\theta)\, V^{\Lambda'}_{a,b}(\theta)\, sh^{q-1}\theta\, ch^{p-1}\theta\, d\theta = \delta(\Lambda-\Lambda')$$

To each discrete series of representations realized on $\mathcal{H}(H^p_q)$ there corresponds the continuous one. Moreover we have one new series of continuous representations of SO(p,q) groups related to a real cone, on which a discrete series cannot be realized (for details see [16]).

The problems related to the eigenfunction expansion realized by the set of considered harmonic functions and interesting connections with the Gelfand triplet of spaces $\phi \subset H \subset \phi'$ are discussed in [21].

8. Discrete Degenerate Representations of the U(p,q) group ($p \geq q > 1$)

The homogeneous spaces given by formulae (2.1) and Table I are suitable for construction of the discrete degenerate representations only in the case of the unimodular unitary groups SU(p,q). To obtain similar representations also for the U(p,q) groups we have to consider manifolds of the type

$$X_+^{2(p+q)-1} = U(p,q)/U(p-1,q) \qquad \text{or}$$

$$X_-^{2(p+q)-1} = U(p,q)/U(p,q-1) \quad . \tag{8.1}$$

It was shown by Rosenfeld ([6], p. 621) that the group U(p,q) acts transitively on these manifolds. In contradistinction to the previous case of SO(p,q) groups we have now two invariant operators on the manifolds $X_+^{2(p+q)-1}$ and $X_-^{2(p+q)-1}$ given by (2.6). Besides the second-order Casimir operator $\Delta(X)$ (2.4) we can still define the first order invariant operator $\hat{M}_{p+q}$

given by the formula

$$\hat{M}_{p+q} = \sum_{i=1}^{p+q} L_i \tag{8.2}$$

where $L_i (i = 1,2,...p+q)$ are the generators of a Cartan subgroup of $U(p,q)$. Thus, in this case, the set of harmonic functions which creates a basis for the representation is given by the set of simultaneous solutions of the equations

$$\Delta(X)\ \psi_M^\lambda(X) = \lambda\psi_M^\lambda(X)$$

$$\hat{M}\ \psi_M^\lambda(X) = \lambda\psi_M^\lambda(X) \quad . \tag{8.3}$$

To obtain an explicit solution of these equations it is necessary to introduce a convenient model of the spaces $X_+^{2(p+q)-1}$ and $X_-^{2(p+q)-1}$, i.e. a manifold which has the same dimension and the same stability group as $X_\pm^{2(p+q)-1}$ and on which the group $U(p,q)$ acts transitively.

The model of $X_\pm^{2(p+q)-1}$ can be realized by the hypersurface in the $(p+q)$-dimensional complex space C^{p+q} which is determined by the equation

$$z^1\bar{z}^1 + z^2\bar{z}^2 + \ldots + z^p\bar{z}^p - z^{p+1}\bar{z}^{p+1} - \ldots - z^{p+q}\bar{z}^{p+q} = \pm 1 \tag{8.4}$$

Besides the space C^{p+q} we shall also consider the flat Minkowski space $M^{2p,2q}$ defined by the relations

$$x^{2k-1} = \mathrm{Re}\, z^k \qquad x^{2k-1},\ x^{2k} \in M^{2p,2q},\ z^k \in C^{p+q}$$

$$x^{2k} = \mathrm{Im}\, z^k \qquad k = 1,2,\ldots p+q \tag{8.5}$$

The solution of the eigenvalue problem is considerably facilitated if a system of coordinates is chosen such that the metric tensor $g_{\alpha\beta}(X_\pm^{2(p+q)-1})$ on the hypersurface $X_\pm^{2(p+q)-1}$ induced by the metric tensor $g_{ab}(M^{2p,2q})$ on $M^{2p,2q}$ is diagonal. As it was shown in a previous paper [8], the so-called biharmonic coordinate system appears to be especially convenient for this purpose. We shall construct it with the help of a recursion prescription: first we construct the coordinate system for the compact sub-manifold satisfying the equation

$$z^1 \bar{z}^1 + \ldots + z^p \bar{z}^p = 1 \quad , \tag{8.6}$$

which is a homogeneous manifold with respect to action of the compact subgroup U(p) of U(p,q). If we suppose that we have constructed the coordinate system for $z''^1, \ldots z''^l$, $l < p$, then the coordinate system for the variables $z'^1, \ldots z'^{l+1}$ is given by

$$\begin{aligned} z'^i &= z''^i \sin \vartheta^{l+1} \qquad & 0 \leq \varphi^i \leq 2\pi \\ z'^{l+1} &= e^{i\varphi^{l+1}} \cos \vartheta^{l+1} \qquad & 0 \leq \vartheta^i \leq \frac{\pi}{2} \end{aligned} \qquad i = 1,2,\ldots,l \tag{8.7}$$

Therefore putting $z^1 = e^{i\varphi^1}$ for p = 1 and applying successively the procedure (8.7) we obtain the coordinate system for the manifold determined by eq. (8.6) for an arbitrary p = 1,2,.. We shall denote the corresponding set of angles by $\omega \equiv \{\varphi^1, \ldots \varphi^p, \vartheta^2, \ldots \vartheta^p\}$.

In the same way, also the coordinate system for the variables $z^{p+1}, \ldots, z^{p+q}$ satisfying the equation

$$z^{p+1}\bar{z}^{p+1} + \ldots + z^{p+q}\bar{z}^{p+q} = 1$$

is constructed. The corresponding set of angles will be denoted by $\tilde{\omega} = \{\tilde{\varphi}^1, \ldots \tilde{\varphi}^q, \tilde{\vartheta}^2, \ldots \tilde{\vartheta}^q\}$. Finally, the complete coordinate system on $X_+^{2(p+q)-1}$ is created by

$$\begin{aligned} z^i &= z'^i \operatorname{ch}\theta \qquad & i = 1,\ldots p \\ z^l &= z'^l \operatorname{sh}\theta \qquad & l = p+1,\ldots p+q \end{aligned} \tag{8.8}$$

The metric tensor $g_{\alpha\beta}(X_+^{2(p+q)-1})$ on the space $X_+^{2(p+q)-1}$ is given by

$$g_{\alpha\beta}\left(X_{+}^{2(p+q)-1}\right) = \sum_{\alpha,\beta=1}^{2(p+q)-1} g_{kl}\left(M^{2p,2q}\right)\partial_\alpha x^k \partial_\beta x^l \qquad (8.9)$$

$$l,k = 1,2,\ldots,2(p+q)$$

where ∂_α, $\alpha = 1,2,\ldots 2(p+q)-1$ denotes partial differentiation with respect to the angles

$$\varphi^1, \varphi^2, \ldots \varphi^p, \tilde{\varphi}^1, \ldots \tilde{\varphi}^q, \vartheta^2, \ldots \vartheta^p, \tilde{\vartheta}^2, \ldots \tilde{\vartheta}^q, \theta$$

Consider first the equations (8.3) on the homogeneous space $X_{+}^{2(p+q)-1}$. Using formulae (2.4) and (8.2) we find that the invariant operators $\Delta(X_{+}^{2(p+q)-1})$ and $\hat{M}(X_{+}^{2(p+q)-1})$ can be expressed in terms of the biharmonic coordinates in the following way

$$\Delta\left(X_{+}^{2(p+q)-1}\right) = -\frac{1}{ch^{2p-1}\theta\, sh^{2q-1}\theta}\frac{\partial}{\partial\theta} ch^{2p-1}\theta\, sh^{2q-1}\theta\frac{\partial}{\partial\theta} + \frac{\Delta(X^{2p-1})}{ch^2\theta} - \frac{\tilde{\Delta}(X^{2q-1})}{sh^2\theta} \qquad (8.10)$$

$$\hat{M}_{p+q} = \sum_{k=1}^{p+q}\frac{\partial}{\partial\varphi^k} = \hat{M}_p + \hat{\tilde{M}}_q, \quad \hat{M}_p = \sum_{l=1}^{p}\frac{\partial}{\partial\varphi^l}, \quad \hat{\tilde{M}}_q = \sum_{l=1}^{q}\frac{\partial}{\partial\tilde{\varphi}^l} \qquad (8.11)$$

where $\Delta(X^{2p-1})$, $\hat{M}_p$ and $\hat{\Delta}(X^{2q-1})$, $\hat{\tilde{M}}_q$ are the invariant operators of the compact unitary group U(p) and U(q) respectively. The eigenfunctions of the operators (X^{2p-1}) and $\hat{M}_p$ for an arbitary U(p) group are given explicitly in the Appendix of [8].

If we represent the simultaneous eigenfunctions of the operators (8.10) and (8.11) as a product of eigenfunctions of $\Delta(X^{2p-1})$ and $\hat{M}_p$ times eigenfunctions of $\tilde{\Delta}(X^{2q-1})$ and $\hat{\tilde{M}}_q$ times an unknown function $\Psi^{\lambda}_{J_p,\tilde{J}_q}(\theta)$ we obtain the following equation for $\Psi^{\lambda}_{J_p,\tilde{J}_q}(\theta)$

$$\left[-\frac{1}{ch^{2p-1}\theta\, sh^{2q-1}\theta}\frac{d}{d\theta}ch^{2p-1}\theta\, sh^{2q-1}\theta\frac{d}{d\theta}-\frac{J_p(J_p+2p-2)}{ch^2\theta}+\right. \tag{8.12}$$

$$\left.\frac{\tilde{J}_q(\tilde{J}_q+2q-2)}{sh^2\theta}-\lambda\right]\psi^{\lambda}_{J_p,\tilde{J}_q}(\theta) = 0$$

where $-J_p(J_p+2p-2)$ and $-\tilde{J}_q(\tilde{J}_q+2q-2)$ are eigenvalues of the operators $\Delta(X^{2p-1})$ and $\tilde{\Delta}(X^{2q-1}_q)$ respectively, with J_p and $\tilde{J}_q$ being certain non-negative integers for $p,q > 1$ (see Appendix of [8]). The simultaneous eigenfunctions of the equations (8.3) are given by

$$Y^{L,J_2,\ldots,J_p,\tilde{J}_2\ldots\tilde{J}_q}_{(\omega,\tilde{\omega},\theta)\; m_1,\ldots,m_p,\tilde{m}_1,\ldots,\tilde{m}_q} = Y^{J_2,\ldots J_p}_{(\omega)\; m_1,\ldots,m_p}\cdot Y^{\tilde{J}_2\ldots\tilde{J}_q}_{(\tilde{\omega})\; \tilde{m}_1,\ldots,\tilde{m}_q}\cdot V^{L}_{J_p,\tilde{J}_q}(\theta) \tag{8.13}$$

where $V^L_{J_p\tilde{J}_q}(\theta)$ are the solutions of equation (8.12) given by [8].

$$V^{L}_{J_p,\tilde{J}_q}(\theta) = \frac{1}{\sqrt{N}}th^{\tilde{J}_q}\theta\, ch^{-(L+2p+2q-2)}\theta\; {}_2F_1\left(\frac{\tilde{J}_q-J_p+L+2q}{2},\frac{\tilde{J}_q+J_p+L+2p+2q-2}{2};\tilde{J}_q+q;th^2\theta\right), \tag{8.14}$$

with

$L = J_p - J_q - 2q - 2n$ $\qquad$ $n = 0,1,2,\ldots$

$\lambda = -L(L+2p+2q-2)$ $\qquad$ M+L even.

The functions Y^J_M ($Y^{\tilde{J}}_{\tilde{M}}$) are simultaneous eigenfunctions of the invariant operators $\Delta(X^{2p-1})$, $\hat{M}_p$ ($\tilde{\Delta}(X^{2q-1})$, $\hat{\tilde{M}}_q$) of the compact U(p) (U(q)), subgroup. These functions can be expressed as a product of the exponential and the usual d-functions in the form (see [20] Appendix).

$$Y^{J_2\dots J_p}_{(\omega)\,m_1,\dots,m_p} = \frac{1}{\sqrt{N_p}} \prod_{l=1}^{p} e^{im_l\varphi^l} \prod_{k=2}^{p} \sin^{2-k}\vartheta^k \, d^{\frac{1}{2}(J_k+k-2)}_{\alpha_k,\beta_k}(2\vartheta^k) \qquad (8.15)$$

The structure of the Hilbert space $\mathcal{H}^L_M(X_+^{2(p+q)-1})$ can be given in the form

$$\mathcal{H}^L_M(X_+) = \sum_{J_p=L+\tilde{J}_q+2q}^{\infty} \sum_{\tilde{J}_q=0}^{\infty} \sum_{M_p=-J_p}^{J_p} \sum_{\tilde{M}_q=-\tilde{J}_q}^{\tilde{J}_q} \mathcal{H}^{L,J_p,\tilde{J}_q}_{M,M_p,\tilde{M}_q}(X_+)\,\delta_{M\,M_p+\tilde{M}_q} \qquad (8.16)$$

$J_p-\tilde{J}_q-L$ even, M_p+J_p even, $\tilde{M}_q+\tilde{J}_q$ even

where $\mathcal{H}^{L\,J_p\,\tilde{J}_q}_{M\,M_p\,\tilde{M}_q}$ is a finite-dimensional subspace in which the irreducible unitary representation of the maximal compact subgroup U(p) × U(q) determined by the invariant numbers J_p, M_p and $\tilde{J}_q$, $\tilde{M}_q$ is realized. The formula (8.16) represents in fact the decomposition of the representation of the U(p,q) group determined by L and M with respect to the representation of its maximal compact subgroup.

We obtain the representations realized in the Hilbert space $\mathcal{H}^L_M(X_-^{2(p+q)-1})$ by formally exchanging J_p, M_p and $\tilde{J}_q$, $\tilde{M}_q$.

In the case of U(p,1) groups the representation space $\mathcal{H}^L_M(X_+^{2p+1})$ is spanned by the set of the harmonic functions of the form

$$Y^{L,J_2,\dots,J_p}_{(\omega,\tilde{\varphi},\theta)\,M,m_1,\dots,m_p,\tilde{m}_1} = \frac{1}{\sqrt{2\pi}} V^L_{J_p\,\tilde{m}_1}(\theta)\, e^{i\tilde{m}_1\tilde{\varphi}}\, Y^{J_2,\dots J_p}_{(\omega)\,m_1,\dots m_p} \qquad (8.17)$$

where

$$V^L_{J_p\,\tilde{m}_1}(\theta) = \frac{1}{\sqrt{N}} \operatorname{th}^{|\tilde{m}_1|}\theta \operatorname{ch}^{-(L+2p)}\theta \; {}_2F_1\left(\frac{|\tilde{m}_1|-J_p+L+2}{2}, \frac{|\tilde{m}_1|+J_p+L+2p}{2}; |\tilde{m}_1|+1; \operatorname{th}^2\theta\right)$$

and the invariant numbers J_p, M_p and $\tilde{m}_1$ of the maximal compact subgroup are restricted by

$$J_p - |\tilde{m}_1| = L + 2 + 2n \qquad n = 0,1,2,\ldots$$

$$M_p = \sum_{i=1}^{p} m_i = J_p - 2n' \qquad n' = 0,1,2,\ldots - J_p$$

The irreducibility of the representations related to the set of the harmonic functions (8.13) or (8.17) was proved in [20]. The unitarity follows from the left-invariance of the measure $d\mu(\omega,\tilde{\omega},\theta)$ on corresponding manifolds [20].

The maximal set of commuting operators is given by the following sequence of operators

$$\Delta[U(p,q)] \quad , \hat{M}_{p,q}$$

$$\Delta[U(p)] \ , \hat{M}_p, \quad \Delta[U(p-1)] \ , \hat{M}_{p-1}, \ldots \Delta[U(2)] \ \hat{M}_2 \ \hat{M}_1$$

$$\tilde{\Delta}[U(q)] \quad \hat{\hat{M}}_q, \quad \tilde{\Delta}[U(q-1)], \quad \hat{\hat{M}}_{q-1}, \ldots \tilde{\Delta}[U(2)] \ \hat{\hat{M}}_2 \ \hat{\hat{M}}_1$$

The series of the continuous degenerate representations of U(p,q) groups determined by a real number $\Lambda (0 \leqslant \Lambda < \infty)$ and by integer M can be also constructed using the above method. It turns out that besides the continuous series of representations which correspond to the discrete series there exists one new continuous series of representations related to a complex one, on which discrete series cannot be realized.

The problems related to the eigenfunctions expansion (e. g. relativistic transition amplitude with U(6,6) symmetry) and interesting connections with Gelfand triplet of spaces $\phi \subset H \subset \phi'$ are discussed in [22].

References

1. See e.g. I. M. Gelfand and M. A. Naimark, Trudi of Math. Steklov Institute 36 (1950)
2. See e.g. M. I. Graev, Trudi of Moscov Math. Soc. 7,335 (1958)

3. I. M. Gelfand, Amer. Math. Soc. Transl. (Ser. 2) 37, 31-34 (1964); I. M. Gelfand and M. I. Graev, Trudi of Moscov Math. Soc. 8, 321 (1959).

4. It turns out that a maximal set of invariant operators does not determine in general an irreducible representation. In ref. [16] it is shown that we must introduce the additional invariant operator with the discrete spectrum which splits a reducible representation space on irreducible parts.

5. See e.g. S. Helgason, chap. IX and X.

6. B. A. Rosenfeld, Dokl. Ak. Nauk. SSSR, 110, 23-27 (1956); A. S. Fedenko, Dokl. Ak. Nauk SSSR 108, 1026-1028 (1956).

7. R. Raczka, N. Limic and J. Niederle preprint ICTP Trieste IC/66/2

8. If the metric tensor $g_{\alpha\beta}(X)$ on the homogeneous space X is induced by the Cartan metric tensor g_{ik} in the Lie algebra R of the group G, then the Laplace-Beltrami operator $\Delta(X)$ is equal to the second order Casimir operator $Q_2 = g_{ik}X^i X^k$ (see [5] chap. X)

9. $SO_o(m,n)$ denotes component of a unity of $SO(m,n)$.

10. Here and elsewhere we use brackets for indices defined as follows

$$\left[\frac{a}{2}\right] = \begin{cases} \frac{a}{2} & \text{if } a = 2r \\ \frac{a-1}{2} & \text{if } a = 2r+1 \end{cases} \qquad \left\{\frac{a}{2}\right\} = \begin{cases} \frac{a}{2} & \text{if } a = 2r \\ \frac{a+1}{2} & \text{if } a = 2r+1 \end{cases} \qquad r=1,2,\ldots$$

11. The measure $d\mu(\Omega) = |\overline{g}(H^p_q)|^{+1/2}\, d\Omega$ is the Riemannian measure which is left invariant under the action of $SO_o(p,q)$ on H^p_q.

12. E. L. Ince, Ordinary differential Equations (Dover Publications 1956).

13. V. Bargmann, Ann. of Math. 48, 568-640 (1947).

14. I. M. Gelfand and M. I. Graev, Trudi of Moscov, Math. Soc. 11 (1962).

15. J. Dixmier, Bull. Soc. Math. France, 89, 9-41 (1961)

16. N. Limic. J. Niederle and R. Raczka, Continuous most degenerate representation of arbitrary noncompact SO(p,q)

groups, ICTP (in print).

17. M. A. Naimark, Linear Representations of the Lorentz group (Pergamon Press 1964), p. 167.

19. B. A. Rosenfeld, Non-Euclidean Geometries, Moscow, 1955.

20. J. Fisher and R. Raczka, preprint ICTP, IC/66/8

21. N. Limic, J. Niederle and R. Raczka, Generalized Fourier transforms related to degenerate representations of SO(p,q) groups, ICTP Trieste, (in print)

22. R. Raczka and J.Fisher, Gelfand-Graev transforms related to degenerate representations of U(p,q) groups, ICTP, Trieste (in print).

TRIPLET MODEL OF ELEMENTARY PARTICLES†

By

W. THIRRING

Institut für Theoretische Physik
Universität Wien

I. Introduction

The model of a fundamental triplet of fractionally charged fermions* (quarks) has been remarkably successful in producing many numbers of direct experimental significance. In this model pseudoscalar and vector SU_3 octets and singulets are pictured as quark-antiquark bound in an s-state. The baryon octet and decuplet are three quarks in an s-state and the forces which hold the quarks together appear to be to a large extent independent of spin and unitary spin. In this SU_6-approximations the above mentioned bosons belong to the 35 and 1 rep-

---

* Their quantum numbers are:

	N	Q	Y	μ/μ_q	T_3	T
p	1/3	2/3	1/3	2/3	1/2	1/2
n	1/3	-1/3	1/3	-1/3	-1/2	1/2
λ	1/3	-1/3	-2/3	-1/3	0	0

(N = Baryon number
Y = Hypercharge
Q = Charge
μ = Magnetic moment)

† Lecture given at V. Internationalen Universitätswochen für Kernphysik, Schladming, 24 February - 9 March 1966.

resentations and the baryons to the 56 representation. Their quark content is as follows

$\underline{1}|$ $\quad X_i = \frac{1}{\sqrt{3}}(\bar{p}p + \bar{n}n + \bar{\lambda}\lambda)$

spins are antiparallel

$$\underline{35}| \quad \begin{array}{llll} \rho^+,\pi^+ = \bar{n}p & & K^{*+},K^+ = \bar{\lambda}p \\ \rho^-,\pi^- = \bar{p}n & & K^{*o},K^o = \bar{\lambda}n \\ \rho^o,\pi^o = \frac{1}{\sqrt{2}}(\bar{p}p-\bar{n}n) & & K^{*-},K^- = \bar{p}\lambda \\ & & \bar{K}^{*o},\bar{K}^o = \bar{n}\lambda \end{array} \qquad (1)$$

$$\omega_i = \frac{1}{\sqrt{2}}(\bar{p}p+\bar{n}n)$$

$$\phi_i = \bar{\lambda}\lambda$$

$$\eta_i = \frac{1}{\sqrt{6}}(2\bar{\lambda}\lambda-\bar{p}p-\bar{n}n)$$

The vector particles correspond to spin parallel and the pseudoscalar particles to spin antiparallel.

$$\underline{56}| \quad \begin{array}{llll} P = ppn & & \Xi^{*-} = n\lambda\lambda \\ N = pnn & & \Xi^{*o} = p\lambda\lambda \\ \Sigma^+ = pp\lambda & & Y^{*-} = nn\lambda \\ \Sigma^o = pn\lambda & & Y^{*o} = pn\lambda \\ \Sigma^- = nn\lambda & & Y^{*+} = pp\lambda \\ \Lambda^o = pn\lambda & & N^{*-} = nnn \\ \Xi^o = p\lambda\lambda & & N^{*o} = nnp \\ \Xi^- = n\lambda\lambda & & N^{*+} = npp \\ \Omega^- = \lambda\lambda\lambda & & N^{*++} = ppp \end{array} \qquad (2)$$

Here in $3/2^+$ particles all spins are parallel and in $1/2^+$ particles one pair of quark spins is antiparallel. The total spin-unitary spin wave function is totally symmetric*. We added a subscript i to some of these bosons since these ideal configurations are mixed by SU_6 breaking terms. Assuming that these are mainly a higher mass for λ than for p and n we see that to first order this induces an equal spacing in the baryon octet and decuplet. The vector meson nonet also has equal spacing whereas the pseudoscalar octet satisfies the Gell-Mann Okubo mass formula. A more exact theory goes as follows. $\bar{\lambda}\lambda$ is heavier by 2Δ than $\bar{p}p$ and $\bar{n}n$. If m_8 and m_1 are the octet and singulet masses for perfect symmetry we find the following masses

$$\begin{aligned} &\rho, \pi : m_8 \\ &K^*, K : m_8 + \Delta \end{aligned} \tag{3}$$

For the T = Y = 0 states we have the mass matrix (in the $\begin{vmatrix} 88 & 81 \\ 18 & 11 \end{vmatrix}$ representation)

$$\begin{vmatrix} m_8 + \frac{4}{3}\Delta & \frac{2\sqrt{2}}{3} I \Delta \\ \frac{2\sqrt{2}}{3} I \Delta & m_1 + \frac{2}{3}\Delta \end{vmatrix} \tag{4}$$

if I is the overlap integral of the 35 and 1 wave functions. The eigenvalues of (4) are

$$m_\pm = \frac{m_8 + m_1}{2} + \Delta \pm \left[\left(\frac{m_8 - m_1}{2} + \frac{\Delta}{3} \right)^2 + \frac{8}{9}\Delta^2 I \right]^{1/2} \tag{5}$$

and the eigenvectors

$$\phi_+ = \phi_8 \cos\alpha + \phi_1 \sin\alpha$$

$$\phi_- = \phi_1 \cos\alpha - \phi_8 \sin\alpha$$

---

* These wave-functions are written out in full in W. Thirring, Acta Physica Austriaca, Suppl. II, 1965.

with

$$tg\alpha = \frac{2\sqrt{2}\ I\ \Delta/3}{\frac{m_8-m_1}{2} + \frac{\Delta}{3} + \left[\left(\frac{m_8-m_1}{2} + \frac{\Delta}{3}\right)^2 + \frac{8}{9}\Delta^2 I^2\right]^{1/2}} \tag{6}$$

For the pseudoscalar particles $m_1-m_8 >> \Delta$ and there is not much octet-singulet mixing. One finds from the observed (masses)^2 $m_8 = 140$, $m_1 = 830$, $\Delta = 122$, $I = 0.65$, $90^\circ-\alpha = 10^\circ$

$$X = X_i \cos 10^\circ + \eta_i \sin 10^\circ \simeq \frac{1}{\sqrt{3}}\left(\bar{\lambda}\lambda(1+\frac{\sqrt{2}}{6}) + \bar{n}n(1-\frac{\sqrt{2}}{12}) + \bar{p}p(1-\frac{\sqrt{2}}{12})\right)$$

$$\eta = -X_i \sin 10^\circ + \eta_i \cos 10^\circ \simeq \frac{1}{\sqrt{6}}\left(\bar{\lambda}\lambda(2-\frac{\sqrt{2}}{6}) - \bar{n}n(1+\frac{\sqrt{2}}{6}) - \bar{p}p(1+\frac{\sqrt{2}}{6})\right) \tag{7}$$

For the vector nonet we have in accordance with SU_6 $m_1-m_8 << \Delta$ and $I \sim 1$. Hence the particles are close to the ideal mixing angle $\mathrm{arctg}\ 1/\sqrt{2} = 35.3^\circ$. One finds* from the masses $m_1 = 810$, $m_8 = 770$, $\Delta = 122$, $\alpha = 40^\circ$

$$\begin{aligned} \phi &= \phi_i \cos 5^\circ + \omega_i \sin 5^\circ \simeq \lambda\bar{\lambda} + \frac{1}{12}\frac{1}{\sqrt{2}}(\bar{p}p+\bar{n}n) \\ \omega &= -\phi_i \sin 5^\circ + \omega_i \cos 5^\circ \simeq \frac{1}{\sqrt{2}}(\bar{p}p+\bar{n}n) - \frac{1}{12}\bar{\lambda}\lambda \end{aligned} \tag{8}$$

In the following we shall calculate from the above reaction rates in the spirit of nonrelativistic nuclear physics except for correction of relativistic kinematics when necessary.

---

* The small ω_i admixture allows the $\phi \to \rho + \pi$ decay and $\alpha = 40^\circ$ is in agreement with the rate of this decay |1|.

II. Strong Interactions

In accordance with our SU_6 assumption we postulate that there is an effective quark-meson interaction of the form

$$L' = q^{+}\left(V + \frac{\vec{\sigma}.\vec{\nabla}}{\mu_o} P\right) q.\sqrt{2}g \tag{9}$$

where q is a six-component (Pauli) spinor for the quarks ($p\uparrow$, $p\downarrow$, $n\uparrow$, $n\downarrow$, $\lambda\uparrow$, $\lambda\downarrow$) and V and P are the field-operator matrices

$$V = \begin{vmatrix} \frac{\omega_i+\rho_o}{2}, & \rho^{+} & K^{*+} \\ \rho^{-} & \frac{\omega_i-\rho_o}{2} & K^{*o} \\ K^{*-} & \overline{K}^{o*} & \phi_i \end{vmatrix}$$

$$P = \begin{vmatrix} \frac{\eta_i}{\sqrt{6}} + \frac{\pi^o}{\sqrt{2}} & \pi^{+} & K^{+} \\ \pi^{-} & \frac{\eta_i}{\sqrt{6}} - \frac{\pi^o}{\sqrt{2}} & K^{o} \\ K^{-} & \overline{K}^{o} & -\sqrt{2/3}\,\eta_i \end{vmatrix} \tag{10}$$

In our nonrelativistic approximation of the vector particles only the zerocomponent enters and one has the usual pseudoscalar $\vec{\sigma}.\vec{\nabla}$ - interaction. μ_o is a mean meson mass and g is the strong interaction constant.

Taking the matrix element of (9) between nucleons we find, f.i.

$$<P|L'|P> = \frac{5}{3} \; \frac{\vec{\sigma}\cdot\vec{\nabla}}{\mu_o} \pi^o g \; + \; \eta - \text{contribution} \tag{11}$$

Thus g is related to the pion nucleon coupling constant $f = \sqrt{4\pi \times 0.08}$ by

$$f = \frac{5}{3} g \frac{\mu_\pi}{\mu_o} \quad . \tag{12}$$

Since there is only one coupling constant the SU_3 ratio F/D is fixed and found to be 2/3. There is no precise experimental evidence on that. However taking matrix elements between $<N|$ and $|N^*>$ of (9) we can also find the $BB^*\pi$ - vertex. From this we calculate with standard formulae from the static model the following widths [2] (Relativistic correction may reduce them up to 20 %).

	theory	experiment
Γ_{N^*}	100 MeV	120 MeV
$\Gamma_{Y^*-\Lambda}$	35 MeV	40 MeV
$\Gamma_{Y^*\Sigma}$	5 MeV	4 MeV
Γ_{Ξ^*}	12 MeV	7,5 MeV

Thus, considering that there is no adjustable parameter in these predictions the fit with the data is pretty good.

Regarding the BBV-vertex one knows experimentally only rough numbers. Better known is the VPP vertex from the decay of the vector mesons. This can also be calculated in our model in the same way by taking $<P|L'|V>$ or $<P|L'|P>$ one finds, f.i. that the $\rho_o\pi^+\pi^-$ vertex is

$$2gV^\mu(PP^+_{,\mu} - P^+P_{,\mu}) \qquad \begin{array}{l} V = \rho \text{ field} \\ P = \pi \text{ field} \end{array} \tag{13}$$

To calculate the decay one has to use the relativistic formula

$$\Gamma_{V\to P+P'} = \frac{g^2}{4\pi}\frac{M_V}{3}\left[1-2\,\frac{m^2+m'^2}{M_V^{\,2}} - \frac{(m^2-m'^2)^2}{M_V^{\,4}}\right]^{3/2} \tag{14}$$

We may use the $\rho\to2\pi$ decay to adjust the still unknown μ_π/μ_0. $\Gamma_{\rho\to2\pi}$= 124 MeV requires $g^2/4\pi = 0.6$ which means $\mu_0 = 4.4\mu_\pi \sim$ $\sim$ 660 MeV, not unreasonable*. Than we predict

	theory	experiment
$\Gamma_{K^*\to K\pi}$	50 MeV	50 ± 2 MeV
$\Gamma_{\phi_i\to 2K}$	3 MeV	2.4 ± .8 MeV

quite impressive an agreement.

The model has many more points of contact with experimental data regarding strong interaction. I shall not discuss them since they involve further dynamical assumptions (dominance of one intermediate state, etc.) and do not allow such a precise test of the theory.

III. Electromagnetic Interactions

After the brilliant success of SU_6 in expressing all static magnetic moments [4] of baryons in terms of μ_q the quark magneton one began to look at transition moments. The N-N^* magnetic transition moment came out by about 1.25 smaller [6] than required by the photoproduction data. This point will have to be reconsidered once low energy pion physics is redone with higher experimental precision. The quark model also predicts the magnitude of the V-P magnetic transition moment once μ_q is taken from the nucleons to be $\mu_P = \frac{2.8e}{2M_P}$. Relativistically the $VP\gamma$-vertex is [7]

* The sceptical reader is referred to [3].

$$2\mu(P_{\varepsilon\mu\nu\sigma\tau}V^{\sigma\tau})_{,\nu}\, A^{\mu} \tag{15}$$

where μ is the transition moment

$$\langle P|\mu_z|V,j_z{=}0\rangle = \mu \quad . \tag{16}$$

(15) gives a decay rate

$$\Gamma_{V\to P+\gamma} = \frac{4}{3}\,\frac{\mu^2}{4\pi}\left(\frac{M_V^2-m_P^2}{2M_V}\right)^3 \tag{17}$$

For the decay $P\to V+\gamma$ one has to multiply (17) by 3 to account for the sum of the polarization of V and exchange M_V and m_P. The transition moments (16) can be easily calculated and one finds

$$\begin{aligned}\Gamma_{\omega\to\pi^0+\gamma} &= 12 \text{ MeV}\\ \Gamma_{\phi\to\eta_r+\gamma} &= 0.25 \text{ MeV}\\ \Gamma_{X_r\to\rho+\gamma} &= 0.17 \text{ MeV}\end{aligned} \tag{18}$$

(Where the subscript r denotes the real particle) Experimentally the situation is as follows [2]. $\Gamma_{\omega\to\pi^0+\gamma}$ agrees within the experimental error of 10 %. For ϕ one knows that $\Gamma_{\phi\to\eta+\gamma}/\Gamma_\phi < 1/8$ whereas (18) predicts $\Gamma_{\phi\to\eta+\gamma}/\Gamma_\phi = 1/13$. The width of X is unknown, <4 MeV, but the observed branching ratio allows us to predict from (18) that $\Gamma_X = 1/2$ MeV.

All the other [1] γ-decays have branching ratios <1 % and hence no significant data can be expected soon.

It should be mentioned here that in the SU_6 approximation where the wave-function factors into a space part and into the SU_6 part we have indicated all baryon form factors should be a universal function of momentum transfer times their zero momentum transfer value. This is in striking agreement with our experimental knowledge. Also, since the N and N^* are L=0 quark states there should be no E_2 -transition moment between them and it is actually <4% of the M_1 moment [6].

Further electromagnetic predictions of the quark model require another parameter namely the spacial extention of the quark antiquark wave function $\psi(x)$. For this we cannot make a precise prediction but putting $|\psi(0)|^2 = 1/r^3$ we will see whether we will get a reasonable value $r \sim 1$ Fermi. Let us first consider the V-γ-vertex. The coupling constant $f_{V\gamma}$ is defined by

$$\langle V|j_\mu(x)|0\rangle = \frac{ef_{V\gamma}}{\sqrt{2M_V\,\Omega}}\, e_\mu e^{ipx} \tag{19}$$

where e_μ is the polarization vector and Ω the normalization volume. Since the electric current j_μ is

$$\sim \frac{2e}{3}\,\bar{p}\gamma_\mu p - \frac{e}{3}\,\bar{n}\gamma_\mu n - \frac{e}{3}\,\bar{\lambda}\gamma_\mu\lambda$$

we see immediately the SU_3 result

$$f_{\rho\gamma} : f_{\omega_i\gamma} : f_{\phi_i\gamma} = 3 : 1 : -\sqrt{2} \tag{20}$$

apart from mass corrections. Writing j_μ out in terms of creation and annihilation operators of quarks we find easily

$$\langle\omega_i, j_3=0|j_3(x)|0\rangle = e\left(\frac{2}{3} - \frac{1}{3}\right)\frac{\psi(0)}{\sqrt{\Omega}}\, e^{ipx} \tag{21}$$

and hence

$$f_{\omega\gamma} = \left[\frac{2m_\omega}{9r^3}\right]^{1/2} \qquad \text{(perhaps} \times 0.8) \tag{22}$$

There may be a slight correction because of the quark anomalous magnetic moment. It has been argued [8] that in a meson the quark will have an effective mass $m_{meson}/2$ and a correspondingly higher magnetic moment. To give it the "observed" moment $\mu_q = \mu_p$ we have to include a $\sigma_{\mu\nu}$-term which gives the 0.8 factor. $f_{\rho\gamma}$ and $f_{\phi\gamma}$ follow than from (20). The experimental evidence on the $f_{V\gamma}$ is confusing. For $\Gamma_{\omega\to e^+e^-}/\Gamma_\omega$ the values 2×10^{-4} [6] and 1×10^{-4} [2] are quoted. From (19) and the well-known electron-positron-loop one finds

$$\Gamma_{V\to e^+e^-} = m_V \frac{4\pi\alpha^2}{3} \frac{f_{V\gamma}^2}{m_V^4} \tag{23}$$

and practically the same for $\Gamma_{V\to\mu^+\mu^-}$. Taking 2×10^{-4} for the ω-ratio we find

$$f_{\omega\gamma} = m_\omega^2/8.2 \tag{24}$$

a value which is also favoured by the isoscalar nuclear form factor. However then (20) predicts to large a value for $f_{\rho\gamma}$. There one has the values [2]

$$\Gamma_{\rho\to e^+e^-}/\Gamma_\rho = 0.65 \times 10^{-4}$$

and [9]

$$\Gamma_{\rho\to\mu^+\mu^-}/\Gamma_\rho = 0.3 \times 10^{-4}$$

and the branching ratio for ρ should be the same as for ω since

$$\Gamma_\rho = 10\Gamma_\omega \quad , \qquad \left(\frac{f_{\rho\gamma}}{f_{\omega\gamma}}\right)^2 \left(\frac{m_\rho}{m_\omega}\right)^3 = 10 .$$

The value 0.3×10^{-4} would give

$$f_{\rho\gamma} = \frac{3m_\rho^2}{20} \tag{25}$$

In the following we shall use as a compromise (20) and

$$f_{\omega\gamma} = m_\omega^2/15 \tag{26}$$

which also predicts a $f_{\phi\gamma}$ which agrees with the (uncertain) evidence from the nucleon form factor. (26) gives $r = 0.8 \times 10^{-13}$cm which is just what one would expect.

Next we will discuss the $P \to 2\gamma$ decay. If one wants to apply the formula (15) for the 2γ-decay of singulet positronium one has to face the question of the qurk mass and its magnetic moment. Introducing the magnetic moment μ

$$\Gamma_{2\gamma} = 4\pi|\psi(0)|^2\left(\frac{e^2}{m}\right)^2 = 16\pi\mu^4\,\frac{m^2}{r^3} \tag{27}$$

we can calculate [10] r from the observed rates. For the π^0-decay, f.i. $\mu = 1/3(2/3-1/3)\mu_q$ and using an effective quark mass $m = \frac{m_{\pi^0}}{2}$ we get $r = 2 \times 10^{-13}$cm for

$$\frac{1}{\Gamma_{\pi^0\to 2\gamma}} = 0.7 \times 10^{-16}\,\text{sec}\,. \tag{28}$$

This is the largest value for $\Gamma_{\pi_0\to 2\gamma}$ one finds in the literature [11] but it should still be larger since the r one gets this way is unreasonably big. The above consideration suffers from the questionable concept of the effective quark mass. However about the same result can be obtained by the following argument. In our nonrelativistic picture the emission of the first γ should just straighten out the two spins, i.e. $\pi^0 \to V+\gamma$ and then the quark-antiquark system can transform into the second

γ. Hence the process should go through $\pi^0 \to V+\gamma \to 2\gamma$ and we have predictions for both vertices so we can calculate the rate. With (15) and (19) we find [7]

$$\Gamma_{\phi\to 2\gamma} = \left(\frac{2\mu f}{4\pi}\right)^2 \frac{\pi}{m_P} \left(\frac{m_P}{m_V}\right)^4 . \tag{29}$$

Inserting the transition moment μ and f from (26) we see that for π^0 the amplitudes for $\pi^0 \to \omega+\gamma \to 2\gamma$ and $\pi^0 \to \rho+\gamma \to 2\gamma$ are equal we get altogether

$$\Gamma_{\pi^0\to 2\gamma} = 4\pi m_{\pi^0} \alpha^2 \left(\frac{2.8 m_\pi}{15\ M_P}\right)^2 = 66\ \text{eV} \tag{30}$$

which is by about a factor 7 bigger than the experimental number (28). Similarly we obtain theoretically

$$\Gamma_{\eta_r\to 2\gamma} = 3\ \text{KeV} \qquad \Gamma_{X_r\to 2\gamma} = 45\ \text{KeV}$$
$$\Gamma_{X_i\to 2\gamma} = 56\ \text{KeV} \qquad \Gamma_{\eta_i\to 2\gamma} = 1.5\ \text{KeV} \tag{31}$$

The value for the η-decay gives a correct η branching ratio if one calculates the $\eta \to 2\pi\gamma$ decay through the corresponding sequence $\eta \to \rho+\gamma \to 2\pi+\gamma$ with (13) and (15) one finds [7]

$$\Gamma_{\eta\to 2\pi\gamma} = \frac{g^2}{4\pi} \frac{(\mu m_\eta)^2}{12\pi^2} \left(\frac{m_\eta}{m_\rho}\right)^2 \frac{m_\eta}{160} . \tag{32}$$

Inserting the corresponding transition moment we obtain

$$\Gamma_{\eta\to 2\pi\gamma} = m_\eta \frac{g^2}{4\pi} \frac{e^2}{4\pi} \left(\frac{2.8 m_\eta}{\sqrt{3} M_P}\right)^2 \left(\frac{m_\eta}{m_\rho}\right)^4 \frac{1}{160\pi} = 390\ \text{eV} \tag{33}$$

and hence a branching ratio

$$\frac{\Gamma_{\eta\to 2\gamma}}{\Gamma_{\eta\to 2\pi\gamma}} = 7.8 \ . \tag{34}$$

This agrees with the data [2] within the 20 % error. For the X the 2γ decay mode has not yet been observed and we predict with (18)

$$\frac{\Gamma_{X\to\rho+\gamma}}{\Gamma_{X\to 2\gamma}} \sim 4 \ . \tag{35}$$

Regarding the discrepancy of the π^0 lifetime one may argue that the π may have an anomalously large r and hence also the virtual ρ and ω in the chain $\pi^0\to V+\gamma\to 2\gamma$. This would decrease $f_{V\gamma}$ and bring the theoretical $\Gamma_{\pi^0\to 2\gamma}$ down. However, if all other predictions of the model, in particular the large η-width ($\sim$8KeV) and the $X\to\rho+\gamma/X\to 2\gamma$ ratio are experimentally verified one may also question the experimental evidence on the π^0 lifetime. (Using (20) for $f_{V\gamma}$ reduces all $P \to 2\gamma$ widths by 2 which may be favoured by experiment.)

IV. Weak Interactions

In this model the weak interactions will be naturally described by the quark transition $p\to n+W^+$, $p\to\lambda+W^+$. In our non-relativistic theory the 0-component of the weak current will be pure vector and the space components pure axial vector. Hence we assume

$$(J_o^+, \vec{J}^+) = \frac{G}{\sqrt{2}} \bar{q}(1,-\vec{\sigma})(E_1\cos\theta+E_2\sin\theta)q \tag{36}$$

where $E_1 p = n$, $E_1\bar{n} = -\bar{p}$, $E_2 p = \lambda$, $E_2\bar{\lambda} = -\bar{p}$, θ is the Cabbibo angle and $GM_P^2 = 10^{-5}$. However, taking matrix elements of (36) between P and N one finds as in (11) a V - 5/3 A interaction instead of the experimental V - 1.2 A. Since there is evidence [12] that the bare quark interaction is pure V-A we conclude that the process of renormalization does not respect SU_6 and the physical quark interaction is

$$(J_o^+, \vec{J}^+) = \frac{G}{\sqrt{2}} \bar{q}(1,-0.7\vec{\sigma})(E_1\cos\theta + E_2\sin\theta)q \qquad (37)$$

which gives V-1.2 A for the N-P-decay. Calculating other matrix elements in the baryon octet we find

$N \to P$: $(V-1.2A)\cos\theta$

$\Sigma^+ \to \Lambda$: $-0.7\sqrt{2/3}\cos\theta$

$\Sigma^- \to \Lambda$: $-0.7\sqrt{2/3}\cos\theta$

$\Lambda \to P$: $\sqrt{3/2}(V-0.7A)\sin\theta$

$\Sigma^- \to N$: $(V+0.23A)\sin\theta$

$\Sigma^0 \to P$: $\sqrt{1/2}(V+0.23A)\sin\theta$

$\Xi^- \to \Sigma^0$: $-\sqrt{1/2}(V-1.2A)\sin\theta$

$\Xi^- \to \Lambda$: $\sqrt{3/2}(V-0.23A)\sin\theta$

$\Xi^0 \to \Sigma^+$: $-(V-1.2A)\sin\theta$ (38)

To check these predictions will keep experimentalists busy for years. Regarding the total rates we find with $\sin^2\theta = 0.058$ the following branching ratios (to the total decay rates)

	theory	experiment
$\Sigma^+ \to \Lambda + e^+ + \nu$	0.19×10^{-4}	$(0.66 \pm .4)10^{-4}$
$\Sigma^- \to \Lambda + e^- + \bar{\nu}$	0.56×10^{-4}	$(0.75 \pm .3)10^{-4}$
$\Lambda \to P + e- + \bar{\nu}$	0.75×10^{-3}	$(0.8 \pm .1)10^{-3}$
$\Sigma^- \to N + e- + \bar{\nu}$	0.94×10^{-3}	$(1.3 \pm .2)10^{-3}$
$\Xi^- \to \Lambda + e^- + \bar{\nu}$	0.54×10^{-3}	$(2 \pm 1)10^{-3}$ (39)

One can also calculate B^*-B-transitions but they cannot compete with strong transitions when they are possible. However, once the $\nu+B\to B^*+e$ rates are experimentally better determined the model has something to say about them. For the Ω^- the most favourable leptonic decay is $\Omega^-\to\Xi^0+e^-+\bar{\nu}$ for which we obtain a branching ratio a little less than one percent. It is clear that (37) reproduces other consequences of the CVC-theory, in particular the $\pi^+\to\pi^0+e^++\nu$ and the $K^+\to\pi^0+e^++\nu$ decay which may be used to determine θ. Regarding the $\pi^+\to\mu^++\nu$, $K^+\to\mu^++\nu$ decays we can in addition to their ratio also estimate the absolute rate. The calculation is analog to the one of the V-γ-vertex. If one defines a c_P by

$$<P|A_\mu^+(x)|0> = \frac{c_P}{\sqrt{m_P\Omega}}\, q_\mu e^{iqx} \tag{40}$$

one can work out

$$\Gamma_{P\to\mu+\nu} = G^2 c_P^2 \frac{m_\mu^2 m_P}{4\pi}\left(1 - \frac{m_\mu^2}{m_P^2}\right)^2 \tag{41}$$

The empirical lifetimes give

$$c_P = m_\pi \cdot 0.7 \quad , \qquad c_K = c_\pi/3.7 \tag{42}$$

Evaluating (40) in the same way as in (21) we find

$$c_\pi \sqrt{m_\pi} = 0.7 \cos\theta \sqrt{2}\ \psi(0)$$

With (42) this gives

$$r = |\psi(0)|^{-2/3} = 1.75 \times 10^{-13}\text{cm} .$$

Here again we get a somewhat large value for the quark wave function in the pion.

In summary one may say that in spite of some failures of the quark model it fits in so many instances so well that this does not seem to be accidental. Of course, not all predictions are independent and some can be obtained from others by weaker assumptions (e.g. only SU_3).

V. Higher Resonances

So far we have been considering only the "classical" particles and resonant states. Now we investigate higher resonances. They find their natural explanation in the quark model as L-excitations. In group-theoretic language this means that our (approximate) invariance group is $O_3 \times U_6$ and we have so far considered the identity representation of O_3. Now we will look for the L = 1 representation. For the $\bar{q}q$ - system we have only the 1 and 35 representation of SU_6 whereas for the qqq the 20 and 70 representations may appear in addition to 56. It may also happen that an $\vec{L}.\vec{S}$-force completely destroys SU_6 for $L \neq 0$.

For the $\bar{q}q$ system the L = 1 states will have positive parity and if we add the quark spin we expect 0^+ (scalar), 1^+ (axial vector) and 2^+ (tensor) particles. More in detail we expect for the SU_6 singulet (s = 0) one axial vector particle and for the 35 a 0^+ nonet (s = 1), a 1^+ nonet (s = 1), a 1^+ octet (s = 0) and a 2^+ nonet (s = 1). The charge conjugation quantum number c for the neutral members can be found from Michel's generalized exclusion principle for positronium.

$$(-)^{\ell}(-)^{s+1} \quad c = -1 \tag{43}$$

For $\ell = 0$ we had c = 1 for the 0^- and c = -1 for the 1^- particles, as is experimentally established. Now we have c = +1 for the 0^+, 1^+ and 2^+ nonets and c = -1 for the 1^+ singulet and octet. It is this prediction regarding parity and* c which is typical for the quark picture since the ordering of increasing energy with angular momentum will come from most reasonable models. The experimental information on all these particles is not yet complete. There are candidates for all these** 3 × 36 particles and they seem to exhaust about most of the

* G is, of course, $C.(-)^I$

** also counting spin directions.

known boson resonances. Best established is the 2^+ nonet with the $A_2(1320)$, $K^*(1410)$, f(1250) and f'(1500). The mass formula gives $m_1=1230$, $m_8=1320$, $\Delta=90\pm15$, $I\approx1$ but there seems to be the difficulty that more recent mass values require I >1. The branching ratios of the various decays seem [1] to be in accord with the nonet assignement. Regarding the scalar and axial vector particles the data is inconclusive so that I shall not discuss them further. Once this has been cleared up it will be an important test for the quark model to see whether the same kind of calculation we indicated for the 0^-, 1^- particles also works here. For the baryons L = 1 will produce a multitude of 1/2, 3/2 and 5/2 states [6]. Actually such states seem to exist beyond the 56-particles but since I do not have precise theoretical and experimental numbers to confront with each other I shall leave this subject.

VI. Forces between Quarks

One of the puzzles of the quark model is why the $\bar{q}q$ and qqq states are the lowest in energy. This must be a peculiar feature of the forces between quarks. Conservatively one would expect the forces between quarks to arise from the exchange of bosons. However, if quarks exist they must be heavy ($M_q \sim 5$GeV) and hence they can take advantage of the strong inner part of a Yukawa potential $e^{-\mu r}/r$ for $\mu \sim 1/2$ GeV. Thus it will not be the exchange of pions which gives the dominant force but all bosons will contribute. If there is some repulsive core which cuts of the 1/r singularity it may also be the motion of the quarks is nonrelativistic around the minimum of the potential yet this minimum is almost as deep as M_q. Thus the quark states may be more like molecular states than light nuclei. Of course a nonrelativistic Schrödinger equation cannot describe the situation correctly for a situation when the total energy approaches zero: Then some disaster is going to happen which will not show up in the Schrödinger equation. However, one may hope that general properties like level ordering can be taken over from our non-relativistic intuation.

Let us see what we may expect if the forces are mainly due to the exchange of the 36 mesons. For the vector particles the force will be nonrelativistically just the (spin-independent) Coulomb force with the Yukawa factor $e^{-\mu r}$. The first qualitative question will be which states are attractive and which repulsive. Exchange of a single vector particle gives an attraction for particle-antiparticle and a repulsion for particle-particle. But since we have many particles the situation appears more complicated. If we write the vector meson matrix V of (10) in the form

$$V = \sum_{i=0}^{8} \lambda_i V^i \tag{44}$$

Their exchange between two quarks will give a potential

$$U_V = \sum_i \lambda_i^{(1)} \lambda_i^{(2)} \frac{e^{-\mu r}}{4\pi r} \tag{45}$$

analog to $\vec{\tau}^{(1)} . \vec{\tau}^{(2)}$ in the nuclear case. The potential generated by exchange of pseudoscalar particles is more complicated because of the

$$(\vec{\sigma}^{(1)} . \vec{\nabla})(\vec{\sigma}^{(2)} . \vec{\nabla}) \frac{e^{-\mu r}}{4\pi r} \qquad \text{term.}$$

This can be decomposed into a tensor force and a central force $\vec{\sigma}^{(1)} . \vec{\sigma}^{(2)}$. Since the tensor force is 0 in s-states we take only the later and add it to (38). The sign comes out correct so that one obtaines a constant plus the quadratic Casimir operator of SU_6 if the 0^+ mesons have an appropriate coupling constant

$$U_{V+P} = \pm \Big(\sum_{i=0}^{8} \lambda_i^{(1)} \lambda_i^{(2)} + \sum_{i=0}^{8} \lambda_i^{(1)} \lambda_i^{(2)} \vec{\sigma}^{(1)} \vec{\sigma}^{(2)} \Big) \frac{e^{-\mu r}}{r} . \tag{46}$$

In order to get something SU_6 invariant we disregarded the fact that the central force from the pseudoscalar mesons is not just $e^{-\mu r}/r$ and that the masses are different. Here the upper sign refers to qq and the lower to $\bar{q}q$. The question now is whether the ( ) is positive or negative in the 35 and 56 representations. This can be answered easily by observing that they contain the

$$\phi = \overset{\uparrow}{\bar{\lambda}}\ \overset{\uparrow}{\lambda} \qquad \text{and the} \qquad \Omega^- = \overset{\uparrow}{\lambda}\ \overset{\uparrow}{\lambda}\ \overset{\uparrow}{\lambda}$$

resp. looking at the explicit form of the vectormeson-matrix we see that the fundamental λ-processes are

$$\lambda \to \lambda + \phi \ , \qquad \lambda \to p + K^{*-} \ , \qquad \lambda \to n + \bar{K}^{*o} \qquad .$$

Hence $\lambda\lambda$ or $\bar{\lambda}\lambda$ can only exchange a ϕ so that $\sum_{i=o}^{8} \langle \lambda_i \lambda_i \rangle > 0$ for 35 or 56. Furthermore since their spins are parallel $\vec{\sigma}^{(1)} \cdot \vec{\sigma}^{(2)} = 1$ so that also the pseudoscalar contribution is > 0. Therefore we come back to our naive conclusion that we get attraction for $\bar{q}q$ and repulsion for qqq. More exactly we learn from group theory that the expectation values for the quadratic Casimir operator

$$\sum_{i=o}^{8} \lambda_i^{(1)} \lambda_i^{(2)} + \sum_{i=o}^{8} \lambda_i^{(1)} \lambda_i^{(2)} \vec{\sigma}^{(1)} \vec{\sigma}^{(2)}$$

has the following relative values in the various representations

1	-10
35	2/7
20	-7
56	+5
70	-1

Thus we are left with the puzzle why the baryons are in the 56 representation. There are several ways out. For the exchange of scalar and axial vector particles between q and q the sign of (46) changes. There the spin independent part comes from the 0^+ particles which give something intrinsically attractive. However, since these bosons are not so well established there is little one can say. One may also argue that threebody forces become important and that is why we see qqq and not qq. Nevertheless it is funny why baryons prefer 56 as we shall discuss in the next section.

VII. The 3-quark States

Here we shall disregard the question why there are attractive forces between three quarks but see how it can come about that the 56 representation is the lowest in energy. Since it is totally symmetric regarding the SU_6-variables it must be totally antisymmetric in space if we want to preserve fermi statistics for the quarks. Of course, usually the totally antisymmetric wave functions are not the lowest ones, in particular the ones with L = 0. However one could imagine that somehow the force is much stronger in the 56 than in the 20 or 70 so that the first antisymmetric state in the 56 potential is below the symmetric state in the 20. This still leaves the puzzle why it is a L = 0-state. To get some feeling for the situation let us consider a quasi-molecular approximation. We imagine that the three quarks perform small oscillations around these equilibrium positions which will be an equilateral triangle. Thus we assume we have an effective Hamiltonian

$$H = \sum_{i=1}^{3} \frac{\vec{p}_i^{\,2}}{2M} + M\omega^2 \sum_{i>k} \left(|x_i - x_k| - R \right)^2 \qquad (47)$$

After getting rid of the center of mass motion we see that

for R = 0 we have a 6-dimensional isotropic oscillator. If R is much larger than the fluctuation of the x_i around their equilibrium positions we have the rotational energy plus three vibrational modes, 2 of which are degenerate. Thus without translations the invariance group is U_6 for R = 0, $U_2 \times O^3 \times P$ for* $R \to \infty$ and $S_3 \times O^3 \times P$ for intermediate distances. Knowing the solution for the two limiting cases one can get a pretty good idea what is going on by connecting the lowest levels of the same quantum numbers. For R = 0 the energy spectrum is

$$E = E_o + \sqrt{3/2}\,\omega . n \ . \tag{48}$$

The first antisymmetric state appears for n = 2, the wave function being

$$\psi_{2a} = (\vec{r}_1 \times \vec{r}_2 + \vec{r}_2 \times \vec{r}_3 + \vec{r}_3 \times \vec{r}_1)\psi_o(r_{ik}) \tag{49}$$

$$r_{ik} = r_i - r_k$$

where the $\vec{r}_i$ are the coordinates of the three quarks. $\psi_o(r_{ik})$ is the gaussian wave function of the ground state and is a totally symmetric 0^+-function.

One sees easily that ψ_{2a} is translational invariant, totally antisymmetric and 1^+. The first totally antisymmetric S-level appears for n = 6 and is of the form

$$\psi_{6a} = (r_{12}^2 - r_{23}^2)(r_{23}^2 - r_{31}^2)(r_{31}^2 - r_{12}^2)\psi_o \tag{50}$$

a similar situation appears for the limit $R \to \infty$ where the energy becomes

$$E = \frac{\ell(\ell+1)}{2MR^2} - \frac{\nu^2}{4MR^2} + \omega\sqrt{3/4}\,(\sqrt{2}\,n_1 + n_2) \ . \tag{51}$$

---

* P is the space reflection

Here ν is the projection of L perpendicular to the plane of the quarks and n_1 and n_2 are the two vibrational quantum numbers. The lowest totally antisymmetric state is again the 1^+-level with $\ell = 1$, $\nu = n_1 = n_2 = 0$ whereas the lowest antisymmetric s-state has quantum numbers $\ell = \nu = n_1 = 0$, $n_2 = 3$. That this has a higher energy corresponds to our general experience that rotational levels are below the vibrational modes. Thus, even granting that the forces in the 56 representation are the strongest it is difficult to see why the lowest state is an s-state.

Summarizing one can say that with conventional ideas the puzzle of the three quark states is hard to understand. How serious one takes this is a matter of private opinion.

I am personally glad that not everything goes smoothly through with the quark model. This will keep this field interesting and we shall not get just a repeatition of nuclear physics.

References

1. S. L. Glashow, R. H. Socolow, Phys. Rev. Letters 15, 329 (1965)
2. For experimental data we use mainly H. A. Rosenfeld et al., Rev. Mod. Phys. 37, 633 (1965)
3. F. Gürsey, A. Pais, L. A. Radicati, Phys. Rev. Letters, 13, 299 (1964)
4. M. Bég, B. W. Lee, A. Pais, Phys. Rev. Letters 13, 514 (1964)
6. R. Dalitz, Resonant States and Strong Interactions, Oxford Conference 1965.
7. L. Brown, P. Singer, Phys. Rev. Letters 8, 460 (1962)
8. Gerasimov, Dubna preprint (1965)
9. J. K. de Pagter et al., Phys. Rev. Letters 16, 35 (1966)
10. J. Kuti, private communication
11. G. Bellettini et al. Phys. Letters 18, 333 (1965)

HIGHER SYMMETRIES[†]

By

F. E. LOW

Massachusetts Institute of Technology
Cambridge

I am going to talk about higher symmetries (and by higher I mean higher than SU(3),symmetries that mix spins in). I will say a few words about my underlying feelings about this and they are that in the long run the symmetry game is not the thing which will provide us with the basic answers. The basic answers will be provided in some kind of dynamics which we will have to understand. Nevertheless, as a phenomenology, as an intermediate step it will be extremly useful to examine these symmetries and they have had an amazing success so far. I am going to talk about theories in which higher symmetries are not an approximation, but hold exactly. Now you may say that this is totally academic and perhaps it is. So: there is something like SU(6) which could be an exact symmetry, and if so does that thing has any resemblance whatever to the usual SU(6) that we read about in the news papers or to the very interesting quark-triplet model that Prof. Thirring discussed. And I believe the answer is no. The answer is that as soon as you ask for something with any trace of an exact symmetry it loses all contact with the model and becomes something quite different. Nevertheless although the thing is different it might still be interesting. My talk will be mostly about the work of other people and only at the end I shall make some extreme and absurd suggestions which you will recognize as my own.

---

† Lecture given at the V. Internationalen Universitätswochen für Kernphysik, Schladming, 24 February - 9 March 1966.

I. Static SU(6)

Let me start with a brief review of the static SU(6) as suggested by Radicati and Gürsey. SU(6) is the algebra of objects Γ_μ (μ not a relativistic index)

$$\Gamma_\mu = \lambda_\alpha \sigma_i \qquad \begin{array}{l} \alpha = 0,1,\ldots 8 \\ i = 0,1,2,3 \\ (\alpha = i = 0 \text{ left out}) \end{array} \tag{1}$$

λ_α are the SU(3) generators, λ_0 is the unit matrix.
σ_i are the SU(2) generators, σ_0 is the unit matrix.

So we have 35 generators of unitary transformations on 6 objects with determinant one. The unitary transformations are then exponentials of arbitrary real combinations of such objects

$$U = \exp i\,(a_\mu\, \Gamma_\mu)\ .$$

In the usual way one wants to associate physical states with the representations of these operators. We will discuss tensor representations. A basic 6-component-vector ξ_μ has the transformation property

$$\xi'_\mu = U_{\mu\nu}\, \xi_\nu\ . \tag{2}$$

The contravariant vector transforms with the complex conjugate U^*

$$\xi'^\mu = U^*_{\mu\nu}\, \xi^\nu\ . \tag{2'}$$

We then can generate arbitrary representations by taking tensor combinations of such objects. Consider the completely symmetric $B_{\{\mu\nu\lambda\}}$. These 56 states then correspond to the baryons: the 8 × 2 nucleon octet ($I = 1/2^+$) and the 10 × 4 decuplet ($I = 3/2^+$). Similarly the low-lying mesons are to belong to the representation M_μ^ν (that is a second rank mixed tensor, whose trace vanishes) which contains $6 \times 6 - 1 = 35$ states. It includes the 8 pseudoscalar pions and the $8 \times 3 = 24$ vector octet ρ's and one ω.(We call the ω the unitary singlet and don't worry about mixing.) The great success of this assignment is that we need only one representation for the low-lying baryons and one for the mesons. Without any further hypothesis one is finished. There is nothing else that can be done.

With one slight further assumption one can get another piece of information. Given the representation one can know the matrix elements of these Γ_μ's between the states. If one assumes that the λ_α's with the appropriate isotopic index times the $\vec{\sigma}$ are the integrated β-decay axial vector currents, then one can derive the axial vector coupling constant renormalization and one finds

$$G_A/G = 5/3 \quad .$$

Experimentally it is 1.2 .

One has to be able to work in motion in order to get answers for coupling constants and S-matrix elements.
Instead of the $\vec{\sigma}$ algebra, one might try a wider algebra, e.g. all the positive parity Dirac operators

$$\sigma_i\left(\frac{1 \pm \beta}{2}\right) \quad . \tag{3}$$

Since $\frac{1 \pm \beta}{2}$ are projection operators the + and - algebra commute with each other, and each algebra by itself is a U(6). Thus the whole algebra is $U(6) \times U(6)$. This algebra might also be used to classify the low-lying states, the baryons are (56,1) and the mesons $(6,\bar{6})$.

A further possibility will be a "chiral group", where instead of $(1 \pm \beta)$ you have $(1 \pm \rho_1) = (1 \pm \gamma_5)$. This algebra is again

$U(6) \times U(6)$. One could also use the widest algebra with all the 16 Dirac matrices. But it is probably much to wide to have anything to do with nature.

Now we have two questions: Can we apply these operators to moving particles and to states with more than one particle? We don't have to know what these operators are in order to derive results. E.g. we can apply three isotopic spinvectors to moving states. We have no physical interpretation of isotopic spin. We simply have three integrals of motion which can be used to classify moving states of one or more particles. Then we can deduce relations between coupling constants and between cross sections without knowing what these operators are. If in addition the generators have physical properties which can be measured in another way then that is even more interesting.

II.

We supplement this by assuming that the objects we are talking about have certain transformation properties. The natural transformation properties are provided for example by the quark-model as suggested by Gell-Mann, that is that to the operator Γ there corresponds an operator in the Hilbert-space $A(\Gamma)$ which is an integral of a current, let's say $J(\Gamma)$. Then all the commutation relations written down can very naturally be reproduced by assuming simply that

$$J(\Gamma) = \psi^{+}(x)\, \Gamma\psi(x) \quad .$$

With the usual canonical scheme (if that makes any sense) all the commutation relations can be reproduced with this Ansatz because one knows that the commutation relations of integrals of this form are the same as the commutation relations of the Γ's themselves. This is excluding such problems as the Schwinger difficulty with the commutator of the charge density and the current and other things associated with the existence of the three dimensional integral.

So $A(\lambda_\alpha \sigma_i)$ would be identified with the three dimensional integral

$$A(\Gamma_{\alpha i}) = \int d\vec{x}\ \psi^+(x)\ \Gamma_{\alpha i}\ \psi(x) \quad . \qquad (4)$$

For i = 0 it is just $\int d\vec{x}\ \psi^+\ \lambda_\alpha\ \psi$ and these are the SU(3) generators. There is obviously no inconsistency in assuming them to be exactly conserved since these are integrals of the time component of conserved currents. Therefore they are invariants and the classification with respect to these operators is very simple.

We consider next

$$\int d\vec{x}\ \psi^+\ \lambda_\alpha\ \sigma_i\ \psi$$

for $i \neq 0$. These operators are in fact the space components of an axial vector, $J^5_{i\alpha}$.

Something else which is of interest taken out of the chiral-algebra is

$$\int d\vec{x}\ \psi^+\ \lambda_\alpha\ \rho_1\ \psi$$

the fourth component of the $J^5_{o\alpha}$ (axial vector).

In terms of the various currents of this type there is the following very interesting suggestion by Gell-Mann: Although the symmetries may not hold exactly, the commutation relations of these integrated currents do hold exactly. That is exceedingly interesting but totally empty unless we can associate the currents with some other physical quantity which we can observe independently.

Naturally since we have a perfectly good vector and axial vector unitary octet both of which are integrated we would like to associate these vectors with the hadronic β-decay current. Let us then ask the question, is it possible for these things to be symmetries of the system in the usual sense, that is, to have something like $A(\lambda_\alpha \sigma_i)$ on baryon states just giving some linear combination of baryons

$$A(\lambda_\alpha \sigma_i)|B>_m = \Sigma|B>_n ((\lambda_\alpha \sigma_i))_{mn} . \qquad (5)$$

Is this possible? Now clearly for the σ_o it is possible because just the generator of unitary symmetry is left. If we can imagine the situation where unitary spin is exactly conserved these are invariants and can be used to classify states with no problem.

For the σ_i the situation is somewhat different and a difficulty was pointed out by Gell-Mann: If (5) is valid then you must have

$$A(\lambda_\alpha \sigma_i)|vac> = 0 \qquad (6)$$

We may now ask, what about the matrix elements of e.g. $J^5_{\mu\alpha}$ between the vacuum state and a pair of baryons

$$<B,\overline{B}|J^5_{\mu\alpha}|vac> \qquad (7)$$

This matrix element is represented by the following graph. It gives the axial vector form factor-crossed (F_A).

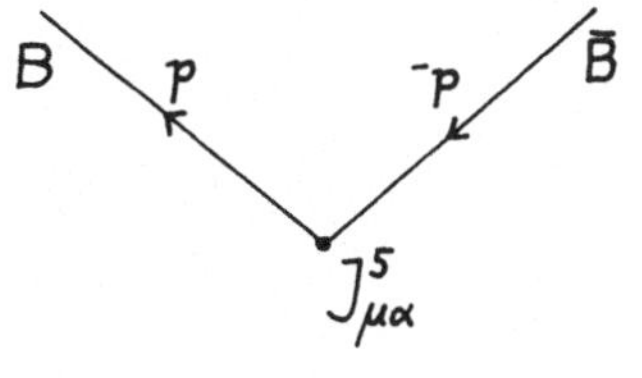

If you take p in the same direction as the vector μ, you

get zero. But if you take p orthogonal to μ you get a contribution to F_A. You evaluate $F_A(t)$ at $t = 4\ (m^2+p^2)$(m is the mass of the baryon).

If $A|vac\rangle = 0$ then also $\langle\vec{p}, -\vec{p}|A|vac\rangle = 0$ and therefore $F_A(t) = 0$ for all p. As an (hopefully) analytic function it vanishes for all t, we have then also $F_A(0) = 0$ but $F_A(0) = = \frac{G_A}{G_V} = \frac{5}{3}$ (for baryons). This is a contradiction with the scheme itself.

If we take $\vec{p}$ in the z direction and $A(\lambda_\alpha \sigma_z)$ then we get 0. So things are better. That does not mean that one of these operators is diagonal on the vacuum.

Which operators can be used to classify states in motion? Let's assume motion in z-direction. A good operator A on a state which is in motion in the z-direction will be elastic and presumably represents the group we are talking about. That is for example the isotopic spin operators on the proton will give the neutron and the proton: that's called elastic. The unitary operators on the nucleon will give N, Λ, Σ, Ξ, but not N^*. In the SU(6) game you have to include N^*. Which of the A's can possibly have this property? They must all have one necessary (not sufficient) property.

That is to say that the matrix element must effectively be an invariant. Because if not, but nevertheless A is diagonal in this sense on the states, then we can calculate the commutation relations and they will only take us within this elastic set of states. The commutation relations are nonlinear $[A, A] = A$. On the left the commutation relations will give us the square of the ratio of $\langle\vec{p}|A|\vec{p}\rangle$. On the right this ratio is either one, if the right hand term is a unitary SU(3) generator or linear: contradiction.

Which A's have these properties? I will give you a mnemonic. Consider $\int\psi^+\psi\, d\vec{x}$, it is an invariant in this sense. It is the integral of the fourth component of a relativistic vector. The integral of an invariant would not work. In particular

$$\rho(p_z) = \frac{\rho(0) + v j_z(0)}{\sqrt{1 - v^2}}$$

v is the velocity corresponding to p_z, $j_z(0) = 0$ (no vector current at rest).

A thing which gets larger by $(1-v^2)^{-1/2}$ is invariant when integrated. If I normalize spinors to be 1, then the matrix element of $u^+(p_z)I\, u(p_z)$ is essentially the integrated charge density, the charge. Invariance essentially means that

$$u_f^+(p_z)\Gamma\, u_i(p_z) = (\chi_f^\times, \Gamma\chi_i) \tag{9}$$

These can be different spin states and of course different isotopic spin states. Now we can simply calculate those operators that commute with $\alpha_z p_z$: For the algebra U(6)×U(6) at the rest, those are the following operators $\sigma_z, \beta\sigma_x, \beta\sigma_y$, Observe that these three operators generate an SU(2) algebra. That is the commutation relations of these are identical to those of SU(2). Therefore the algebra generated by

$$A(\lambda_\alpha\ \sigma_z),\ A(\lambda_\alpha\beta\sigma_x),\ A(\lambda_\alpha\beta\sigma_y)\ .$$

is another SU(6) algebra called $SU(6)_W$. But for the algebra SU(6) at rest we get another algebra

$$A(\lambda_\alpha),\ A(\lambda_\alpha\sigma_z) \qquad \text{or} \qquad A(\lambda_\alpha \frac{1 \pm \sigma_z}{2})$$

and this is therefore U(3) × U(3).

The $U(6)_W$ can be used to classify states in one dimensional motion. If we consider scattering experiments (two dimensional motion in a plane), there is a perpendicular direction which leaves $\beta\sigma_\perp$. And that's just the perpendicular component of spin. If you quantize spin perpendicular to the plane of an event the spin is conserved. That has been studied by Amati and Alles. They have shown that the U(3) × U(3) based on $\beta\sigma_\perp$ is in contradiction with two particle unitarity in two different channels used in the Mandelstam sense,that is to say s and t. If you consider scattering in the s-channel and impose this U(3) × U(3), it gives you restrictions on amplitudes. Now cross this amplitude to the t-channel then it is in contradiction with unitarity. This was proved for quark-quark scattering. Gell-Mann argued: quarks are very heavy and therefore there is no elastic channel in the crossed channel. Nevertheless it seems unlikely that if you can prove it for the simplest nontrivial case that it should work for a larger representation. All the outlined proofs of inconsistency have to do with highly relativistic phenomena.

III. Coleman's Counterargument

Let us consider the linear group U(3) × U(3) of one dimensional motion, where there was no contradiction with unitarity. But there is an argument by Coleman:

We suppose that

$$\int J_z^5(\vec{x},t)\; d\vec{x}\,|1,p_z\rangle$$

where $|1,p_z\rangle$ is a one-particle state, has no inelastic matrix element; this means

$$\langle n,p_z|\; \int J_z^5(\vec{x},t)\; d\vec{x}\,|1,p_z\rangle = 0 \qquad (10)$$

$|n,p_z\rangle$ is a many particle state. Equation (10) is equivalent to

$$\langle n,p_z|J_z^5(x)|1,p_z\rangle = 0$$

or covariantly

$$<n|\{\Box J_\mu^5(x) - \frac{\partial}{\partial x^\mu}\frac{\partial}{\partial x_\nu} J_\nu^5(x)\} .P^\mu|1> = <n|Q_\mu P^\mu|1> = 0 \tag{11}$$

(Proof: We can find a coordinate system $\vec{p}_n = \vec{p}_1 = \vec{k} . p .$ (z-direction)) Then equation (11) becomes

$$(E_n - E_1)^2 \, p<n,\vec{k}.p|J_z^5(x)|1,\vec{k}p> = 0$$

Consider a wave packet of $|1,p_z>$ and a n-wave packet of $<1', p_z'| + <n'|$. Now we localize $|1>$ and $|1'>$ in x and p space with smooth functions such that $p_n - p_1$ is time like. We take the position of the wave packet far from the variable x in $J^5(x)$ and get

$$<1'|P_\mu|1> <n'|Q^\mu|vac> = 0$$

or

$$<n'| \Box J_\mu^5(x) - \frac{\partial}{\partial x^\mu}\frac{\partial}{\partial x_\nu} J_\nu^5(x)|vac> = 0$$

Taking the curl we obtain

$$<n'| \Box(\partial^\mu J_\nu^5(x) - \partial^\nu J_\mu^5(x))|vac> =$$

$$= M_{n'}^2 <n'|\partial^\mu J_\nu^5 - \partial^\nu J_\mu^5|vac> = 0 \qquad \text{if } M_{n'}^2 \neq 0 .$$

Since n' is arbitrary,

$$\partial^\mu J_\nu^5 - \partial^\nu J_\mu^5|vac> = 0$$

and since any local operator which annihilates the vacuum is zero, therefore

$$\partial^\mu J_\nu^5 - \partial^\nu J_\mu^5 = 0$$

For $\mu = z$ and $\nu = 0$ we find that

$$\int J_z^5(x)\, d\vec{x}$$

is a constant of motion. This is in contradiction, because at the beginning it was assumed that the axial vector current is not conserved.

The attempt to make moving SU(6), that is the U(3) × U(3) subgroup of SU(6) into a true symmetry for finite velocity particles, fails as shown. That is to say that the generators of the form $A(\lambda_\alpha \sigma_z)$ on one particle states $|p_z, m\rangle$ give elastic and inelastic states and therefore one cannot use the A's to generate the symmetry in the usual sense.

IV.

There is an attempt to cure this problem which I will briefly review as an exercise to find the proof that it is false, which was also shown by Coleman. However it is interesting in the following way:

Instead of trying to assume (5) which we have shown to be inconsistent with locality and relativistic invariance, we make a different assumption. We have seen that (5) led us to disconnected diagrams because it was possible to localize the nucleon and the field far away from each other. Therefore the best thing to do is to try to tie these two things together so that they cannot be moved apart. This is done in the following way: Instead of the A acting on a nucleon state giving a nucleon state, you say that the A, commuted with a creation or an annihilation operator of an "in"- or "out" field $\psi_m^{in,out}(x)$ would give a similar sum.

$$[A(\Gamma_i), \psi_m^{in,\, out}(x)] \simeq \sum_n A_{nm}\, \psi_n^{in,out}(x) \quad .$$

By Fourier transformation you can reduce this to statements

about the momentum space operators $a_m^{in,out}(p_z)$, $a_m^{+in,out}(p_z)$ (motion in the z-direction):

$$[A(\Gamma_i), \begin{matrix} a_m \\ a_m^+ \end{matrix}] = \pm [A(\Gamma_i)]_{mn} \begin{pmatrix} a_n \\ a_n^+ \end{pmatrix} \qquad (12)$$

Essentially the particles then would generate in this sense the representations of the group but the generator would be tied to the particle variable itself and you would not be able to move them apart.

Basically what it amounts to is saying that the matrix element

$<n|A(\Gamma_i)|1> = 0$, graphically 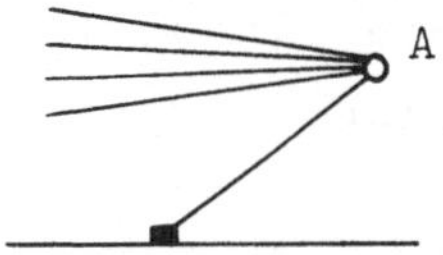 ,

except for disconnected diagrams, for example you would say that this 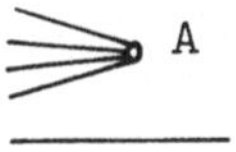

is not zero. This corresponds simply to dropping disconnected diagrams.

First we will show that the commutation relations for the A's are saturated by the one particle states and it is a simple matter to check that. We use equation (12) and calculate one particle matrix elements of the A's,

$${}_{out}<p,m'|A(\Gamma_i)|p,m>_{in} = {}_{out}<p,m'|A(\Gamma_i)\, a_m^{+in}|vac> ,$$

therefore

$${}_{out}<p,m'|A(\Gamma_i)|p,m">_{out}\ {}_{out}<p,m"|A(\Gamma_j)|p,m>_{in} =$$

$$= {}_{out}<p,m'|A(\Gamma_i)|p,m">_{out} \Big[{}_{out}<p,m"|A^j_{mn}|p,n>_{in} +$$

$$+ {}_{out}<p,m"|a_m^{+in} A(\Gamma_j)|vac> \Big]$$

The "in" and "out" one particle states are the same, so that the "in" creation operator can be taken over and it annihilates the one particle state $|p,m''>_{out}$,

$$_{out}<p,m''|a_m^{+in}A(\Gamma_j)|vac> = \delta_{mm''}<vac|A(\Gamma_j)|vac> = 0$$

for one particle states. This doesn't work for many particle states, they are disconnected diagrams and these are explicitely not zero. So the commutation rules are saturated only by one particle states.

Again, this approach was destroyed by Coleman by a very similar construction to the previous one.

Consider

$$<n|[(\Box J_\mu^5 - \frac{\partial}{\partial x^\mu}\frac{\partial}{\partial x_\nu} J_\nu^5)P^\mu , a_m^{out}(\vec{p})]|vac> , \qquad (13)$$

n is inelastic.

This is the covariant version of

$$p_z(E_n+E_m)^2<n|[J_z^5 , a_m^{out}(\vec{p})]|vac> .$$

$$<n|[J_z^5 , a_m^{out}(\vec{p})]|vac> = (J_z^5)_{m\mu}<n|a_\mu^{out}|vac> = 0 .$$

so that

$$<n|[Q_\mu P^\mu , a^{out}(\vec{p})]|vac> = 0$$

or

$$<n + \vec{p}|Q_\mu P^\mu|vac> = 0 .$$

If you have at least four identical particles in the final

state, you can use symmetry to show that the matrix element vanishes for four different momenta. This is sufficient to show that $Q_\mu = 0$ to a state with four or more identical particles.

Now suppose that it doesn't vanish to some state (e.g. a two fermion state) and consider

 Q_μ, and bring in another particle 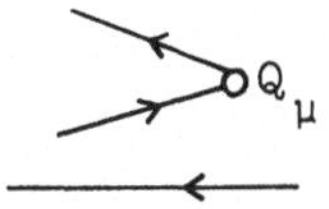

in such a way, that scattering from one particle to three particles will occur.

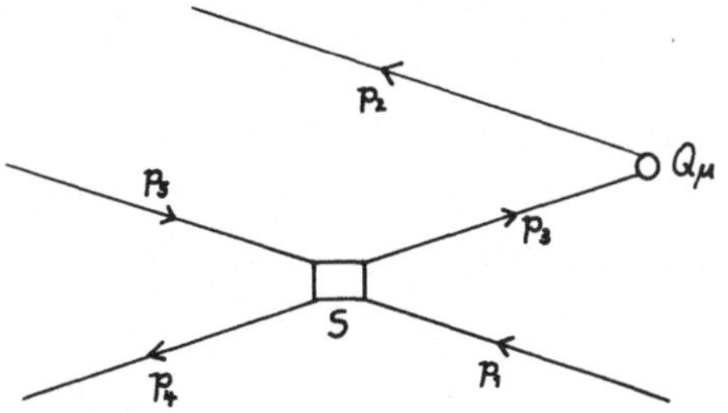

If there is any crossing analyticity then you can turn the p_1 line around to get a vacuum to four particles transition.

Now, if scattering actually takes place you get

$$\langle \bar{p}_1, p_2, p_4, p_5 | Q_\mu | vac \rangle \neq 0$$

and this gives a contradiction again: Q_μ has to vanish to any state, leading to the previous results. The other possibility without contradiction is S = 1 (no scattering), so that this scheme gives at best a relativistic SU(6) type free particle theory.

V.

Now there is a new suggestion, which has been put forward

by many people: Fubini and Furlan (who suggested it but did not carry it out immediately), Adler, Schnitzer, Gerstein and Lee (who used it with great success). Gell-Mann and Dashen have now just published a preprint in which they discuss this scheme in great detail.

The rule is: you work at p_z equals infinity, and this seems to cure a large number of problems.

These people were not interested in curing precisely this problem, we have found. However, if you look at the paper of Adler on the axial vector coupling constant, you will see that disconnected diagrams are ignored. In the paper of Weisberger they are not ignored. Weisberger writes down two types of Feynman diagrams: connected graphs and disconnected graphs. When taking the square of the matrix elements there are squares of connected graphs, squares of disconnected graphs and cross-terms. The totally disconnected terms can be ignored since they cancel out in the commutator. The cross-terms are dominated by a one-pion state, the other contributions are strongly damped, and so on. It is a rather elaborate procedure. - Adler never tells you about disconnected diagrams. He goes directly to $p_z \to$ infinity. At that point he basically assumes that he has only connected diagrams to work with.

Something like this was already known to Adler when he published that famous paper. These people want to consider a slightly different aspect. Gell-Mann and Dashen, however, were specifically interested in curing this trouble of the disconnected diagrams.

What happens to disconnected diagrams when you work at infinity:

Suppose we have

$$\langle p_z|A(\Gamma_i)|n\rangle\langle n|A(\Gamma_j)|p_z\rangle$$

where $|p_z\rangle$ is a one particle state. Let's consider then disconnected contributions; this means:

|n> equals $|p_z$ + another state n'> and furthermore the matrix element is equal to a δ-function of p_z times the matrix element from the vacuum to the state |n'>

$$\langle vac|A(\Gamma_i)|n'\rangle\langle p_z+n'|A(\Gamma_j)|p_z\rangle$$

Now what happens? This is a sum over n' and the momentum of n' is something like zero. The point is that the state $|p_z+n'\rangle$ of course has an infinite mass. That is, mass of $|p_z+n'\rangle$ goes to infinity, $m^2 = (M_n+E(p_z))^2-p_z^2 = M_n^2+ M^2+ + 2E(p_z)M_n \to \infty$ as $p_z \to \infty$.

The basic idea here is that if in the physics of strongly interacting particles infinite mass - or momentum-transfer-matrix elements vanish sufficient rapidly for the purpose at hand, then you have two different classes of matrix elements, if you work at infinity: finite momentum-transfer matrix elements evaluated at infinite velocity (but that doesn't matter because one can always transform back, and find out what they are) and infinite mass matrix elements, which simply go to zero.

Now the question is: what are the operators, which will have this property. Which $A(\Gamma_j)$ are such that infinite mass matrix elements might (because we do not really know!) vanish?

Fubini, Walecka and Segré categorized operators into two classes, which they call "good" and "bad" operators. They can be defined in this context in two different ways.

First definition: Do commutation relations leak? Remember we have seen that the relativistic transformation properties of $A(\lambda_\alpha\sigma_z)$ are such that they don't leak. In fact we found that these things act as if they were invariants for transformations in directions parallel to σ_z. Therefore you might guess that the $A(\lambda_\alpha\sigma_z)$ should be good operators.

On the other hand there is leakage for operators like

$A(\lambda_\alpha \sigma_y)$, they leak like $\frac{m}{E} = \sqrt{1 - v^2}$.

As v goes to 1, leakage is total for $A(\lambda_\alpha \sigma_y)$, but there is no leakage for $A(\lambda_\alpha \sigma_z)$ for any v.

The second consideration is the following:
Previously we calculated the matrix element

$$\langle p_z | A(\Gamma_i) | p_z \rangle = u^+(p_z) \Gamma_i u(p_z)$$

Now you consider the matrix element to a pair state (for example a positron of momentum $-p_z$ and an electron of momentum $+p_z$)

$$\langle p_z, -p_z | A(\Gamma_i) | vac \rangle \quad .$$

Do you get an infinite mass matrix element?

$$\langle p_z, -p_z | A(\Gamma_i) | vac \rangle = u^+_{(+)}(p_z) \Gamma_i \; u_{(-)}(p_z) \rightarrow$$

$$\rightarrow \frac{1-\alpha_z}{\sqrt{2}} \cdot \Gamma_i \frac{1+\alpha_z}{\sqrt{2}} \quad , \text{ as } p_z \rightarrow \infty \quad . \tag{14}$$

For σ_z, which commutes with the Dirac operators, this is zero. The same thing which gave you no leakage in the one-particle to one-particle transitions gives you zero in the case of vacuum to two particles.

On the other hand σ_y does not give zero at all, even as p_z goes to infinity.

This leads to the second definition:

$$\begin{aligned} &\text{good operator, if } [\Gamma_i, \alpha_z]_- = 0 \\ &\text{bad operator, if } \{\Gamma_i, {}_z\}_+ = 0 \end{aligned} \tag{15}$$

Now we have the division into good and bad operators and we would like to assume that for good operators the problems at infinity go away; one works only with good operators.

An important point to add is that this is a weaker assumption than the earlier one, where you assumed that there are good generators which work at all momenta. Now we have a very special restriction: only if we work at infinity we can use the symmetry and working at infinity in this context is not a question of Lorentz transformations. These two things are different and lead to different results.

VI. Consequences

Now let me give one of the consequences of the theory at $p_z \to \infty$, which is rather drastic.

The symmetry we are using at $p_z \to \infty$ is

$$U^{(+)}(3) \times U^{(-)}(3) \qquad ,$$

where $U^{(\pm)}$ is given by the generators $A^{\pm} = \int \psi^+ \lambda_\alpha (\frac{1 \pm \sigma_z}{2}) \psi d\vec{x}$, or equivalently $\lambda_\alpha (\frac{1 \pm \rho_1}{2})$. Of course, $A^+ + A^- = A(\lambda_\alpha)$, which are the old SU(3) generators.

$A^+ - A^-$ is something else, for which you don't immediately have a physical significance. It is $A(\lambda_\alpha \sigma_z)$. For λ_0 you have $A(\sigma_z) = \int \psi^+ \sigma_z \psi d\vec{x}$, which you might call quark spin in a quark model.

Let us talk about the assignment for the baryons with respect to this group. First we have a (10,1). This contains 10 helicity 3/2 states, because the 10 is completely symmetric in its three quark indices. (1,10) then has helicity $\lambda = -3/2$. For the helicity 1/2 states we have a (6,3) with 18 states, the decuplet and the octet. (3,6) would be $\lambda = -1/2$. This was the conventional assignment. On can say one thing on transformation properties. Consider the action of the operator $P\, R_y(\pi)$, parity times rotation about y-axis by an angle π, on these matrix elements. This operator takes the momentum into itself, because P reflects the momentum and $R_y(\pi)$ brings it back again. P doesn't affect the spin, but $R_y(\pi)$ does. Therefore spin and helicity changes sign, $\lambda \to -\lambda$ under this operation and it takes $U^{(+)}$ into $U^{(-)}$.

If λ corresponds to a representation (m,n), $-\lambda$ must correspond to (n,m).

Let us consider the following. We want to calculate the matrix elements of "good" operators only. To calculate magnetic moments, I cannot use the current j_μ, rather I have to consider the so called "good" operator $\rho(\vec{x})$. We calculate $\rho(\vec{x})$ between states moving in the z-direction with infinite momenta $\vec{p} - \vec{q}/2$ and $\vec{p} + \vec{q}/2$, respectively.

$$\langle\vec{p}-\vec{q}/2|\rho(\vec{x})|\vec{p}+\vec{q}/2\rangle \quad , \quad \vec{p} \to \infty \quad , \tag{16}$$

and for convenience with $\vec{p} \cdot \vec{q} = 0$.

We assume the local commutation relations of $\rho(\vec{x})$ with the generators.

$$\rho(\vec{x}) = \int \psi^+ \lambda_Q 1 \psi \, d\vec{x} = \int \psi^+ \lambda_Q \left(\frac{1+\sigma_z}{2} + \frac{1-\sigma_z}{2}\right) \psi \, dx \quad ,$$

with $\lambda_Q = \lambda_3 + \lambda_8/\sqrt{3}$. Therefore under commutation with A, ρ transforms like (8,1) and (1,8).

Let us see where the magnetic moments come from in this matrix element (16), which is of finite momentum transfer $t = -\vec{q}^2$. First I use Lorentz invariance. We have to write a four vector current.

$$j_\mu = i \bar{u} \gamma_\mu u[e F_1(t) + i\sigma_{\mu\nu} \cdot q_\nu \, \delta\mu . F_2(t)] \quad ,$$

with F_1, F_2... Hofstadter form factors, normalized to one,

$\delta\mu$... anomalous magnetic moment.

I simply calculate the expectation value for $\rho(\vec{x})$ at infinite momentum, $p_z \to \infty$ and q finite and it approaches

$$\langle|\rho|\rangle = e\, F_1(\vec{q}^2)(\chi_f,\chi_i) - (\chi_f, \sigma_z \vec{\sigma}\cdot\vec{q}\,\delta\mu F_2(\vec{q}^2)\chi_i) \quad .$$

In this limit one can get the matrix element of the magnetic moment, if you allow a helicity flip, since $\sigma_z\vec{\sigma}.\vec{q}$ flips the helicity. We are now at zero momentum basically, so we are talking about normal Pauli spin matrices.

Now we are completely finished. We want to go from (6,3) to (3,6) by (8,1) + (1,8) and it doesn't go, because the "1" on the "3" gives "3" and you have to match it with the "6" and the same with $1 \times 6 = 6 \neq 3$!

So you cannot do it. There is no overlap and therefore the anomalous magnetic moment is zero. This is rather unsatisfactory from an experimental point of view.

Suppose we have fixed this up. What do we need? We need a matrix element $\langle(m,n)|(1,8)+(8,1)|(n,m)\rangle$. In order to get $\delta\mu \neq 0$, we must have $m = n$.

Now we have to consider representations of the form (m,m) and we furthermore ask these to contain octet-like representations. The lowest one of these is (8,8) for the baryons, built up by three quarks.

We might look at this (8,8) and see what kind of states it contains. We decompose it according to SU(3) itself. This gives

$$1 + 8 + 8' + 10 + \overline{10} + 27$$

Let me just write down particles and see how far we can go.

$J_z = 1/2(?)$

$\underline{1}$	$\underline{8}$	$\underline{10}$	$\underline{8'}$	$\overline{\underline{10}}$	$\underline{27}$
$Y_o^*(1405)$	N	Δ	D(1512)	N'(1420)	$S_{1/2,1/2}(1680)$
			$Y_o^*(1520)$		$S_{1/2,3/2}(1680)$
			$\Xi^*(1818)$		
			Y^*(over 2000 by sum rule)		
$(1/2^-)$	$(1/2^+)$	$(3/2^+)$	$(3/2^-)$	$(1/2^+)$	

$J_z = 3/2(?)$

$(6,3) = \underline{10} \quad \underline{8}$

Remarks: It is amusing that all strangeness zero states with $J \leqslant 3/2$ fit so neatly in this scheme, especially the lastly detected ones. The missing Y = 2 and Y = 0 states will have to be studied.

There is no rule against mixing of parities in this game, because I am working at infinite momentum and $\sigma_z = \rho_1$.

In the $\underline{8'}$ there may be the $Y_1^*(1660)$, but probably it has J = 5/2. According to the mass formula Y_1^* would come at 2000. It was reported that both $S_{1/2,1/2}$ and $S_{1/2,3/2}$ were seen, hiding under the old 900 MeV nucleon resonance.

The interesting thing about this scheme is that it is inconsistent with exact symmetry. Presumably one has to break it in two ways. The basis of inconsistency is the following: what kind of transitions can the operator $J^5_{o\alpha}$ make? Consider it between states of opposite parity (at $p_z \to \infty$)

$$\langle p_z = \infty, -|J^5_{o\alpha}|p_z = \infty, +\rangle \quad =$$

$$= \frac{\langle p_z = 0,-|J^5_{0\alpha}|p_z = 0,+\rangle + v\langle p_z = 0,-|J^5_{z\alpha}|p_z = 0,+\rangle}{\sqrt{1 - v^2}}$$

Basically we have to look at the numerator as $v \to 1$:

a. For parity conserving transitions only the second term gives a contribution, $\Delta J = 1$.

b. For parity violating transitions $J^5_{z\alpha}$ acts like a vector current and there is no such current at $p_z = 0$. There is thus no change of angular momentum, $\Delta J = 0$.

E.g.: We have an $\underline{8}(3/2^-)$ and an $\underline{8}(1/2^+)$. The transition matrix element is parity violating and cannot lead to $\Delta J \neq 0$. That is the inconsistency.

How to cure?

The first thing to do is to mix the representations of SU(3), so that in fact the $1/2^+$ state is not pure and so on.

If you have mass differences then the symmetry at infinite momentum is not so simply connected with the symmetry at rest by Lorentz transformations. There has to be some kind of relation between mass differences and matrix elements of the generators. That is a way of possibility such a theory has to be worked out.

RELATIVISTIC INVARIANCE AND LOCAL FIELDS†

By

P. T. Matthews

Imperial College
London

I. Introduction

These lectures were supposed to be about $\tilde{U}(12)$ [1]. I shall only mention $\tilde{U}(12)$ towards the end. What I wish to do is to study in more general terms the relationship between internal symmetries and relativistic invariance, and then to discuss in this light the achievements and also the limitations of $\tilde{U}(12)$ and related theories. The work I am reporting was done in collaboration with Professor G. Feldman. Much of it is already in the literature. In particular, a great deal of what I have to say comes very directly from extremely elegant and informative papers of Weinberg [2].

In elementary particle physics we are interested in relativistic quantum mechanics. The literature on this subject can be separated into two main categories, which hardly seem to make contact with each other. The first category is quantum field theory, which builds directly on the work of Dirac. It has been developed by a series of heuristic jumps based on physical insight and inspired guess work. It is amazingly successful leading quantitatively to an explanation of the Lamb shift and electron magnetic moment. Even in situations where perturbation theory is not applicable, it gives rise to many very fruitful notions - the relation between spin and

---

† Lecture given at the V. Internationalen Universitätswochen für Kernphysik, Schladming, 24 February - 9 March 1966.

statistics, crossing symmetry, the substitution law and the CTP theorem.

The second category of papers is based on a purely group theoretic approach, associated primarily with the name of Wigner. This approach is extremely logical and appears to be relatively sterile. One parameterises the allowed particle representations in terms of mass and spin, but one never even meets an anti-particle, let alone the other notions listed above. For this reason many physicists have tended to ignore the work of the Wigner school.

However with the coming of SU(3) and the mixing of spin with SU(3) through SU(6) [3], it has become clear that it is really necessary to understand relativistic quantum mechanics from this purely group theoretic point of view, and to bridge the gap which seems to exist between the field theorists and the group theorists.

The crux of the matter is the role of the homogeneous Lorentz group. The group theorists remind us, quite rightly, that this is a non-compact group; that its finite representations are non-unitary; that they have nothing to do with the unitary representations of the Poincaré (inhomogeneous Lorentz) group, to which the physical states must belong. On the other hand the familiar fields ψ_α, A_μ, $F_{\mu\nu}$ are each finite representations of the homogeneous Lorentz group, as denoted by the suffix, and the introduction of these finite representations of the homogeneous group seems to be the crucial step in the whole development. Our first task is to sort out this apparent paradox, and then to see how the vital extra assumption of causality, when combined with Poincaré Invariance leads to anti-particles and the important related properties.

II. Poincaré Invariance and Auxiliary Operators

Our problem is to construct states and S-matrix elements which are invariant under the transformations of the Poincaré group

$$U(\ell,\eta) = \exp[-i(P_\mu \ell^\mu + 1/2\ \eta^{\mu\nu} J_{\mu\nu})] \ . \tag{2.1}$$

A pure Lorentz transformation, which on a four vector p_μ gives ($\mu,\nu = 0,1,2,3$, with metric $(1, -1, -1, -1)$)

$$p'_\mu = \Lambda^{-1\nu}_{\mu}\ p_\nu \quad , \tag{2.2}$$

may be specified in terms of $\underline{v}$ - the velocity of the transformation. This corresponds to

$$U(\eta) = \exp[-i\ \underline{\eta}\ .\ \underline{K}] \quad , \tag{2.3}$$

where

$$K_i = J_{oi} \quad , \tag{2.4}$$

$$\hat{\underline{\eta}} = \hat{\underline{v}} \quad , \tag{2.5}$$

$\cosh \eta = \gamma$, $\sinh \eta = \gamma|v|$, $\gamma = (1 - v^2)^{-1/2}$.

We shall consistently use Λ or η to specify a pure Lorentz transformation which takes $p \to p'$.

The infinitesimal generators of the Poincaré group satisfy the commutation relations

$$[J_{\mu\nu}, J_{\pi\rho}] = i(-g_{\mu\pi}\ J_{\nu\rho} - g_{\nu\rho}\ J_{\mu\pi} + g_{\mu\rho}\ J_{\nu\pi} + g_{\nu\pi}\ J_{\mu\rho}) \ ,$$

$$[P_\mu\ , J_{\pi\rho}] = i(g_{\mu\pi}\ P_\rho - g_{\mu\rho}\ P_\pi) \ ,$$

$$[P_\mu\ , P_\nu\ ] = \quad 0 \ . \tag{2.6}$$

We may also consider the parity operator R which satisfies

$$\{R,P_i\ \} = 0, \qquad [R,\ P_o] = 0 \ ,$$

$$\{R, J_{oi}\} = 0, \qquad [R, J_{ij}] = 0. \tag{2.7}$$

The states of an irreducible unitary representation are specified by constructing a complete commuting set from these operators. We may first specify

$$\langle P_\mu^2 \rangle = m^2 \quad , \tag{2.8}$$

which is a property of the multiplet (or representation) and then

$$\langle \underline{P} \rangle = \underline{p} \quad , \tag{2.9}$$

which partially specifies a state. Given m^2 and $\underline{p}$ (equivalent to p_μ), we may still consider eigenstates of those $J_{\mu\nu}$, which leave a particular p_μ invariant (These, by definition, are the infinitesimal generators of the "little group"). In the rest frame

$$p_\mu = m_\mu = (m, \underline{0}) \quad , \tag{2.10}$$

the little group is clearly the group of rotations with generators J_{ij}. From these we may specify $\langle J^2 \rangle$ and $\langle J_3 \rangle$, which for single particle states gives the spin s and spin component s_3. We thus have particle multiplets determined by m^2, s, and states

$$|m^2, s; \underline{p}, s_3 \rangle \equiv |p, s\rangle \quad , \tag{2.11}$$

the final expressing being simply an abbreviation. We are still free to specify the parity, but we leave this for the moment.

Before considering the transformation properties of these states it is convenient to introduce the boost. This is defined to be the Lorentz transformation which transforms a rest frame state, $|m,s\rangle$, into a moving state $|p,s\rangle$;

$$|p,s\rangle = N \exp\left[-i\underline{\varepsilon}(p) \cdot \underline{K}\right]|m,s\rangle \tag{2.12}$$

where

$$P_\mu |p,s\rangle = p_\mu |p,s\rangle \quad . \tag{2.13}$$

Using only the general commutation relations (2.6) it can be shown that (2.13) is satisfied provided

$$\cosh|\varepsilon| = \frac{p_o}{m}, \ \sinh|\varepsilon| = \frac{|\underline{p}|}{m} \ , \ \hat{\underline{\varepsilon}} = \hat{\underline{p}} \quad . \tag{2.14}$$

Under a pure Lorentz transformation $\underline{\eta}$, a general state $|p,s\rangle$ is transformed to

$$U(\eta)|p,s\rangle = e^{-i\underline{\eta}\cdot\underline{K}}|p,s\rangle \tag{2.15}$$

$$\equiv e^{-i\underline{\varepsilon}'\cdot\underline{K}} \left(e^{i\underline{\varepsilon}'\cdot\underline{K}} \, e^{-i\underline{\eta}\cdot\underline{K}} \, e^{-i\underline{\varepsilon}\cdot\underline{K}}\right) |m, s\rangle \ , \tag{2.16}$$

where

$$\underline{\varepsilon}' \equiv \underline{\varepsilon}\,(p') \tag{2.17}$$

Relation (2.16) is identically equal to (2.15). By the definition of the boost (2.12), the effect of the three exponential factors in brackets is successively to take $m \to p$, $p \to p'$, $p' \to m$. Thus they take a rest frame state to a rest frame state, and together induce only a rotation of the spin in the rest frame. Thus (2.16) may be written

$$U(\eta)|p,s\rangle = e^{-i\underline{\varepsilon}'\cdot\underline{K}} \,|m,s'\rangle \langle m,s'|D(p)|m,s\rangle =$$

$$= |p',s'\rangle \langle m,s'|D(p)|m,s\rangle \tag{2.18}$$

where D(p) is the, socalled, Wigner rotation induced by the factor in brackets in (2.16).

$$<m,s'|D(p)|m,s> =$$

$$<m,s'|e^{i\underline{\varepsilon}'.\underline{K}}\ e^{-i\underline{\eta}.\underline{K}}\ e^{-i\underline{\varepsilon}.\underline{K}}|m,s> . \qquad (2.19)$$

Under translations

$$U(\ell)|p,s> = e^{iP_\mu \ell^\mu}|p,s> = e^{ip_\mu \ell^\mu}|p,s> . \qquad (2.20)$$

These transformation properties may be expressed in terms of ordinary Fock annihilation and creation operators on a non-degenerate vacuum state $>_o$,

$$a^+(p,s)>_o \equiv |p,s> \qquad (2.21)$$

where, with covariant normalization,

$$[a(p,s),\ a^+(p',s')]_\pm\ 2\pi\ \theta(p_o)\ \delta(p^2-m^2) = (2\pi)^4\ \delta^4(p-p')\delta_{ss'} . \qquad (2.22)$$

Then by (2.18) and (2.21)

$$U(\eta)\ a(p,s)\ U^{-1}(\eta) = <m,s|D^{-1}(p)|m,s'>\ a(p',s') =$$

$$= <m,s|e^{i\underline{\varepsilon}.\underline{K}}\ e^{i\underline{\eta}.\underline{K}}\ e^{-i\underline{\varepsilon}'.\underline{K}}|m,s'>\ \ a(p',s'), \qquad (2.23)$$

and by (2.20),

$$U(\ell)\ a(p,s)\ U^{-1}(\ell) = e^{-ip_\mu \ell^\mu}\ a(p,s) . \qquad (2.24)$$

The transformation properties of these operators under Lorentz transformations are complicated because the Wigner rotation, D(p), depends not only on $\underline{\eta}$, the parameter of the Lorentz transformation but also on the momentum of the particular state being transformed.

To construct operators with simple transformation properties, we must first develop explicit representations for the three factors appearing in the Wigner rotation (2.19). To do this we must introduce an auxiliary group which must contain the Lorentz boosts $\underline{K}$ and the spin operators J_{ij} as subgroups. The simplest group to take is the homogeneous Lorentz group, and we may use any representation, which includes the spin s in its decomposition. It is again simplest to use finite representations $|\alpha\rangle$, which are consequently non-unitary. It is in this auxiliary capacity that the finite non-unitary representations of the homogeneous Lorentz group enter the theory. Thus

$$\langle m,s|D(p)|m,s'\rangle = \langle m,s|\alpha\rangle \langle\alpha|e^{i\underline{\varepsilon}\cdot\underline{K}}|\beta\rangle \langle\beta|e^{i\underline{\eta}\cdot\underline{K}}|\delta\rangle \times$$

$$\times \langle\delta|e^{i\underline{\varepsilon}'\cdot\underline{K}}|\gamma\rangle \langle\gamma|m,s'\rangle \quad . \tag{2.25}$$

Now we define the auxiliary operator

$$A_\alpha(p) \equiv \langle\alpha|e^{-i\underline{\varepsilon}\cdot\underline{K}}|\beta\rangle \langle\beta|m,s\rangle\, a(p,s) \equiv u_\alpha(p)^s\, a(p,s). \tag{2.26}$$

Then by (2.23) and (2.25)

$$U(\eta)A_\alpha(p)U^{-1}(\eta) = u_\alpha(p)^s\, U(\eta)\, a(p,s)\, U^{-1}(\eta)$$

$$= (e^{i\underline{\eta}\cdot\underline{K}})_\alpha^\beta\, A_\beta(p') \quad . \tag{2.27}$$

This is just what is wanted. In the transformation of $A_\alpha(p)$, p is replaced by p', and the transformation of the index is a matrix transformation parameterised by $\underline{\eta}$ alone. The factor $u_\alpha(p)^s$ is a generalised spinor, and the relation (2.25), which defines these spinors as an explicit non-unitary representation of the boost operator is the crucial algebraic link, between operator fields and the group theoretic analysis.

The dual operator, denoted by an upper suffix, by definition transforms contravariantly, so

$$A^{\alpha}(p) \equiv a^{+}(p,s) <s,m|\beta> <\beta|e^{i\underline{\varepsilon}.\underline{K}}|\alpha> \equiv a^{+}(p,s)\ u_{s}(p)^{\alpha} \ . \tag{2.28}$$

In these non-unitary representations

$$<\beta|K|\alpha> \equiv K^{\alpha}_{\beta} \quad , \tag{2.29}$$

is no longer a hermitian matrix. (It is, in fact, anti-hermitian), so that $A^{\alpha}(p)$ is not equal to $(A_{\alpha}(p))^{+}$. The relationship between the dual and the hermitian conjugate has to be evaluated for each particular representation. For the Dirac case (see below) $A_{\alpha}(p) = \psi_{\alpha}(p)$;

$$A^{\alpha}(p) \equiv \psi^{\alpha}(p) = (\psi^{+}(p)\gamma_{o})^{\alpha} \ . \tag{2.30}$$

Under translations

$$A_{\alpha}(p) \to e^{-ip_{\mu}\ell^{\mu}}\ A_{\alpha}(p) \ , \tag{2.31}$$

$$A^{\alpha}(p) \to e^{ip_{\mu}\ell^{\mu}}\ A^{\alpha}(p) \quad . \tag{2.32}$$

With these auxiliary operators, it is a very simple matter to construct Poincaré invariants. All we have to do is to take a product of the appropriate operators and saturate the indices (in the sense that each index must appear as part of a dummy summation between upper and lower indices). Thus a simple form is

$$T = \int A^{\alpha}(p_1)\ A^{\beta}(p_2)\ \ldots\ A_{\alpha}(p_3)\ A_{\beta}(p_4)\ \times$$
$$\times \quad \delta(p_1+p_2+\ \ldots\ -p_3-p_4)\ d^4(p_1)\ \ldots\ d^4p_4 \ . \tag{2.33}$$

The saturation of upper and lower indices ensures that the factors $e^{i\underline{\eta}.\underline{K}}$ all just cancel out. The δ-function is required to remove the phase factors, which arise from translations, and shows clearly how energy-momentum conservation is related to displacement invariance. This particular form is also invariant under the purely index transformations of the auxiliary group.

Since under Lorentz transformations

$$p_\mu A_\alpha(q) \rightarrow p_\mu (e^{i\underline{\varepsilon}.\underline{K}})^\beta_\alpha \, A_\beta(p') \quad =$$

$$= \Lambda^\nu_\mu p'_\nu \, (e^{i\underline{\eta}.\underline{K}})^\beta_\alpha \, A_\beta(p') \;, \tag{2.34}$$

p_μ transforms effectively as a four-vector when it appears explicitly as a factor in such a density. If we take $|\alpha>$ to be the Dirac ((1/2, 0) + (0, 1/2)) representation for spin 1/2 particles we have

$$(K_i)^\beta_\alpha = (\sigma_{oi})^\beta_\alpha \tag{2.35}$$

and four vector and pseudo-scalar densities can be constructed from

$$\psi^\alpha(p)\; (\gamma_\mu)^\beta_\alpha \; \psi_\beta(q), \; \psi^\alpha(p)\; (\gamma_5)^\beta_\alpha \;\; \psi_\beta(q) \tag{2.36}$$

respectively. The most general Poincaré invariant involving such operators is then the form -

$$T = \int \psi^\alpha(p_1)\ldots \; t(\gamma_\mu p^\mu, \gamma_5)^{\beta\ldots}_{\alpha\ldots} \; \ldots\psi_\beta(p_2)$$

$$\delta^4(p_1 + \;\ldots\; - p_2\ldots)\; d^4p_1\ldots \; d^4(p_2) \;. \tag{2.37}$$

III. Causality and Local Fields

Notice that the problem of constructing Poincaré invariants, which may, in particular, be Poincaré invariant S-operators, has been solved without introducing anti-particles or any of the general properties associated with them such as crossing symmetry and CTP invariance. These concepts arise from the notion of local fields, to which we now turn.

For a consistent relativistic quantum theory it is essential that the mutual disturbances associated with observations do not travel faster than light. A sufficient condition for this (though not obviously a necessary one) is to require that the theory is constructed from local field operators, which satisfy causal (anti) commutation relations

$$[\psi_\alpha(x),\ \psi_\beta(y)]_\pm = 0, (x-y)^2 < 0. \tag{3.1}$$

We shall refer to this as the causality condition. It is this requirement of auxiliary local field operators in configuration space, which has all the fruitful consequences, mentioned above.

To construct $\psi_\alpha(x)$ from $A_\alpha(p)$, we take a Fourier transform, but it will prove necessary to have terms corresponding to both positive and negative frequencies. Given any representation of a Lie algebra of the form

$$[J_i, J_j] = i f_{ijk} J_k \ , \tag{3.2}$$

there is always another representation

$$J' = -J^T \ . \tag{3.3}$$

The corresponding transformations are

$$U' = \exp\,[i\varepsilon J'] = \exp[-i\varepsilon J^T] = (U^{-1})^T. \tag{3.4}$$

If the algebra is that of three dimensional rotations these are related by a matrix B. Thus

$$<m,s|D(p)|m,s'> = <m,s'|B^{-1}D^{-1}(p)\ B|m,s> . \qquad (3.5)$$

Then

$$U(\eta)\ a^{+}(p,s)\ U^{-1}(\eta) = <m,s|B^{-1}D^{-1}(p)\ B|m,s'>\ a^{+}(p',s'), \qquad (3.6)$$

and we can introduce an alternative auxiliary field

$$\tilde{A}_{\alpha}(p) \equiv <\alpha|e^{-i\underline{\varepsilon}.\underline{K}}|\beta> <\beta|B|m,s>\ a^{+}(p,s) \equiv \tilde{v}_{\alpha}(p)^{s}\ a^{+}(p,s). \qquad (3.7)$$

Under pure Lorentz transformations

$$U(\eta)\ \tilde{A}_{\alpha}(p)\ U^{-1}(\eta) = (e^{i\underline{\eta}.\underline{K}})_{\alpha}^{\beta}\ \tilde{A}_{\beta}(p), \qquad (3.8)$$

which is the same as (2.26). However under translations

$$\tilde{A}_{\alpha}(p) \to e^{ip_{\mu}\ell^{\mu}}\ \tilde{A}_{\alpha}(p) \qquad (3.9)$$

which is opposite to (2.31).

We can now define a field

$$\psi_{\alpha}(x) \equiv \int (A_{\alpha}(p)\ e^{-ipx} + \tilde{B}_{\alpha}(p)\ e^{ipx}) \times$$

$$\times\ \theta(P_{o})2\pi\ \delta(p^{2}-m^{2})\ \frac{d^{4}p}{(2\pi)^{4}} . \qquad (3.10)$$

To allow for later developments we have introduced a second particle of mass $\underline{m}$ and spin $\underline{s}$, with auxiliary operator

$$\tilde{B}_{\alpha}(p) = \tilde{v}_{\alpha}(p)^{\bar{s}}\ b^{+}(p,\bar{s}) . \qquad (3.11)$$

The variable $\bar{s}$ has the same range of values as s, but the constant spinor $<\alpha|m,\bar{s}>$ may have different properties from $<\alpha|m,s>$.

If we now construct the (anti) commutator, using (3.10) and (2.21)

$$[\psi_\alpha(x), \psi_\beta^+(y)]_\pm = \int \left(u_\alpha(p)^s (u_\beta(p)^s)^* e^{-ip(x-y)} \mp \right.$$

$$\left. \mp \tilde{v}_\alpha(p)^{\bar{s}} (\tilde{v}_\beta(p)^{\bar{s}})^* e^{ip(x-y)}\right) \theta(p_o) 2\pi \, \delta(p^2-m^2) \frac{d^4p}{(2\pi)^4} . \tag{3.12}$$

In order that the causality relation be satisfied we must have

$$[\psi_\alpha(x), \psi_\beta^+(y)]_\pm = f_\alpha^\beta(\partial) \int (e^{-ip(x-y)} - e^{ip(x-y)}) \times$$

$$\times 2\pi \, \theta(p_o) \, \delta(p^2-m^2) \frac{d^4p}{(2\pi)^4} \equiv f_\alpha^\beta(\partial) \, \Delta(x-y) . \tag{3.13}$$

The crucial feature is the minus sign appearing between the two exponentials in the penultimate expression. Since one term arises from the a-particles and the other from the b-particles (anti-particles), it is clearly necessary to include the latter to satisfy the causality condition. The causality relation further leads to the usual relation between spin and statistics, and to the relative parities of particle and anti-particle. We illustrate this by considering spin zero and spin 1/2 (in the Dirac representation).

For the case of spin zero, everything is extremely simple. The state $|p,s\rangle$ is the trivial representation of the little group, and $|\alpha\rangle$ can be taken as the trivial representation of the homogeneous Lorentz group. Thus

$$u = \tilde{v} = 1 . \tag{3.14}$$

and

$$A(p) = a(p), \; \tilde{B}(p) = b^+(p) . \tag{3.15}$$

Assuming Bose statistics (i.e. commutation relations for the operators a and b) we arrive immediately at the causal relation

$$[\phi(x), \phi(y)] = \Delta(x-y) . \tag{3.16}$$

For causality to be satisfied it is not necessary to make any statement about parity, but if we make the natural requirement that the local field has simple transformation properties

$$R\phi(x)\,R^{-1} = \pm\,\phi(x_0, -\underline{x}) \quad , \tag{3.17}$$

then spin zero particles and anti-particles must have the same intrinsic parity.

To discuss spin 1/2, we use the reducible (1/2,0) + + (0, 1/2) representation of the auxiliary group, which implies that α is four valued. If we use Dirac matrices

$$\{\gamma_\mu, \gamma_\nu\} = 2g_{\mu\nu}, \qquad \sigma_{\mu\nu} = \frac{i}{2}\,[\gamma_\mu, \gamma_\nu] \tag{3.18}$$

and

$$\gamma_5 = -\,i\,\gamma_0\gamma_1\gamma_2\gamma_3 \tag{3.19}$$

then, from (2.26) and (3.7)

$$\begin{aligned} u_\alpha(p)^s &= (e^{-i\varepsilon_i\sigma_{oi}/2})^\delta_\alpha\,\langle\delta|m,s\rangle \\ u_\alpha(p)^{s^*} &= \langle m,s|\delta\rangle\,(e^{-i\varepsilon_i\sigma_{oi}/2})^\alpha_\delta \\ \tilde{v}_\alpha(p)^{\bar{s}} &= (e^{-i\varepsilon_i\sigma_{oi}/2})^\delta_\alpha\,\langle\delta|B|m,\bar{s}\rangle \\ \tilde{v}_\alpha(p)^{\bar{s}^*} &= \langle m,\bar{s}|B^{-1}|\delta\rangle\,(e^{-i\varepsilon_i\sigma_{oi}/2})^\alpha_\delta \quad . \end{aligned} \tag{3.20}$$

Also (2.14) and (3.18) give

$$e^{-i\varepsilon_i\sigma_{oi}} = \frac{(\gamma p)}{m}\,\gamma_0 \quad , \tag{3.21}$$

and the matrix γ_0 satisfies the relations (2.7) and represents the parity operator. In order to satisfy causality, it is necessary to assume anti-commutation relations. The evaluation of the spin sums uu^* and $\tilde{v}\tilde{v}^*$, which appear in (3.12), depends on what conditions are placed on the constant spinors $\langle\alpha|m,s\rangle$ and $\langle\alpha|m,\bar{s}\rangle$.

If there are as many particles $|m,s\rangle$ as there are components $A_\alpha(o)$, then

$$\langle\alpha|m,s\rangle = \delta^s_\alpha \quad , \tag{3.22}$$

and we have parity doubling

$$\gamma_o|m,s_{1,2}\rangle = +\,|m,s_{1,2}\rangle \quad , \tag{3.23}$$

$$\gamma_o|m,s_{3,4}\rangle = -\,|m,s_{3,4}\rangle \quad ,$$

and similarly for the anti-particles. Then using (3.20) and (3.21) to evaluate (3.12)

$$\{\psi_\alpha(x),\ \psi^+_\beta(y)\} = (i\gamma\partial\gamma_o)^\beta_\alpha\ \Delta(x-y) \quad . \tag{3.25}$$

Apart from the Klein-Gordon equation there is no restriction on $A_\alpha(p)$ or $\tilde{B}_\alpha(p)$ and thus no "equation of motion". As far as causality is concerned there is no need to introduce anti-particles. We can even identify the particles with the anti-particles and take $b(p,\bar{s}) \equiv a(p,s)$.

The conventional theory is obtained by restricting the parity of the particles by the relation

$$(\gamma_o)^\beta_\alpha\ \langle\beta|m,s\rangle = \langle\alpha|\ m,s\rangle \ . \tag{3.26}$$

One finds that it is only possible to satisfy causality if the anti-particles are taken to have the opposite parity

$$(\gamma_o)^\beta_\alpha\ \langle\beta|m,\bar{s}\rangle = -\ \langle\alpha|m,\bar{s}\rangle \ . \tag{3.27}$$

This is an important consequence of the causality requirement. One now has for the sums appearing in (3.12), using (3.21),

$$(uu^*) = e^{-i\varepsilon_i\sigma_{oi}/2} \left(\frac{1+\gamma_o}{2} \right) e^{-i\varepsilon_i\sigma_{oi}/2} = \frac{(\gamma p + m)}{2} \gamma_o \quad (3.28)$$

$$(\tilde{v}\tilde{v}^*) = e^{-i\varepsilon_i\sigma_{oi}/2} \left(\frac{1-\gamma_o}{2} \right) e^{-i\varepsilon_i\sigma_{oi}/2} = \frac{(\gamma p - m)}{2} \gamma_o \quad (3.29)$$

giving the causal commutator

$$\{\psi_\alpha(x), \psi^+_\beta(y)\} = ((\gamma\partial + m)\gamma_o)^\beta_\alpha \, \Delta(x-y), \quad (3.30)$$

and the equation for the field

$$(\gamma\partial + m)\, \psi(x) = 0 \quad . \quad (3.31)$$

There are a number of features of this analysis which illustrate quite general points. To satisfy causality we have found it necessary to introduce anti-particles and to assume commutation relations (Boson statistics) for integer spin and anti-commutation relations (Fermi statistics) for half integer spin. The parity of particle and anti-particle must be chosen the same for Bosons, but opposite for Fermions. The parity type restrictions (3.26) and (3.27) reduce the number of independent components of the field $A_\alpha(p)$ to the number of independent operators a(p,s). Any "equations of motions" such as (3.31) (i.e. other than the Klein-Gordon equation) are just the boosts of these restrictions.

Finally, even if the auxiliary group is restricted to be the homogeneous Lorentz group, there is considerable freedom of choice for the representation $|\alpha\rangle$ for any given spin s. It is this freedom that accounts for the various field theories for higher spin - Bhabha, Rarita-Schwinger, Proca, Kemmer, etc.

IV. Higher Symmetries

Local fields which combine particles of different spin may

easily be constructed by taking the auxiliary group to be larger than the homogeneous Lorentz group. This we may take U(2,2) - or $\tilde{U}(4)$. The Dirac representation may be regarded as the basic representation of this group, with generators in this representation

$$\Gamma_r = 1,\ i\gamma_5,\ \gamma_\mu,\ \gamma_5\gamma_\mu,\ \sigma_{\nu\nu} \tag{4.1}$$

such that

$$\gamma_o\ \Gamma_r^+\ \gamma_o = \Gamma_r\ .$$

The formalism of spin 1/2 is then as in the previous section. But we may also define a spin 0-1 combination with auxiliary operator

$$A_\alpha^\beta(p) = (u(p)^s)_\alpha^\beta\ a(p,s)\quad , \tag{4.2}$$

where

$$(u(p)^s)_\alpha^\beta = \langle{}_\alpha^\beta|e^{-i\underline{\varepsilon}\cdot\underline{K}}|{}_\pi^\rho\rangle\ \langle{}_\pi^\rho|m,s\rangle \quad =$$
$$= (e^{-i\varepsilon_i\sigma_{oi}/2})_\alpha^\pi\ \langle{}_\pi^\rho|m,s\rangle\ (e^{i\varepsilon_i\sigma_{oi}/2})_\rho^\beta\ . \tag{4.3}$$

The upper suffices in the kets refer to the dual - or minus transpose - representation (See 3.3). Similarly

$$\tilde{B}_\alpha^\beta(p) = \langle{}_\alpha^\beta|e^{-i\underline{\varepsilon}\cdot\underline{K}}|{}_\pi^\rho\rangle\ \langle{}_\pi^\rho|B|m,\bar{s}\rangle\ b^+(p,\bar{s}) \equiv$$
$$\equiv (\tilde{v}(p)^{\bar{s}})_\alpha^\beta\ b^+(p,s)\ , \tag{4.4}$$

and from these we can construct a local field $\phi_\alpha^\beta(x)$.

Without restricting the spin we can impose a variety of conditions on the constant spinor $\langle{}_\alpha^\beta|m,s\rangle$. The most restrictive specify the parities of the "quark" content, i.e. for negative parity of the whole multiplet, we can have

$$(\gamma_0)^{\pi}_{\alpha} <^{\beta}_{\pi}|m,s> = <^{\beta}_{\alpha}|m,s> \quad (4.5)$$

$$<^{\rho}_{\alpha}|m,s> (\gamma_0)^{\beta}_{\rho} = -<^{\beta}_{\alpha}|m,s> \quad . \quad (4.6)$$

To satisfy causality the "quark" parities of $|m,\overline{s}>$ must be taken opposite to the above. Then there is a causal commutation relation and the field satisfies the equations of motion which are the boost of these restrictions, namely,

$$(i\gamma\partial-m)^{\pi}_{\alpha} \phi^{\beta}_{\pi}(x) = 0 \quad (4.7)$$

$$\phi^{\rho}_{\alpha}(x) (i\gamma\partial+m)^{\beta}_{\rho} = 0 \ . \quad (4.8)$$

As far as spin and parity are concerned these are just the spin 0-1 fields proposed by Salam, Delbourgo and Strathdee [1]. They are those fields which one constructs from a single quark and its anti-quark.

These extended multiplets must belong to some extended little group. This can be determined by considering those transformations of the auxiliary group which leave invariant the rest frame projection of the free particle energy operator.

$$ma^{+}(o,s)\ a(o,s) = m(A_{\alpha}(o))^{+} <\alpha|m,s> <m,s|\beta> A_{\beta}(o) \ .$$

If the restrictions on $<\alpha|m,s>$ are always of the parity type considered above $<\alpha|m,s> <m,s|\beta>$ only involved γ_0 and the little group generators are those Γ_r (4.1) which commute with γ_0, namely,

$$1,\gamma_0,\sigma_i,\gamma_5\gamma_i,\gamma_0\sigma_i \quad . \quad (4.10)$$

These generate $U(2) \times U(2)$.

It is very easy to extend these fields to include the outer product with any internal symmetry S (such as SU(3)). One simply constructs fields

$$\psi_{\alpha,s}(x) \tag{4.11}$$

where α specifies a representation of the space-time auxiliary group - e.g. homogeneous Lorentz $\mathcal{L}$ - and s defines a representation of S. The auxiliary group is then $S \times \mathcal{L}$; the little group is $S \times U(2)$. Poincaré and internal group invariant densities are constructed by saturating independently suffices referring to basic representations in the two groups.

Since the auxiliary group and the internal symmetry group S are both by definition groups of pure index transformation, it is very simple, in the construction of local fields, to extend the auxiliary group from the outer product of the two groups to some wider group. Thus if (4.11) is the basis representation of $U(2,2) \times SU(3)$, so that $\alpha = 1,\ldots,4$; $s = 1,2,3$, we can consider a local field

$$\psi_X \equiv \psi_{\alpha,s} \qquad X = 1,\ldots,12 \tag{4.12}$$

which transforms as the basic representation of the extended group $U(6,6)$ - or $\tilde{U}(12)$ - such fields correspond to particle multiplets of the extended little group $U(6) \times U(6)$ and thus reproduce the observed correlation between spin and SU(3) properties of the particles. It was the achievement of $\tilde{U}(12)$ theorists to show how to construct, and decompose, such fields. Any theory based on them certainly has all the well known advantages of a local field theory - antiparticles, crossing, CTP invariance - but it is very importand to notice that it is the auxiliary group, not the Poincaré group, which has been extended to combine in an intimate way with the internal symmetry. There has been no extension of the Poincaré group.

The question arises what use can be made of these fields to built further restrictions into the theory. The most natural requirement is that S-matrix elements should be index U(6,6) invariant. We are thus restricted to forms such as (compare with (2.37))

$$T = \int \psi^X(x)\ldots\psi^Y(y)\ldots\psi_X(w)\ldots\psi_Y(z)\ d^4x\ldots d^4y\ldots d^4w\ldots d^4z. \tag{4.13}$$

which is consistent with $P \times SU(3)$ invariance, but consideralby more restrictive leading to $U(6) \times U(6)$ invariant interactions in the static limit.

However, as was stated in the original papers, it is not possible to construct a theory which leads to such elements, because this will inevitably involve spin sums and propagators which will introduce factors represented by t_α^β in (2.37) which are excluded in (4.13).

Even if we do not require a 'theory' but impose the form (4.13) as a prescription for T matrix elements we are in conflict with unitarity. This is because

$$\mathrm{Im}\ T = T\ \rho\ T^+ \quad . \tag{4.14}$$

If T and T^+ have the form (4.13), the spin sums involved in the factor ρ will introduce into Im T factors which are excluded by the rule.

It would appear that the local U(6,6) fields can at best be used to construct vertex parts (which are not in conflict with physical unitarity) and that all further dynamical considerations must be incorporated through techniques such as dispersion relations and peripheral model, in which the normal input is the single particle amplitudes and the three particle couplings.

Acknowledgements

The author would like to thank Professor Urban for the hospitality extended to him at the Schladming School. A more detailed account of the above work will be published elsewhere in collaboration with Professor G. Feldman.

References

1. A. Salam, R. Delbourgo, J. Strathdee, Proc. Roy. Soc. A.284, 146 (1965); B. Sakita and K. Wali, Phys. Rev. 139, B1355 (1965).
2. S. Weinberg, Phys. Rev. 133, B1318 (1964); 139, B597 (1965).
3. F. Gürsey and L. A. Radicati, Phys. Rev. Letters 13, 173 (1964); B. Sakita, Phys. Rev. Letters 13, 643 (1964); Phys. Rev. 136, B1756 (1964).

WHAT IF ANYTHING IS SU(6) SYMMETRY?†

By

P. STICHEL

Physikalisches Staatsinstitut der Universität Hamburg
and Deutsches Elektronen-Synchrotron (DESY),Hamburg

In this seminar I want to talk about the violation of some basic physical concepts caused by applying formal group theoretical methods to elementary particle physics. What I want to present is a critical review of the so called SU(6) symmetry. I will neither discuss group theoretical details nor numbers and experiments. What I want to talk about are the basic ideas and the main defects of the different approaches within the SU(6) business. In particular, I will discuss the connections between the different approaches which are of a logical, quasilogical or nonlogical nature. As we have learned in Prof. Källén's seminar talk, such connections may best be illustrated by a diagram consisting of boxes and lines. So let us start with the SU(6) box

SU(6)

Now we will look at this box by using a microscope in order to discover its fine structure. What we will see is illustrated by the following diagram.

You see different boxes connected by unbroken, broken or forbidden lines expressing the logical, quasilogical or nonlogical nature of the considered connection.

† Lecture given at the V. Internationalen Universitätswochen für Kernphysik, Schladming, 24 February - 9 March 1966.

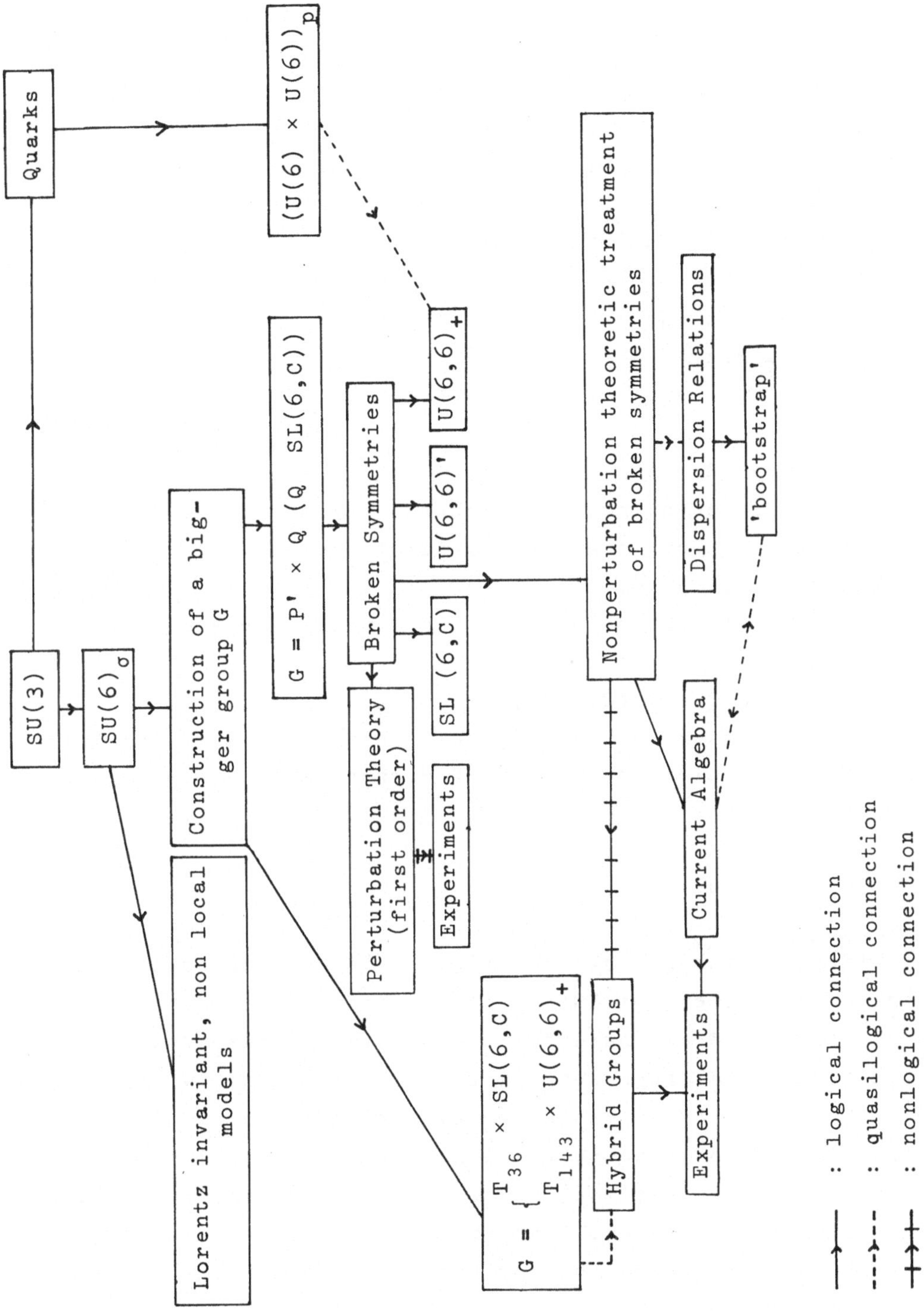
Quarks
(U(6) × U(6))p
SU(3)
SU(6)σ
Construction of a bigger group G
G = P' × Q (Q SL(6,C))
Broken Symmetries
U(6,6)+
U(6,6)'
SL (6,C)
Perturbation Theory (first order)
Experiments
Nonperturbation theoretic treatment of broken symmetries
Dispersion Relations
'bootstrap'
Current Algebra
Lorentz invariant, non local models
G = { T36 × SL(6,C) T143 × U(6,6)+
Hybrid Groups
Experiments
: logical connection
: quasilogical connection
: nonlogical connection

Before going into the details let me make two statements which will be the main conclusions of this talk:

Statement 1:

There does not exist any 'SU(6)-symmetric' theory of strong interaction which fulfills all of the following requirements

a. Relativistic invariance
b. Locality (Crossing symmetry),
c. Completeness of the physical states (Unitarity of the S-matrix)
c. $S \neq 1$

Statement 2:

The approach of current algebras may be considered as a dynamical concept within the general framework of axiomatic quantum field theory.

Now I will discuss some of the details contained in the diagram:

I. Static SU(6) ($SU(6)_\sigma$):

Roughly described this is an extension of the nonrelativistic supermultiplet theory of Wigner (applied in nuclear physics) by the substitution

$$SU(2)_I \rightarrow SU(3)$$

(Strictly speaking this description is not correct, because the SU(4) content of $SU(6)_\sigma$ is different from $SU(4)_{Wigner}$. This follows from the different physical interpretation of the fundamental representation of SU(4) in the two cases.)

Therefore, $SU(6)_\sigma$ is a nonrelativistic theory of spin independence of the interaction between elementary particles. I remind you of the meaning of spin independence: Consider for example NN forces, then spin independence means that you may have $\vec{\sigma}_1 \cdot \vec{\sigma}_2$ terms, but no tensor forces or spin orbit couplings.

Defects of SU(6) :

a. Nonrelativistic theory

b. Important p-wave couplings are forbidden:

$N \not\rightarrow N + \pi, \quad N^* \not\rightarrow N + \pi, \ldots$

Consider the NNπ vertex. In the nonrelativistic limit it is proportional to $X_2^+(\vec{\sigma} \cdot \vec{q})X_1$, i.e. it contains explicitly a spin-orbit coupling which fails to be SU(2) invariant.

II. Relativistic invariant, nonlocal models (Radicati, Gürsey [1], Schroer [2])

This is an immediate extension of the concept of spin independence to relativistic theories. This is a possible marriage between SU(6) and the Poincaré group P without requiring the construction of a larger group G. The SU(2) content of SU(6) in this case is Wigner's little group for time-like p_μ with $p^2 > 0$. At first this little group is only defined in the subspace of fixed p_μ. But what we want is an operator representation within the whole Hilbert space H. It turns out that such an operator representation may be constructed only within a subspace for a fixed particle number n H_n. Therefore we need a theory in which the particle number is conserved. This is necessarily a nonlocal theory: Consider a local field operator A(x) acting on the vacuum $A(x)|0\rangle$. In nontrivial theories this state always contains components with arbitrary particle numbers!

In the following consideration I restrict myself to the SU(2) spin-content of SU(6)

H_1 :

The one-particle subspace contains particles of mass M. Then the operators of our SU(2) are defined as follows:

$$S_i^{(1)} = \frac{1}{M} a_i^{\mu} (\Lambda_P^{-1}) \Gamma_\mu^{(1)}$$

where P is an operator. If P is a c number, then Λ_P is a special Lorentztransformation which transforms p rotational-free from the rest system into the moving system. It is identical with the 'boost' transformation introduced by Prof. Matthews in his lecture.

$\Gamma_\mu^{(1)}$ is the representation of the Bargman-Wigner spin operator in H_1, i.e.

$$\Gamma_\mu^{(1)} \equiv \frac{1}{2i} \varepsilon_{\mu\nu\rho\sigma} M_{(1)}^{\nu\rho} P_{(1)}^{\sigma}$$

Then we obtain explicitly for the $S_i^{(1)}$

$$S_i^{(1)} = \frac{1}{M} \left(\Gamma_i^{(1)} - \frac{i\,\Gamma_o^{(1)} P_i^{(1)}}{M + P_o^{(1)}} \right)$$

The operators $M_{\mu\nu}^{(1)}$, $P^{(1)}$, $S_i^{(1)}$ do not constitute a closed commutator algebra. For this consider the following counter example:

$$\left[M_{oj}^{(1)} , S_l^{(1)} \right] = i \frac{\delta_{jl} \vec{S}^{(1)} \cdot \vec{P}^{(1)} - S_j^{(1)} P_l^{(1)}}{M + P_o^{(1)}}$$

It may be shown that the smallest closed commutator algebra containing $M_{\mu\nu}^{(1)}$, $P^{(1)}$ and $S_i^{(1)}$ consists of an infinite number of linearly independent elements.

H_n:

I will not give an explicit construction of the $\vec{S}$ for $n \geq 2$. It should only be mentioned that the $\vec{S}_{(n)}$ are in general functions of the representation of the $M_{\mu\nu}$ and P in all n one-particle subspaces, i.e.

$$\vec{S}_{(n)} = \vec{S}_{(n)} (M_{\mu\nu}^{(1)}, \ldots M_{\mu\nu}^{(n)} , P_\sigma^{(1)}, \ldots P_\sigma^{(n)})$$

This means: In each H_n we have a different SU(6) group! The relativistic generalization of spin-independence of the interaction may then be formulated as usual

$$[H^{(n)} , \vec{S}^{(n)}] = 0$$

where $H^{(n)}$ is the Hamilton-operator acting in H_n.

It may turn out that such nonlocal models will have some physical interest if one looks for a relativistic generalization of the static quark model discussed in the lectures by Prof. Thirring.

III. Construction of a larger group G

In the following we denote by $\overline{P}$ the covering group of the connected part of the Poincaré group. We have now to discuss two cases.

IIIa. $\overline{P} \subset G$.

From a naive point of view one would look for a group G with the following properties:

1. $\overline{P} \subset G$
2. T is an invariant subgroup of G (T = translations)
3. The action of G on T preserves the Minkowski metric (i.e., if $g \in G$, then $gP^{\mu}P_{\mu}g^{-1} = P^{\mu}P_{\mu}$).

This is an extension of requirement 2.

4. The little group in G/T for a time-like translation is SU(6) (this means: we can classify the states for a given p according to SU(6)).

Michel and other people have shown that such a group does not exist, because according to requirements 1,2, and 3, the little group cannot be a simple one.

So let us drop requirement 3. (This seems to be the only possibility.) Therefore the number of momenta has to be enlarged. According to Rühl the smallest group with the required properties is

$$G = T_{36} \times SL(6,C)$$

Because of T_{36} we are in a larger Hilbert space, and this is the crucial point: the physical Hilbertspace is not an invariant subspace with respect to G. Therefore S, i.e. the operator which transforms the physical outgoing states into the physical incoming states, is not a unitary operator.

The restriction to so-called hybrid groups for colinear or coplanar processes does not prevent unitarity violation. This has been explicitly demonstrated for qq and $q\bar{q}$ scattering by Alles and Amati. It turns out that unitarity and crossing symmetry only allow the trivial S-matrix in this case.

IIIb. Group extension.

$\bar{P} = G/Q$

Q = internal symmetry group which is an invariant subgroup of G. For central extensions it has been shown (Michel et al.) that

$$G = \bar{P} \times Q$$

If $Q \supset SU(6)$ then $Q \supset SL(6,C)$

This $\bar{P}$ must not necessarily be the physical Poincaré group, because group theory only tells us that this $\bar{P}$ is isomorphic to the physical $\bar{P}$, so let us write

$$G = \bar{P}' \times Q \quad .$$

To interpret this G as a relativistic generalization of the concept of spin independence, one may argue, according to Michel:

Suppose there exists a unitary representation of G in the physical Hilbert space with the following prescription:

Consider any many-particle state in momentum space

$|p_1 \ldots p_n \, , \, \eta_1 \ldots \eta_n\rangle$

where the η_i are the internal degrees of freedom (spin, unitary spin), whereas $\overline{P}'$ only acts on the momenta p_1 and Q acts only on the η_i.

Now what is the physical Poincaré group $\overline{P}$? $\overline{P}$ consists of $\overline{P}'$ and the SL(2,C) content of SL(6,C).

IV. Broken Symmetries

It turns out that the mentioned prescription given for G is in contradiction to any particle interpretation of field theory. This means: G has no unitary representation in the physical Hilbert space.

To show this let us first consider old fashioned Lagrangian field theory:

$$L = L_o + L_{int}$$

If the theory contains spin 1/2 particles, then L_o has the form

$$L_o = \overline{\psi}(i \, \gamma\partial + M)\psi + \ldots$$

This L_o cannot be invariant with respect to $\overline{P}'$ because of the presence of γ_μ, which is not transformed by $\overline{P}'$. But the L_{int} may be formally invariant with respect to G.

Let us consider some examples within the Quark-model (restricting ourselves to four-fermion interactions without derivatives):

$$Q = SL(6,C) \quad : \quad L_{int} = g_1 \, (\overline{\psi}\psi)^2 + g_2 \, (\overline{\psi}\gamma_5\psi)$$

$$Q = U(6,6)_+ \quad : \quad L_{int} = g \, (\overline{\psi}\psi)^2$$

$$Q = U(6,6)' \quad : \quad L_{int} = g' \, (\overline{\psi}\gamma_5\psi)^2$$

It turns out, that this formal invariance of the interaction Lagrangian is only a game with the insertion of certain functions depending on γ-matrices and Gell-Mann's λ_i. But this game has nothing to do with a symmetry about which we speak only if there exists an unitary representation of the underlying symmetry group in the physical Hilbert space.

This game with γ-matrices and λ_i leads to relations between certain unrenormalized coupling constants. Some theoretists like to compare the predictions for the ratios of these unrenormalized coupling constants with the ratios of renormalized coupling constants as measured by experimental physicists. In some cases one gets a rather good agreement! Now we know that in general first order perturbation theory does not work for strong interaction physics. This means, if we get agreement with experiments by this game we have a situation which is by no means understood. It may be that it is possible to understand such partial successes by careful study of broken symmetries within the framework of current algebras. I want to make one further remark on this game: It turns out that the calculation of effective vertices in first order perturbation theory as mentioned violates the original idea of spin independence, because of the spin-orbit coupling terms contained in the free field (Wigner-Bargman) equations.

V. Current Algebra

We have seen that the Lagrangian formulation of broken symmetries may fulfill all our requirements, with the exception of the symmetry requirement itself. But nevertheless we have certain relations between unrenormalized coupling constants.

Therefore the question arises: Is there any nonperturbation theoretic formulation of broken symmetries containing the mentioned ratios between unrenormalized coupling constants in a certain approximation? Indeed, such a scheme exists, namely the approach of equal-time current commutators (for the details I remind you of the lectures given by Prof. Moffat). It turns out that one obtains the same ratios between coupling constants as in first order perturbation theory, if one restricts oneself to the one-particle contributions within the dispersion theoretical treatment of current commutation relations according to Fubini, Furlan and Rossetti. The many-particle contributions if they are physically interpretable are a numerical measure of the renormalization effects (as an example I refer to the Adler-Weisberger relation). In cases where the many-particle contributions are not physically interpretable two different approximation schemes may be discussed:

1. The approximation of the two (resp. three)-particle contributions by the relevant resonances*

2. The extension of the dispersion-theoretic one-particle contribution to a local one-particle contribution, whereby the latter contains the full contribution of the one particle intermediate state in the commutator. In such an approach the full vertex structure of the one-particle matrix elements of currents is taken into account rather than just the coupling constants**.

Finally, I only want to mention that all problems connected with the dependence of certain results on current commutators on the frame of reference may be solved by considering only local approximation on local decompositions respectively of matrix elements of current commutators [3]

VI. Miscellaneous Remarks

There are some boxes in the diagram which I have not yet discussed.

---

* I am grateful to the audience for mentioning this possibility.

**Some work along this line is in progress.

Dispersion relations:

In this context the usual N/D job is meant, whereby as an input certain one-particle exchange diagrams are taken as driving forces. The ratios of coupling constants at the vertices of the OPE diagram may be taken from the first-order perturbation results of broken symmetries.

'Bootstrap':

If one does not like bare coupling constants to start with one may try to determine renormalized coupling constants, if possible, self-consistently. So, the bootstrap we are speaking about in this context is the bootstrap in a definite technical sense and not the general bootstrap philosophy (which seems to be empty from the point of view of general local field theory!)

References

1. Proceedings of the seminar on high-energy physics and elementary particle physics, Trieste, 3 May-30 June 1965, and all the literature quoted there.
2. Seminar on high-energy physics and elementary particle physics, Trieste, 3 May - 30 June 1965, paper SMR 2/72 (presented by B. Schroer).
3. B. Schroer and P. Stichel: 'Current commutation relations in the framework of general quantum field theory' (Hamburg preprint).

SPIN AND LORENTZ EXTENSIONS OF INTRINSIC SYMMETRY GROUPS†*

By

H. D. DOEBNER and G. C. HEGERFELDT

Institut für Theoretische Physik (I)
Marburg

Introduction

The well known problem of combining intrinsic symmetry and space time groups in a non-trivial way, i.e. not in the form of the direct product, has led to two negative theorems by Michel [1], McGlinn [2], and others, stating that under quite general assumptions, without enlarging the group one always gets the direct product of the respective Lie algebras. To obtain new results an enlargement of the group seems to be necessary. For a special case Gürsey and Radicati [3] proposed a successful model which combines SU(3) and SU(2), for particles at rest the spin group, as commuting subgroups in the larger group SU(6). Relativistic extensions of this model have been investigated by Salam [4], Fulton and Wess [5] and others.

The idea of Gürsey and Radicati is not such that it could easily be carried over to other types of intrinsic symmetry groups, as for instance to the repeatedly discussed G_2, Sp(6) and B_3.

In this lecture we give, starting from three heuristic physically plausible postulates, a purely group theoretic method

†Lecture given at the V. Internationalen Universitätswochen für Kernphysik, Schladming, 24 February - 9 March 1966.
* Supported in part by the Bundesministerium for wissenschaftliche Forschung, Bad Godesberg, Germany.

to construct a unique spin and Lorentz extension for a large class of intrinsic symmetry groups without using a model Lagrangian or other field theoretic techniques. This method was first applied to the spin extension of G_2 [6]. The complete evaluation of the postulates and the solution of mathematical problems related to them were facilitated by results of Dynkin [7] and Malcev [8].

1. Principles

The embedding of an intrinsic symmetry group G together with the Lorentz group L in a Lorentz extension $\overline{G}$ of G in such a way that G and $\overline{L}$, the covering of the Lorentz group, are commuting subgroups of $\overline{G}$, possesses, without further restrictions, infinitely many solutions. In order to formulate physically plausible requirements and to construct $\overline{G}$ we shall consider a certain class of intrinsic symmetry groups and proceed in two steps:

I. Embedding of G and SU(2) in a larger group $\tilde{G}$ (spin extension of G)

II. Construction of a group $\overline{G}$ which contains $\tilde{G}$ and $\overline{L}$.

We first concentrate on step I, proposing three heuristic principles as postulates and using them to construct $\tilde{G}$. We shall confine ourselves to connected Lie groups.

Let the compact Lie group G be an intrinsic symmetry group which is assumed to commute with the Lorentz group L. Within a G-multiplet, all particles have the same spin so that one has not only an irreducible representation of G but also one of SU(2), the covering group of the rotation group, for particles at rest characterising the spin. Therefore, the physically realised multiplets should correspond to a subset of the set of irreducible representations of the direct product $G \times SU(2)$.

Many of the symmetry groups currently being discussed possess the property that all representations (d,s) of $G \times SU(2)$ which are physically realised by multiplets with spin s, can be obtained from Kronecker products of the form

$$(d,\tfrac{1}{2}) \times \ldots \times (d,\tfrac{1}{2}) \times (\bar{d},\tfrac{1}{2}) \times \ldots \times (\bar{d},\tfrac{1}{2}) \qquad (1)$$

where d is an irreducible representation of G with the dimension as low as possible and $\bar{d}$ is the complex conjugate representation. When this holds we say that G × SU(2), or G, admits of a generalised quark model. Associating with d formal spin 1/2 particles ("quarks") one can interpret (1) in such a way that the particles of the respective higher multiplets are bound states of quarks and antiquarks, as for example in SU(3) with d as the 3-dimensional fundamental representation of SU(3) [9].

We shall now restrict ourselves to such intrinsic symmetry groups. The dimension of d will be denoted by N, and the image of G by G_Q, i.e. $d(G) = G_Q$. The representation $(d,\frac{1}{2})$ then has the dimension 2N.

To get a first restriction on a spin extension $\tilde{G}$ of G, we demand that $\tilde{G}$ is compact and that the formal spin $\frac{1}{2}$ particles of the quark representation $\tilde{d}$ belong to a representation $\tilde{d}$ of $\tilde{G}$ which does not contain further particles, i.e. d must be 2 N-dimensional. In addition, $G_Q \times SU(2)$ should be a proper subgroup of $\tilde{d}(G)$ since otherwise one would have a spin extension in the form of the direct product, so

$$\tilde{d}(\tilde{G}) \supset G_Q \dot{\times} SU(2) \qquad (2)$$

The compactness assumption ensures the existence of finite dimensional unitary representations of $\tilde{G}$ and will later allow the interpretation of $\tilde{G}$ as little group of a Lorentz extension. Postulate (3) guarantees that the irreducible representations of G contain all physically relevant representations of G × SU(3). Indeed, according to (1) the latter are obtained from Kronecker products of $G_Q \times SU(2)$ and $\bar{G}_Q \times SU(2)$. Therefore one only needs to reduce

$$\tilde{d} \times \ldots \times \tilde{d} \times \bar{\tilde{d}} \times \ldots \times \bar{\tilde{d}} \tag{3}$$

to irreducible representations of $\tilde{G}$ and then restrict these to $G_Q \times SU(2)$. With G not admitting of a quark model this need not be so. Take for instance the isospin group SU(2), then SU(4) contains SU(2) × SU(2). One gets all irreducible representations of SU(4) by reduction of Kronecker products of the form SU(4) ×...× SU(4), the SU(2) × SU(2) content of which is obtained from (SU(2) × SU(2)) ×...× (SU(2) × SU(2)), i.e. from

$$(\tfrac{1}{2}, \tfrac{1}{2}) \times \ldots \times (\tfrac{1}{2}, \tfrac{1}{2}) \quad . \tag{4}$$

In (4) there are only representations with spin and isospin being either both integer or half-integer. Hence physical representations like $(\frac{1}{2}, 0)$ are not contained in representations of SU(4).

In view of (3) we can identify $\tilde{G}$ with $\tilde{d}(\tilde{G})$ and take $\tilde{G}$ as a linear or matrix group.

The Lie algebra G° of a compact Lie group G has the form

$$G^{\circ} = H_1^{\circ} + \ldots + H_n^{\circ} + K_1^{\circ} + \ldots + K_m^{\circ} \tag{5}$$

(H_i° simple compact and K_j° one-dimensional Lie algebras)

With G as an intrinsic symmetry group, a K_j° leads to an additive quantum number which is completely independent from the rest of the group, as e.g. the baryon number B from SU(3). In a spin extension, this characteristic independence should be retained so that for the following construction these independent quantum numbers can be neglected and the K_j° in (5) be dropped. Then G° and G become semisimple, and $G_Q = d(G)$ consists of unimodular matrices.

The requirement (2) for $\tilde{G}$ can be realised in many different ways, for $G \simeq G_Q = SU(3)$ one has for instance

$$\tilde{G} = U(3) \times SU(2) \tag{6}$$

as a possible solution. One would prefer, however, to couple G and SU(2) in such a way that $\tilde{G}$ cannot be decomposed into two factors one of which contains G_Q while the other contains SU(2). That is, if $H_1 \supset G_Q$, $H_2 \supset SU(2)$, then

$$\tilde{G} \neq H_1 \times H_2 \qquad (7)$$

Whenever this does not hold, as for instance in (6), we call $\tilde{G}$ a trivial spin extension. Condition (7) is easily satisfied by demanding that $G_Q \times SU(2)$ is contained in a simple subgroup of $\tilde{G}$. The necessity of this requirement can be proved in the case of a simple symmetry group G:

Theorem: Let the compact Lie group $\tilde{G}$ satisfy (2) and be a non-trivial spin extension of the simple Lie group G with the quark representation G_Q. Then $G_Q \times SU(2)$ is contained in a simple subgroup of $\tilde{G} = \tilde{d}(\tilde{G})$.

For a proof cf. [10]. Obviously one should not choose $\tilde{G}$ too large. This means that the number of generators and the rank of $\tilde{G}$ should be as small as possible. If one takes, however, for $\tilde{G}$ a group which contains G × SU(2) and which possesses a minimal number of generators, one collides with the quark model. SU(3) × SU(2), for instance, is contained in SU(6) and also in SU(5), in a different form though. But since the quark model guarantees the correct G × SU(2) - content of the representations of $\tilde{G}$ one should not abandon it but take the minimality requirement into account by choosing the spin extension obtained from a quark model as small as possible. So in theorem 1 one should take the simple subgroup which also contains $G_Q \times SU(2)$ instead of the possibly larger group $\tilde{G}$ as spin extension. We see that in order to fulfill (7), i.e. to exclude trivial spin extensions, one may as well require $\tilde{G}$ to be simple.

The conditions on G and the above requirements on $\tilde{G}$ may be summarized as follows:

Let the compact Lie group G be an intrinsic symmetry group which admits of a quark model (cf. (2)) with the irreducible group $d(G) = G_Q$ of $N\times N$-matrices (According to the remarks to equation (5) we can take G to be semisimple and G_Q to be unimodular).

Postulates for a spin extension $\tilde{G}$ of G:

a) $\tilde{G}$ is simple and compact.

b) $\tilde{G}$ is a group of $2N\times 2N$-matrices which contains $G_Q \times SU(2)$.

c) $\tilde{G}$ is minimal among the groups satisfying a), b).

We remark $\tilde{G}$ is called minimal with respect to a) b) if it has no proper subgroup satisfying a) b).

2. The Main Theorem

At first glance, the postulates a),b),c) might seem to allow for many solutions for a given G_Q and one could be tempted to look for further restrictions. This, however, is not necessary owing to the following:

Main theorem:

a) Let the irreducible group G_Q of unimodular $N\times N$-matrices be different from the lowest spinor representation of $B_n \simeq O(2n+1)$, with $n \geqslant 3$, i.e. not given by the DYNKIN diagram [11]

 $\simeq B_n \quad , \; n \geqslant 3 \quad ,$ (8)

Then the spin extension $\tilde{G}$ is uniquely determined by the postulates a) b) c), specifically one has for $N \geqslant 3$:

(9)

$$\tilde{G} = \begin{cases} SU(2N) \text{ if } G_Q \text{ has no bilinear invariant} \\ Sp(2N) \text{ if } G_Q \text{ has a symmetric bilinear inv.} \\ O(2N) \text{ if } G_Q \text{ has an antisymm. bilinear inv.} \end{cases}$$

For $N = 2$ one has

$$\tilde{G} = SU(4) \quad . \tag{10}$$

b) Let G_Q be given by the lowest spinor representation of B_n,

$n \geq 3$, i.e. by (9).

1. n even: $\tilde{G}$ is uniquely determined by a)b)c), and is given by the lowest spinor representation of $D_{n+2} \simeq O(2n+4)$,

$\tilde{G} =$  $\simeq D_{n+2}$ (11)

2. n odd: There are two solutions $\tilde{G}_1$ and $\tilde{G}_2$ satisfying a)b)c) namely

$\tilde{G}_1 =$  $\simeq D_{n+2}$ (12)

and

$$\tilde{G}_2 = \begin{cases} Sp(2^{n+1}) & \text{for } \frac{1}{2}\, n\, (n+1) \text{ even} \\ O(2^{n+1}) & \text{for } \frac{1}{2}\, n\, (n+1) \text{ odd.} \end{cases} \quad (13)$$

A proof of this theorem and methods to determine the nature of the bilinear invariant of a group as well as applications to SP(6), G_2 , and B_3 will be published elsewhere [10]. We note that for G = SU(3) one gets the familiar result $\tilde{G}$ = SU(6) since SU(3) possesses no bilinear invariant.

3. Comparison with Another Method

Recently, a spin extension of the intrinsic symmetry group Sp(6) [12] was proposed [13] which was obtained by a procedure already indicated in [14]:

1. Completion of the lowest non-trivial irreducible representation of the Lie algebras of SU(2) and Sp(6) to associative algebras.

2. Tensor multiplication of the two algebras and transiti- to a simple algebra by omitting the centre.

In step 1 the authors find as completion the Lie algebras of U(2) and U(6) respectively.

Tensor multiplication leads to the algebra of U(12) from which the simple Lie algebra of SU(12) is obtained by omitting the multiples of the unit matrix. In contrast to this, a) b) c) lead to O(12) which is a subgroup of SU(12) and much smaller so that SU(12) as spin extension seems less satisfying. Moreover, considering the following inclusions:

$$SU(12) \supset SU(6) \times SU(2) \supset Sp(6) \times SU(2) \qquad (1)$$

one notes that SU(12) introduces additional purely intrinsic symmetry since $SU(6) \supset Sp(6)$ is not connected with the spin group SU(2).

In view of the discussion of (1.4) it is clear that a generalisation of the above procedure to other symmetry group might lead to groups whose representations do not contain the physically relevant representations of the internal symmetry group. One may wonder, however, if this procedure could not lead, in some cases, to smaller groups than a) b) c), assuming of course that one starts with the same representation of the intrinsic symmetry group. This is not so, as will be shown now.

Applying 1. to SU(2) and to some intrinsic symmetry group G, one gets irreducible (complex)associative matrix algebras which, according to a theorem of Burnside (cf. [15], p. 92), are the algebras A_N of all $N \times N$-matrices and A_2 respectively for SU(2), where N is the dimension of the irreducibly representation of G. The tensor product of A_N with A_2 then gives the algebra of all $2N \times 2N$-matrices. Dropping the centre of A_{2N} leads to the simple Lie algebra of SU(2N). Comparing this with the main theorem in section 2 one gets the following theorem:

The group G determined by the postulates a) b) c) is smaller than the group obtained from 1. and 2. whenever the irreducible representation of G used in the construction has a bilinear invariant; otherwise the results are equal.

4. Lorentz Extension

Turning to the question of enlarging the spin extension $\tilde{G}$ to a group $\overline{G}$ in such a way that not only SU(2) but also the covering $\overline{L}$ of the inhomogeneous Lorentz group is contained in $\overline{G}$ as a subgroup, one may for example proceed in two steps:

a) Embedding of $\tilde{G}$ and SL(2,C), the homogeneous part of $\overline{L}$, in a homogeneous Lorentz extension $\overline{G}_n$.

b) Adjoining the translations.

In the first step, we carry over to $\tilde{G}$ a procedure already used for SU(6) [16] by considering the groups H of 2N×2N-matrices which contain $\tilde{G}$ and SL(2,C) in such a way that their intersection corresponds to the covering SU(2) of the rotation group.

By (1.2), SU(2) was embedded in $\tilde{G}$ in the form $1_N \times SU(2)$, the Kronecker product of the N×N unit matrix 1_N with SU(2), so that SL(2,C) should lie in H as $1_N \times SL(2,C)$. We therefore demand

$$\tilde{G} \subset H \quad , \qquad 1_N \times SL(2,C) \subset H \quad ,$$

$$\tilde{G} \cap 1_N \times SL(2,c) = 1_N \times SU(2) \tag{1}$$

and take for $\overline{G}_n$ the smallest group H satisfying (1). If two groups H_1, H_2 satisfy (1), so does their intersection. Therefore $\overline{G}_n$ is uniquely determined by this requirement. One can now prove [10]:

Theorem 1:

The smallest group of 2N×2N-matrices satisfying (1) is the complex extension $\tilde{G}_C$ of $\tilde{G}$

$$\overline{G}_n = \tilde{G}_C \tag{2}$$

For the case of $\tilde{G}$ = SU(6), theorem 1 gives the familiar result $\overline{G}_n$ = SL(6,C).

In order to take the translation group T_4 into account, physical interpretation suggests the requirement that T_4 should be invariant under $\overline{G}_n$ and transforms under Lorentz transformations in the usual way. In general, however, this is impossible as is well known for SL(6,C) [5]. For let g be an element of $\overline{G}_n$ and t be an element of an arbitrary translation group T. Invariance of T then means that gtg^{-1} is again in T. The mapping

$$\alpha(g) : t \to gtg^{-1} \tag{3}$$

of T onto T is an automorphism of T, and the mapping $g \to \alpha(g)$ is a homomorphism of $\overline{G}_n$ into the group of automorphisms of T. This group, however, is just given by the set of non-singular linear transformations of T, bearing in mind that T is also real vector space.

Hence $g \to \alpha(g)$ is a representation of $\overline{G}_n$ by linear transformations of this real vector space T. Now, $\overline{G}_n$ will in general not have a real 4-dimensional representation in which the representation matrices of $1_N \times SL(2,C)$ act as Lorentz transformations. A possible way out of this difficulty is to embed T_4 in a larger translation group T and to demand the invariance of T under $\overline{G}_n$, naturally choosing T as small as possible. The dimension of T is uniquely determined by $\overline{G}_n$. The familiar 36-dimensional translation group T constructed for SL(6,C), is a special case of the following result [10].

Theorem 2: For $\overline{G}_n = \tilde{G}_C = SL(2N,C), O(2N,C), Sp(2N,C)$, the smallest possible translation group T is $(2N)^2$ -dimensional.

One of us (G.C.H.)gratefully acknowledges the financial support of the Bundesministerium für wissenschaftliche Forschung, Bad Godesberg, Germany.

References

1. L. Michel, Phys. Rev. 137B, 405 (1965).
2. W. D. McGlinn, Phys. Rev. Lett. 12, 467 (1964).
3. F. Gürsey, L. A. Radicati, Phys. Rev. Lett. 13, 173 (1964).
4. R. Delbourgo, A. Salam, J. Strathdee, Preprint IC/65/1.
5. T. Fulton, J. Wess, Phys. Lett. 14, 57 (1965)
6. H. D. Doebner, G. C. Hegerfeldt, Spin Extension of G_2, Preprint Marburg 1965.
7. E. B. Dynkin, Am. Math. Soc. Transl., Ser. 2.6 (1957), 111-378.
8. A. I. Malcev, Am. Math. Soc. Transl., No. 33 (1950).
9. M. Gell-Mann, Y. Ne'emann, The Eightfold Way, Benjamin Inc. New York, 1964.
10. H. D. Doebner, G. C. Hegerfeldt, Group Embedding of Space-Time and Intrinsic Symmetry Groups, to be published.
11. H. Joos, Schladming Lectures 1964, Acta Physica Austriaca Suppl. 1.
12. H. Bacry, J. Nuyts, L. van Hove, Phys. Lett. 9, 279 (1964).
13. D. G. Fakirov, B. W. Struminsky, I. T. Todorov, Phys. Lett. 17 (1965), 342.
14. V. G. Kadyshevsky, A. N. Tavkhelidze, I. T. Todorov, Phys. Lett. 15, 180 (1965).
15. H. Weyl, The Classical Groups, Princeton 1946.
16. L. Michel, B. Sakita, Anales de l'Institut Henri Poincaré, 1965, in Press.

RELATIVISTIC SU(6)[†]

By

C. FRONSDAL

International Centre for Theoretical Physics
Trieste

I. Introduction

I shall begin by discussing:

a. The socalled difficulties of making SU(6) relativistic.

b. The limitations on relativistic SU(6) theories.

c. The benefits of a relativistic SU(6) theory.

Afterwards I shall present the mathematical tools that are essential, and lastly the applications that have been made to date.

a. The literature, even the very recent one, is full of oblique references to "the difficulties of reconciling SU(6) with Lorentz invariance". It has long been recognized that a relativistic SU(6) must somehow involve a noncompact group; such groups are notorious because all unitary irreducible (nontrivial) representations are infinite dimensional. First it was hoped, but in vain, that this noncompactness was associated with the Lorentz part of a bigger group only. When it became clear that the group that labels particles at rest is also noncompact one was faced by the following dilemma: Either to use finite, and hence non-unitary, representations [1], or to use unitary, and hence infinite, representations. Neither course presents a real difficulty. In the first case,

† Lecture given at the V. Internationalen Universitätswochen für Kernphysik, Schladming, 24 February - 9 March 1966.

however, in order to obtain a physically interpretable theory, it is necessary to invent a system of field equations, whose solutions do not incorporate the high degree of symmetry that had been postulated ab initio. There is therefore a practical difficulty, rather than a difficulty in principle, because the high SU(6) including symmetry is not a property of physical scattering amplitudes. This is true even if all multiplets are mass-degenerate. In the second case (unitary infinite dimensional representations) there is no difficulties what so ever, even though there exists at least one opinionated writer who regards the infinity of the multiplets as catastrophic [2]. I hope to show that it is, instead, of the greatest benefit.

b. The group SU(6) includes elements of the internal group SU(3) and of the external spin group in an essentially mixed way; that is, SU(6) is not a direct product of SU(3) and the spin group. To make it relativistic we have to consider the spin group as a subgroup of the entire Poincaré group P, and construct a larger group G that contains P and SU(3) in an essentially mixed way. There are very strong limitations on how this can be done. These limitations are expressed by the statements

(I) G = P.S (semidirect product)

(II) S is a group that commutes with translations.

The proof has two parts. The first part is so well known that I shall only give the result, namely [3]: If the mass spectrum of an irreducible representation of G has at least one discrete point, then it contains that point only. In other words, irreducible multiplets of G contain states with exactly the same mass. Here is a sketch of the second part of the proof of (I) and (II): [4]

Assume that a representation D of G exists, in which the spectrum of the energy is bounded below. Let H be the Hilbert space in which D is realized, and let H_o be the subspace in which the energy has its minimum value (rest-states). Let G_o

be the stability group of H_o (little group including space translations). Then the states of H_o are labeled by the vectors of an irreducible representation D_o of G_o, and may thus be denoted $|\alpha,m\rangle$, where m is the value of the energy (fixed in H_o) and $\alpha = 1,2,\ldots$ Suppose that G has k infinitesimal generators $L_1 \ldots L_k$ that do not commute with the energy and are linearly independent modulo G_o. Then we construct a k-parameter family of states

$$|\alpha,m,\varepsilon_1\ldots\varepsilon_k\rangle = (1+i\varepsilon_k L_k)|\alpha,m\rangle$$

Now the essential idea is that the states of H are completely determined by the states in the rest system plus the velocity; i.e., by α and three parameters v_1,v_2,v_3. Consequently, the states just constructed are interpretable only if k = 3, and the L_k must be the infinitesimal accellerations. That is, every generator of G that is not in G_o is a pure Lorentz transformation (modulo G_o). Since G_o commutes with both the energy and the momentum, there follows that the only elements in G that do not commute with the translations are the Lorentz transformations. The group G therefore consists of P, a set S of transformations that commutes with the translations (clearly S is a group), and products PS and SP.

Now it is trivial to prove that $PSP^{-1} = S$, which completes the proof.

The structure defined by (I) and (II) constitutes the limitations on relativistic theories in general. To obtain theories of the SU(6) type, in which the spin group and SU(3) are joined together in G_o, one has to impose two more conditions on the group S, namely:

(III) S contains SU(6) as a subgroup.

(IV) S contains a subgroup isomorphic to the Lorentz group and this subgroup does not commute with Lorentz transformations.

The first of these conditions is obvious; the second is necessary to keep our theory from being trivial.

c. Now it seems that we are left with a theory that is essentially trivial. Let $L_{\mu\nu}$ be the generators of the Lorentz group and let $s_{\mu\nu}$ be the generators of SL(2,C)—we use this name for that subgroup of S that is isomorphic, but of course not identical to the Lorentz group. Since $L_{\mu\nu}$ do not commute with $s_{\mu\nu}$ there is only one possible set of commutation relations; these may be written most simply as follows:

$$[L_{\mu\nu}-s_{\mu\nu},s_{\lambda\rho}] = 0 \tag{1}$$

In simple cases, including the simplest (S = SL(6,C)), this implies that the subalgebra $L_{\mu\nu}-s_{\mu\nu}$ commutes with all elements of S. If therefore we define a group P', isomorphic to the Poincaré group P, as the group generated by this subgroup, plus the translations, then the structure (I) can be written simply

$$G = P' \times S \text{ (direct product)}$$

In this way we seem to have been reduced to the trivial case. I hope to dispell this impression in the next lecture. At this point I only make a few clarifying remarks.

It is important to realize that $P' \neq P$. It is possible to make use of the simple structure $P' \times S$ for the purpose of developing the mathematics and the calculational aids, but in the end it is necessary to decompose the Hilbert space according to irreducible representations of the real Poincaré group P, for these are directly associated with the physical states.(It is a curious fact that, if the mass spectrum were actually completely degenerate, then we might as well define elementary particles as irreducible representations of P'. The point is that the degeneracy is lifted in such a way that the eigenstates of mass are associated with irreducible representations of P.)

To clarify the difference between P and P' further, it is helpful to construct the representations of G [5]. This may be done in exactly the same way that Wigner found the representations of P. In fact, let D be an irreducible representation of G in a space H, and let H_p be the subspace of all states with momentum p. Let G_p be that subgroup of G that leaves H_p invariant (leaving out the translations); this is the direct generalization of Wigner's little group. In our case

$$G_p = L_p \, . \, S \text{ (Semidirect product)}$$

where L_p is Wigner's little group. For simplicity, let $\vec{p} = 0$, then

$$G_o = \{L_{ij}\} \, . \, S$$

where $\{L_{ij}\}$ is the group generated by L_{ij}, i,j = 1,2,3, i.e. the ordinary rotation group. As we have seen, $\{L_{ij} - s_{ij}\}$ commutes with S, so

$$G_o = \{L_{ij} - s_{ij}\} \times S \text{ (direct product)} .$$

Following Wigner, we may find all unitary irreducible representations of G by induction from those of G_o. Since G_o is a direct product, its irreducible representations are given by one irreducible representation of the rotation group $\{L_{ij} - s_{ij}\}$ and one irreducible representation of S. Now the essential SU(6) idea is that the spin (i.e. L_{ij}) is a subgroup of S... at least in the rest system. Therefore we may take the choice of representation defined by

$$L_{ij} - s_{ij} = 0 \tag{2}$$

as the essence of the SU(6) idea. It remains, then, to choose an irreducible representation of S.

Returning to an arbitrary frame, L_p is the group generated by [6]

$$w_\mu = \varepsilon_{\mu\nu\lambda\rho} L^{\nu\lambda} p^\rho \quad . \tag{3}$$

The image in S is now the group generated by

$$\tilde{w}_\mu = \varepsilon_{\mu\nu\lambda\rho} s^{\nu\lambda} p^\rho \quad . \tag{4}$$

and the general form of (2) is

$$w_\mu - \tilde{w}_\mu = 0 \quad . \tag{5}$$

It is interesting to note that this rule is equivalent to taking only spin - 0 representations of P'. Hence this is the "orbital group". This separation of spin from orbital, or coordinate, transformations occurs in several other fields of physics as well.

The simplest choice of S that is consistant with conditions (I) ... (IV) is the group SL(6,C). We shall find the physical states by reducing an infinite unitary representation of SL(6,C) according to a maximal compact subgroup. This is isomorphic to SU(6). However, SU(6) may be pulled out of SL(6,C) in many different ways, just like L_p for every timelike momentum, is a maximal compact subgroup of the Lorentz group. It is clear that the relevant subgroup, which we call $SU(6)_p$, is that subgroup that contains the "spin"-group generated by $\tilde{w}_\mu$. This group is identical with the "colinear" group SU(6,W). Reduction of a representation of G according to P consists of (i) picking a definite momentum p, and (ii) reducing H_p according to $SU(6)_p$. The main reason the whole thing is nontrivial and is in fact fraught with physical content, is that the projection operators that project onto an $SU(6)_p$ irreducible multiplet is a highly nonlinear function of p; that is, the projection operators are nonlocal.

Our plan is now, first, to study the irreducible representations of SL(6,C), then the reduction according to $SU(6)_p$ with all its physical applications.

II. Some Representations Of SL(n,C)

In this hour we shall

a. Show the connection of SL(n,C) with SU(n)

b. Construct some representations of SU(n)

c. Construct some representations of SL(n,C)

d. Investigate the integrability of these representations.

Afterwards we shall assign particles to some of these representations, construct the Yukawa coupling, and consider other applications.

a. The group SL(n,C) is the group of n-dimensional complex unimodular matrices. The associated algebra (same name) is the algebra of n-dimensional complex traceless matrices. The largest compact subalgebra is SU(n). Let λ_A^B (A,B = 1,...n) be the generators of SU(n) in the n-dimensional, fundamental representation of SU(n) by traceless hermitian matrices ($\sum_A \lambda_A^A = 0$). Let

$$\lambda'^B_A = i\lambda_A^B$$

and consider $\{\lambda_A^B, \lambda'^B_A\}$ as the generator of a real Lie algebra. This is SL(n,C). The commutation relations are

$$[\lambda_A^B, \lambda_C^D] = i(\delta_C^B \lambda_A^D - \delta_A^D \lambda_C^B)$$

$$[\lambda'^B_A, \lambda_C^D] = i(\delta_C^B \lambda'^D_A - \delta_A^D \lambda'^B_C)$$

$$[\lambda'^B_A, \lambda'^D_C] = -i(\delta_C^B \lambda_A^D - \delta_A^D \lambda_C^B)$$

The problem of constructing the representations of this algebra is considerably simplified when we notice that the operators

$$M_A^B = \frac{1}{2} (\lambda_A^B + i\lambda'^B_A)$$

commute with the operators

$$N_A^B = \frac{1}{2} (\lambda_A^B - i\lambda'^B_A) \quad .$$

The commutation relations satisfied by the M_A^B or by the N_A^B, are the same as those satisfied by the λ_A^B .

Suppose that an irreducible representation D of SL(n,C) is given, then it is clear that the operators M_A^B form an irreducible representation D_1 of SU(n), and the operators N_A^B form an irreducible representation D_2 of SU(n). Conversely, every pair D_1, D_2 of irreducible representations of SU(n) determine an irreducible representation of SL(n,C).

Suppose that D is unitary; then the operators λ_A^B and λ'^B_A are hermitian. From this it follows that the M_A^B, N_A^B are not hermitian, instead

$$(M_A^B)^+ = N_A^B \quad .$$

The two representations D_1, D_2 of SU(n) are therefore not unitary, but complex conjugates of each other [6,7] :

$$D_2 \text{ is equivalent to } D_1^* \; .$$

Our problem is thus reduced to finding pairs of mutually conjugate representations of SU(n).

b. It would not be hard to discuss all the representations of SU(n) [7], but it is convenient to limit ourselves to the simplest ones. The first fundamental irreducible representation is n-dimensional and given by

$$M_A^B \to P_A^B \, , \; (P_A^B)_C^D = \delta_C^B \, \delta_A^D - \frac{1}{n} \, \delta_A^B \, \delta_C^D \quad . \tag{6}$$

Let ξ_A $(A = 1,\ldots n)$ be a system of basis vectors for this representation. Then we may write

$$M_A^B = \xi_A \frac{\partial}{\partial \xi_B} - \frac{1}{n} \delta_A^B \, \xi_C \frac{\partial}{\partial \xi_C} \quad .$$

A special family of irreducible representations is induced on the set of monomials of degree M_1 in the ξ_A :

$$\xi_1^{a_1} \, \xi_2^{a_2}, \ldots \xi_n^{a_n} \quad , \quad \Sigma a_n = N \quad .$$

If M_1 is a positive integer, and a_1, a_2,... are positive integers less than an equal to M_1, then there is only a finite set of these monomials, and they span a finite irreducible representation of SU(n). Instead of this notation we may write

$$\Psi_{A_1 \ldots A_{M_1}} = \xi_{A_1} \, \xi_{A_2} \ldots \xi_{A_{M_1}}$$

where the tensor $\Psi_{A_1 \ldots A_{M_1}}$ is, by construction, symmetric in all its indices. We label this irreducible representation $D_1(M_1,0,\ldots)$. Viewed as a set of monomials of fixed degree M_1, the basis vectors are easily generalized to the case when M_1, and one or more among the a_i, are arbitrary complex numbers. Then one sees immediately that repeated application of the M_A^B give ever new values of the a_i ; hence the representation is infinite dimensional [8]. This is clearly what we need in order to construct unitary representations of SL(n,C). One may doubt the meaning of the tensor notation in that case, but I shall show that it is very useful. Essentially we shall use the tensor method in the following way [7]: calculate useful quantities, like matrix elements or coupling coefficients, for integer but otherwise arbitrary values of M_1; then continue the result analytically to that value of M_1 that corresponds to unitary representations of SL(n,C).

In addition to the n-dimensional representation (6), associated with the n "quarks" ξ_A we have to consider the n-dimensional representation

$$M_A^B \to -(P_A^B)^T$$

which we may associate with n "antiquarks"

$$M_A^B = -\xi^B \frac{\partial}{\partial \xi^A} + \frac{1}{n}\delta_A^B \xi^C \frac{\partial}{\partial \xi^C} \quad . \tag{7}$$

This latter representation is inequivalent to the former one, except in the special case n = 2. Again we may construct monomial representations, or tensors:

$$\psi^{A_1 \ldots A_{M_{n-1}}} = \xi^{A_1} \xi^{A_2} \ldots \xi^{A_{M_{n-1}}} \quad .$$

This representation we call $D_1(\ldots,0,M_{n-1})$.

The most general representation of SU(n) that we shall need is

$$\psi^{B_1 \ldots B_{M_{n-1}}}_{A_1 \ldots A_{M_1}} = \xi_{A_1} \ldots \xi_{A_{M_1}} \xi^{B_1} \ldots \xi^{M_{n-1}} \quad .$$

This is irreducible if the tensor is traceless; that is, if

$$\delta^{A_1}_{B_1} \psi^{B_1 \ldots B_{M_{n-1}}}_{A_1 \ldots A_{M_1}} = 0 \quad ,$$

which is automatically satisfied if

$$\xi_A \; \xi^A = 0 \quad .$$

This representation we call $D_1(M_1,\ldots M_{n-1})$. Both M_1 and M_{n-1} may be complex numbers.

c. The above construction of representations of the SU(n) algebra generated by the M_A^B may be repeated for the N_A^B... Since the two algebras commute, it is necessary to distinguish the variables, or the tensor indices, on which the one or the other acts. Our convention shall be to put dots over all indices that refer to N_A^B ... one quickely realizes that this should include the indices on the generators as well. Let us call $D_2(N_1,\ldots N_{n-1})$ the representation whose basis is the tensor

$$\psi^{\dot{B}_1\ldots\dot{B}_{N_{n-1}}}_{\dot{A}_1\ldots\dot{A}_{N_1}} = \xi_{\dot{A}_1}\ldots\xi_{\dot{A}_{N_1}}\;\xi^{\dot{B}_1}\ldots\xi^{\dot{B}_{N_{n-1}}}$$

Having thus found representations of both the M_A^B and the $N_{\dot{A}}^{\dot{B}}$ we have ipso facto a representation of SL(n,C) on the basis

$$\psi^{C_1\ldots C_{M_{n-1}},\;\dot{D}_1\ldots\dot{D}_{N_{n-1}}}_{A_1\ldots A_{M_1},\;\dot{B}_1\ldots\dot{B}_{N_1}} = \xi_{A_1}\ldots\xi^{C_1}\ldots\xi_{\dot{B}_1}\ldots\xi^{\dot{D}_1}\ldots \qquad (8)$$

this representation we call $D(M_1,\ldots M_{n-1},N_1,\ldots N_{n-1})$. It is not the most general representation, but it suffices for our purpose. Not all of these representations are unitary [7] , but here we shall limit the discussion of unitarity to those special representations that we shall assign to elementary particles.

d. We have found representations of the algebra of SL(n,C); can they be "exponentiated" to give representations of the group? Consider first the compact subgroup. Our representations may be reduced into a sum of irreducible representations of the compact subgroup, and obviously we must require that each of these is integrable. This is simple: an irreducible representation of a compact algebra is integrable if and only if it is finite. In fact, Harish - Chandra has shown that, if every irreducible representation of the compact subgroup that occurs is finite, then the whole representation is integrable and determines a representation of SL(n,C).

It is obviously important to reduce our tensor according to the compact subgroup; fortunately this is relatively easy, in principle at least. We shall carry it our for the most important cases.

III. Multiplet Assignment, Yukawa Coupling

Next we shall

a. Select an SL(6,C) representation for baryons.

b. Reduce this according to $SU(6)_p$.

c. Reduce the product of baryon and antibaryon representations.

d. Determine the SL(6,C) representation for mesons.

This will lead up to the first numerical success of the theory. Afterwards I shall discuss applications to scattering, weak and electromagnetic interactions, and mass formulae.

a. Here our first criterion is that of simplicity; and the simplest possibility is to take, for baryons, the representation whose basis is the tensor

$$\psi^{\dot{B}_1 \ldots \dot{B}_N}_{A_1 \ldots A_{N+k}} \quad ;$$

with N and k still to be determined. We shall see that k must be 3 and N = -9/2.

Consider first the reduction by the compact subgroup. The

generators are

$$\lambda_A^B = M_A^B + N_{\dot{A}}^{\dot{B}} \quad .$$

The M_A^B act on the undotted indices, while the $N_{\dot{A}}^{\dot{B}}$ act in the same way on the dotted ones. The compact subgroup, therefore, acts indiscriminatly on all the indices. We may therefore ignore the dots for the present, and consider our tensor as a regular SU(6) tensor. As such it is not irreducible; for even though it is symmetric in upper indices and symmetric in lower indices, it is not traceless. The reduction therefore consists of using the operation of contraction on an upper and a lower index. Clearly the smallest SU(6) representation that can be obtained is

$$\tilde{\Psi}_{A_{N+1}\ldots A_{N+k}} \equiv \Psi_{A_1\ldots A_{N+k}}^{A_1\ldots A_N} \quad . \tag{9}$$

This must be a finite representation of SU(6); hence k must be a positive integer. Moreover, it must be the 56-dimensional representation of SU(6), hence k = 3.

Unitarity how restricts the value of N to -9/2 + iρ [5,7], where ρ is real. However, a stronger restriction follows from demanding that the 56 have the same parity. In fact the parity operator P must satisfy the relation [5,7,10]

$$PM_A^B = N_{\dot{A}}^{\dot{B}} \, P$$

Thus P must commute with the SU(6) generators (and hence be the same for all the particles in each SU(6) multiplet). If we apply the noncompact generator $M_A^B - N_{\dot{A}}^{\dot{B}}$ to (9), then we find

$$(M_A^B - N_{\dot{A}}^{\dot{B}})\Psi_{A_1 A_2 A_3} = -(N-1)\tilde{\Psi}^B_{AA_1 A_2 A_3} + (N+\tfrac{9}{2}) \cdot \;[56] \tag{10}$$

where $\tilde{\Psi}^{B}_{AA_1A_2A_3}$ is the traceless part of

$$\Psi^{\dot{B}\dot{A}_1 \ldots \dot{A}_N}_{AA_1 \ldots A_{N+3}} \tag{11}$$

This is an irreducible 700 dimensional representation of SU(6). If P is to anticommute with $(M-N)^B_A$, than this multiplet must have parity opposite to that of the 56. In (10), the quantity [56] is a linear combination of the components of the 56; these have the same parity as the 56, hence the coefficient N + 9/2 must vanish. Thus the existence of a parity operator is consistent with the unitarity condition $N = -9/2 + i\rho$, and demands that we take $\rho = 0$.

b. We have already defined two of the irreducible SU(6) subspaces, by (9) and by the traceless part of (11). We are here reducing with respect to the rest-system SU(6) = $SU(6)_o$. The complete reduction is clearly of the form

$$\Psi^{\dot{B}_1 \ldots \dot{B}_N}_{A_1 \ldots A_{N+k}} = S \sum_{t=o}^{\infty} \beta\,(N,t)\ \tilde{\Psi}^{B_1 \ldots B_t}_{A_1 \ldots A_{t+k}}\ \delta^{B_{t+1}}_{A_{t+k+1}} \ldots \delta^{B_N}_{A_{N+k}} \tag{12}$$

where S stands for symmetrization in upper indices, and in lower indices, $\tilde{\Psi}^{B_1 \ldots B_t}_{A_1 \ldots A_{t+k}}$ is the traceless part of the (N-t)-fold trace of the big tensor and the coefficients $\beta(N,t)$ are to be determined. We find

$$\beta(N,t) = \frac{(2t+k+n-1)!}{t!(t+k)!(N-t)!(t+n+k+N-1)!} \ . \tag{13}$$

If N is positive integer, then $\beta(N,t)$ vanishes for $t>N$; in the case of unitary representations, with $N = -(k+n)/2 = -9/2$, (13) reduces to

$$\beta(t) = (-)^t \frac{(2t+k+n-1)!}{t!(t+k)!} \ . \tag{14}$$

Since we know how the generators M_A^B and N_A^B act on left side of (12) we can calculate directly the transformation properties of the finite SU(6) tensors $\tilde{\Psi}^{\cdots}_{\cdots}$. Under compact transformations we have of course the usual SU(6) formula:

$$(M_A^B + N_{\dot{A}}^{\dot{B}})\, \tilde{\Psi}^{B_1 \ldots B_t}_{A_1 \ldots A_{t+k}} = - \sum_{s=1}^{t} \delta_A^{B_s}\, \Psi^{BB_1 \ldots B_t}_{A_1 \ldots A_{t+k}} + $$

$$+ \sum_{s=1}^{t+1} \delta^B_{A_s}\, \tilde{\Psi}^{B_1 \ldots B_t}_{AA_1 \ldots, \ldots A_{t+k}} -$$

$$- \frac{k}{n}\, \delta_A^B\, \tilde{\Psi}^{B_1 \ldots B_t}_{A_1 \ldots A_{t+k}}$$

Under noncompact transformations the result is a bit more complicated, namely [7,11]

$$(M_A^B - N_{\dot{A}}^{\dot{B}})\, \tilde{\Psi}^{B_1 \ldots B_t}_{A_1 \ldots A_{t+k}} = -(2t+k+n)\, \tilde{\Psi}^{BB_1 \ldots B_t}_{AA_1 \ldots A_{t+k}} -$$

$$- \frac{t(t+k)}{(2t+k+n-1)(2t+k+n-2)}\, S\Big[\delta_A^B\, \delta^{B_1}_{A_1} \tilde{\Psi}^{B_2 \ldots B_t}_{A_2 \ldots A_{t+k}} +$$

$$+ (t-1)\, \delta^B_A \delta^{B_1}_{A_1} \delta^{B_2}_{A_2} \tilde{\Psi}^{BB_3 \ldots B_t}_{\ldots A_{t+k}} + (t+k-1)\, \delta^{B}_{A_1}\, \delta^{B_1}_{A_2} \tilde{\Psi}^{B_2 \ldots B_t}_{AA_3 \ldots A_{t+k}} -$$

$$- \frac{(t-1)(t+k-1)}{2t+k+n-3}\, \delta^{B_1}_{A_1} \delta^{B_2}_{A_2}\, \tilde{\Psi}^{BB_3 \ldots B_t}_{AA_3 \ldots A_{t+k}} - (2t+k+n-2) \times$$

$$\times\, \delta_A^{B_1}\, \delta^B_{A_1} \tilde{\Psi}^{B_2 \ldots B_t}_{A_2 \ldots A_{t+k}} \Big] . \qquad (15)$$

Here there is no longer any reference to tensors with complex numbers of indices.

The invariant norm may be determined in various ways. The most direct, if not the most elegant, method is to write down the most general SU(6) invariant:

$$\sum_{t=o}^{\infty} \gamma(t)\tilde{\Psi}^{*A_1\ldots A_{t+k}}_{B_1\ldots B_t} \Psi^{B_1\ldots B_t}_{A_1\ldots A_{t+k}} \tag{16}$$

apply $(M-N)^B_A$, and determine the coefficients $\gamma(t)$ by requiring that the result be zero. The solution is that (16) is invariant if [7,11]

$$\gamma(t) = \frac{(2t+k+n-1)!}{t!(t+k)!} \quad . \tag{17}$$

As we see, (16) is positive definite. We have now completed our reduction of the formal tensor language to an ordinary notation. We have reduced according to static SU(6), but it is trivial to generalize to arbitrary $SU(6)_p$. All we have to do is to replace everywhere

$$\delta^{\dot{B}}_{A} \text{ by } \frac{1}{m}\, p^{\dot{B}}_{A} = \frac{1}{m}(p_o\delta^{\beta}_{\alpha} - \vec{p}.\vec{\sigma}^{\beta}_{\alpha})\delta^{b}_{a}$$

$$\delta^{B}_{\dot{A}} \text{ by } \frac{1}{m}\, p^{B}_{\dot{A}} = \frac{1}{m}(p_o\delta^{\beta}_{\alpha} + \vec{p}.\vec{\sigma}^{\beta}_{\alpha})\delta^{b}_{a} \tag{18}$$

The compact generators of $SU(6)_p$ are

$$p^{\dot{B}}_{A}\, N^{\dot{C}}_{\dot{B}} + M^{B}_{A}\, p^{\dot{C}}_{B} = \lambda^{\dot{C}}_{A}(p) \tag{19}$$

and the corresponding noncompact generators are

$$-p^{\dot{B}}_{A}\, N^{\dot{C}}_{\dot{B}} + M^{B}_{A}\, p^{\dot{C}}_{B} = i\lambda'^{\dot{C}}_{A}(p) \quad . \tag{20}$$

c. We now turn to a study of the "currents" that can be formed out of $\Psi^{\cdots}_{\cdots}$ and $\Psi^{*\cdots}_{\cdots}$, and ask if there exists a system of currents that can be coupled to the mesons. There are two possibilities: either the meson SL(6,C) representation contains an SU(6) singlet, or else it does not. Here I shall discuss only the first possibility, because I think that this is by far the simplest. In fact it is doubtful whether a representation for mesons can be constructed that does not contain an SU(6) singlet, and which is selfconjugate.

We assume, then that the meson representation is irreducible, and that it decomposes into a set of SU(6) multiplets that include a singlet meson, κ say, a 35 dimensional representation, $\pi^{-B}_{\ A}$ say, and others. The invariant Yukawa coupling will then have the form

$$I = J\kappa + J^{-B}_{\ A}\,\pi^{-A}_{\ B} + J^{+B}_{\ A}\,\pi^{+A}_{\ B} + \ldots \tag{21}$$

where the currents J, $J^{\pm}$,... are linear in Ψ and linear in Ψ^*. The first of these is invariant under SU(6), and must therefore have the form

$$J = \sum_{t=0}^{\infty} \gamma(t)\, j^Q_t (\Psi^*\Psi)_{t,t} \tag{22}$$

where

$$(\Psi^*\Psi)_{t,t} = \tilde{\Psi}^{*A_1\ldots A_{t+k}}_{\quad B_1\ldots B_t}\ \Psi^{B_1\ldots B_t}_{\quad A_1\ldots A_{t+k}}$$

Since the quantity (21) is invariant, the set of currents transform among themselves according to an irreducible representation that is contragradient to that of the mesons. We use this fact to calculate the coefficients j^Q_t in (22), [7].

Two of the Casimir operators of SL(n,C) are

$$Q_M = M^B_A\, M^A_B \qquad \text{and} \qquad Q_N = N^{\dot{B}}_{\dot{A}}\, N^{\dot{A}}_{\dot{B}}\ .$$

Applied to the vectors of an irreducible representation these operators are multiplets of the unit operator. For any representation of SL(6,C) that contains the one-dimensional representation of SU(6) we have $M_A^B + N_{\dot{A}}^{\dot{B}} = 0$ on that subspace, and thus

$$Q_M = Q_N = Q \text{ , say.}$$

Suppose now that Ψ and Ψ^* are coupled to an irreducible representation D that contains the trivial representation of the compact subgroup, then the current J satisfies

$$Q_M J = Q_N J = QJ$$

where Q is a pure number. We have used (15) to evaluate $Q_M J$ explicitly; the result is the following recursion relation for the coefficients

$$(t+n-1)(t+k+n-1)(2t+k+n)j^Q_{t+1} + t(t+k)(2t+k+n-2)j^Q_{t-1} =$$

$$= \left[2t(t+k+n-1) + (n-1)(k+n)+2Q\right](2t+k+n-1)j^Q_t \quad . \qquad (23)$$

In the special case Q = 0, J is the SL(6,C) invariant (16), and (23) has the solution $j^Q_t = 1$. When Q = 2n = 12 the complete solution is

$$j^Q_t = 1 + \frac{4nt}{n-1} \; \frac{t+k+n-1}{(k+n)^2} \qquad (24)$$

The general solution for arbitrary Q is not yet known, Notice, however, that the current J is completely determined. That means that the choice of representation for the mesons is known except for the value of Q.

d. Having thus determined J we may apply the noncompact generators to it and obtain other currents that must also be coupled to mesons [7]. In this way the meson representation is also mapped out in detail. The result is that the mesons must belong to the representation whose tensorial basis is

$$\phi^{C_1 \ldots C_M \dot{D}_1 \ldots \dot{D}_M}_{A_1 \ldots A_M \dot{B}_1 \ldots \dot{B}_M} \qquad (25)$$

That this representation can be coupled to the baryon currents is actually obvious if one is satisfied with a formal construction [5]. But now we also know that this is the only one. The simplest SU(6) multiplets that are contained in the reduction of this representation are: the singlet κ^+ (positive parity), the usual 35-plet π^{-B}_{A}, a positive parity 35-plet π^{+B}_{A}, and so on. (It is not possible to do without the second 35-plet)

As far as unitarity of the meson representation is concerned, we have to require that either

$$M = -\frac{5}{2} + i\lambda \quad , \qquad \lambda \text{ real}$$

or

$$-3 < M < -2 \; .$$

But the meson representation has to be self-conjugate (because $\bar{\pi}^0 = \pi^0$), and this means that M must be real. The value of the Casimir operators is related to M by

$$Q = 2M(M+n-1)$$

and we find

$$-12.5 < Q < -12 \quad .$$

It is worth writing down the first terms of the final expressions for the currents J, $J^{\pm}$:

$$J = \Psi^{*ABC}\,\Psi_{ABC} \qquad (26)$$

$$J^{-B}_{A} = \left[-\frac{n+1}{n-1}\,Q\right]^{1/2}\left[\tilde{\Psi}^{*CDE}\tilde{\Psi}^{B}_{ACDE} + \tilde{\Psi}^{*BCDE}_{A}\,\tilde{\Psi}_{CDE}\right] + \ldots \qquad (27)$$

$$J^{+B}_{A} = \frac{k}{k+n}\left[-\frac{n+1}{n-1}\,Q \Big/ \left(-1-\frac{2Q}{n(n-2)}\right)\right]^{1/2} \bar{\Psi}^{BCD}\,\Psi_{ACD} + \ldots \qquad (28)$$

This result has been used to evaluate the physical vertex function to lowest order in the momenta.

IV. Applications

In this section of the notes we study

a. The strong vertex

b. Scattering

c. Form factors

Mass formulae are discussed separately in the following joint work with G. Bisiacchi.

a. The result (21), (26), (27), (28) for the strong vertex does not say anything about the interactions between physical particles unless all three particles are at rest; for the tensors that occur there are SU(6) tensors, rather than $SU(6)_p$ tensors. Note, in fact that the pseudoscalar meson current J^{-B}_{A} does not contain a $(\overline{56}, 56)$ term. For simplicity, suppose that the mesons are at rest, and let p, p' be the momenta of the two baryons. Then there remains only to express $\tilde{\Psi}^{\cdots}_{\cdots}$ in terms of $SU(6)_p$ tensors and $\tilde{\Psi}^{*\cdots}_{\cdots}$ in terms of $SU(6)_{p'}$ tensors. We have seen above how such objects are defined; by either using these definitions, or making Lorentz transformations, it is quite straightforward to calculate (21) to lowest order in momenta [12]. The result is

$$I = 60\sqrt{-6/7Q}\,(\psi^{*}\psi)\kappa + \{4(\psi^{*}\psi)^{B}_{A}\,(\pi^{-C}_{\ B}\, v^{A}_{C} + \pi^{-A}_{\ C}\, v^{C}_{B}) -$$

$$- 2(\overline{\psi}\psi)^{AB}_{CD}\,\pi^{-C}_{\ A}\, v^{D}_{B} + 3(\psi^{*}\psi)\pi^{-B}_{\ A}\, v^{A}_{B} + 5(\overline{\psi}\psi)^{B}_{A} \times$$

$$\times (\pi^{-C}_{\ B}\, w^{A}_{C} - \pi^{-A}_{\ C}\, w^{C}_{B})\} + \tfrac{2}{3}[6/(-24-2Q)]^{1/2}(\overline{\psi}\psi)^{B}_{A} \times$$

$$\times \pi^{+A}_{\ B} + \ldots$$

Part of the results, but not all, are beclouded by the fact that some processes described by this vertex function are only virtual; for these processes it is not quite clear what masses to use in (18). The following results are, nevertheless, unambiguous:

(i) for pseudoscalar mesons, $D/F = 9/5 = 1.8$

(ii) for the vector meson octet, $F/D = 25/3 >> 1$.

The first result is particularly attractive, since the PCAC hypothesis predicts the same D/F for weak as for strong interactions. The fact that the vector mesons have essentially a pure F-coupling is also in accord with current ideas.

b. About scattering I have only this to say, that in colinear processes, in which only two momenta p and p' intervene, the relevant $SU(6)_p$ subgroups have an intersection which is $SU(3) \times SU(3)$. All the results of this colinear group are therefore included in the theory: in particular, the most successful among the Johnson-Treiman relations.

Turning now from the strong interactions to electromagnetic interactions, it is necessary to point out a missing link in the theory as it has developed until now. In the case of fermions we should very much like to see a first order wave equation; this we have not been able to do in a satisfactory way. Perhaps this has something to do with Professor Matthews conjecture (made in yesterday's discussion), that it may not be easy to develop a local field with anticommutators. Anyhow, sooner or later we shall have to construct first order field equations, or obviate the need for them. In the meantime we hope to approximate the final version of the theory by using a Klein-Gordon type of equation for all particles, including fermions.

c. The SL(6,C) invariant for baryons may be written

$$\overline{\Psi}^{A_1 \ldots A_{N+k}}_{\dot{B}_1 \ldots \dot{B}_N} \; \Psi^{\dot{B}_1 \ldots \dot{B}_N}_{A_1 \ldots A_{N+k}} \qquad (29)$$

where

$$\bar{\Psi}^{A_1 \cdots}_{\dot{B}_1 \cdots} = (\psi^{\dot{B}_1 \cdots}_{A_1 \cdots})^* \beta$$

and the matrix β is defined by

$$\tilde{\tilde{\Psi}}^{A_1 \cdots A_{t+k}}_{B_1 \cdots B_t} = \psi^{*A_1 \cdots A_{t+k}}_{B_1 \cdots B_t} \cdot (-)^t$$

Now let us remember the part P' of our group G. This is simple; just let $\Psi^{\cdots}_{\cdots}$ depend on p_μ , and make the representation irreducible by writing

$$(p^2-m^2)\ \Psi^{\cdots}_{\cdots}(p) = 0\ . \qquad (30)$$

(Remember that the essence of SU(6) is that no spin is associated with P'.)

A perfectly possible electromagnetic interaction is obtained by replacing

$$p_\mu \to p_\mu - eA_\mu$$

in this equation, although this is not the physically correct electromagnetic interaction for baryons. An equation that coincides with (30) in the absence of A_μ may be constructed from the generators (19) and (20)

$$[\lambda^B_A\ \lambda^A_B - \lambda'^B_A\ \lambda'^A_B - m^2]\psi = 0\ . \qquad (31)$$

This gives rise to a different electromagnetic interaction; I do not yet know whether it is the correct one. Nevertheless, for illustration only, let us consider

$$[(p-eA)^2 - m^2]\Psi = 0\ .$$

The electromagnetic current, according to this equation, is

$$\overline{\Psi}^{\cdots}_{\cdots}(x)\ \overleftrightarrow{\partial}_{\mu}\Psi^{\cdots}_{\cdots}(x)$$

or in p-space

$$J_{\mu}(q) = \int d^4p\ \overline{\Psi}^{\cdots}_{\cdots}(p+q)(2p+q)_{\mu}\ \Psi^{\cdots}_{\cdots}(p). \tag{32}$$

I think that it is with a good deal of justification that I refer to this interaction as a local interaction. I shall defer, for just a moment, a discussion of the more interesting properties of this type of local interaction, in order to say something about weak currents.

Let us suppose that the weak interaction currents are related to the compact generators, for example

$$\overline{\Psi}^{\cdots}_{\cdots}(pM+Np)^{\dot{B}}_{A}\ \Psi^{\cdots}_{\cdots} \tag{33}$$

Here is already something striking: these quantities are four-vectors; thus we have to do with vector currents, without having to require this as an independent postulate. Not only that, but these four-vectors have mixed parity: in the rest system of p_{μ}, the fourth component belongs to a vector while the space-components belong to a pseudo-vector. Of course we could separate the currents into a vector and a pseudovector part, but they would not separately be local. Thus we see that the assumption of a local interaction leads us to a $V \pm A$ interaction. This part of the theory is brand new, so we have not yet had time to investigate the prediction for G_A/G_V when mass breaking is taken into account. There is also some uncertainty in interpreting (33), for we don't know what momentum vector p_{μ} is.

To illustrate what are the most exciting aspects of the theory, and one that depends crucially in the infinite dimensionality of the representations, let us consider the simple electric current (32). We want to discover the electric form factor of the 56 baryons, so we must expand $\Psi(p)$ and $\overline{\Psi}(p+q)$ into $SU(6)_p$ and $SU(6)_{p+q}$ multiplets respectively, and retain only the 56-dimensional part:

$$J_\mu\Big|_{\overline{56},56} \sim \Psi^{*ABC}(p')\; p'^{E_1}_{\dot{D}_1}\cdots p'^{E_N}_{\dot{D}_N}(2p+q)_\mu\; S p_{E_1}^{\dot{D}_1}\cdots p_{E_N}^{\dot{D}_N}\;\Psi_{ABC}(p)$$

where $p' = p+q$ and S stands for symmetrization with respect to the set $A,B,C,E_1\ldots E_N$.

We hope soon to be able to evaluate this sum [13]. In the meantime I shall give the result for $k = 0$ and $n = 2$ (that is, we consider SL(2,C) instead of SL(6,C)). In that case

$$J_\mu\Big|_{\overline{1},1} = E(p^2,p'^2,q^2)(2p+q)_\mu\;\Psi^*(p+q)\;\Psi(q)$$

where the subscript $\overline{1},1$ means that we have projected out the SU(2) singlet, that is, we have calculated the electric form factor of a spin-zero meson. We find

$$E(p^2,p'^2,q^2) \sim \left(\frac{p^2p'^2}{\Delta}\right)^{1/2}\frac{1}{N+1}\sin h\left[\frac{N+1}{2}\ln\frac{p^2+p'^2-q^2+\sqrt{\Delta}}{p^2+p'^2-q^2-\sqrt{\Delta}}\right]$$

with

$$\Delta = p^4+p'^4+q^4-2p^2p'^2-2p^2q^2-2p'^2q^2 \quad .$$

The number N+1 in the model corresponds to our number N+9/2 in the realistic case. This function is a polynomial if N is an integer, but unitarity requires that $-1 < N < 0$ or $N+1$ = pure imaginary. It has the following interesting properties:

(i) As $-q^2 \to \infty$ for fixed p^2,p'^2, E goes to zero like $(-q^2)^{\mathrm{Re}N}$.

(ii) On the principal sheet of the logarithm the function is analytic in q^2. The logarithmic branch-point lies at infinity; except for that the function is very similar to what one finds when we calculate the triangle diagram in ordinary field theory. In fact, on every sheet except the principal one there is a singularity at $\Delta = 0$. This is a well-known non-Landau singularity of local field theories.

(iii) The function is real, and symmetric in $N + 1 \to -(N+1)$, which reflects the fact that two representations that are related in this way are equivalent.

(iv) By taking N close to zero one may let E approach a constant for low q^2, and arbitrarity well. This means that wide angle scattering will be affected.

References

1. R. Delbourgo, A. Salam and J. Strathdee, Proc. Roy. Soc. A284, 146 (1965)
   W. Rühl, Nuovo Cimento 37, 301 and 319 (1965)
   T. Fulton and J. Wess, Phys. Lett. 14, 57 and 334 (1965)
   M. A. B. Bég and A. Pais, Phys. Rev. Lett. 14, 267(1965)
   Bardakci, Cornwall, Freund and Lee, Phys. Rev. Lett. 14, 48 (1965)
2. S. Coleman, Phys. Rev. 138, B1262 (1965)
3. L. O'Raifeartaigh, Phys. Rev. Lett. 14, 575 (1965)
4. P. Budini and C. Fronsdal, Phys. Rev. Lett. 14, 968 (1965)
5. See Ref. 4 and C. Fronsdal, "Relativistic Symmetries", Proceedings of the Seminar on Elementary Particles and High Energy Physics, Trieste (1965), published by IAEA (1965).
6. G. Bisiacchi and C. Fronsdal, Nuovo Cimento 41, 35 (1966)
7. C. Fronsdal, "Representations of SL(n,C)", UCLA Preprint, submitted to J. Math.Phys. (January 1966)
8. This generation of infinite dimensional representations by

analytic continuation was explained in detail for the simplest of all noncompact groups by A. O. Barut and C. Fronsdal, Proc. Roy. Soc. A287, 532 (1965)

9. Generalized tensors were explained in detail in C. Fronsdal, "The Theory of Representations of Non-Compact Lie-Algebras", Proceedings of the Seminar on High Energy Physics and Elementary Particles, Trieste (1965), published by IAEA, 1965.

10. W. Rühl, "The Parity Transformation for a Symmetry Model Which Uses Unitary Representations of the Group SL(6,C)", CERN Preprint (1965)

11. A. Salam and J. Strathdee, ICTP Preprint 65/79, Trieste (1965)

12. C. Fronsdal and R. White, "The Strong Vertex According to SL(6,C)", ICTP Preprint IC/66/20, Trieste

13. Work carried out in collaboration with G. Cocho, H. Rashid and R. White.

ATTEMPTS AT UNDERSTANDING THE MASS SPECTRA OF ELEMENTARY PARTICLES†

By

G. BISIACCHI and C. FRONSDAL

International Centre for Theoretical Physics

Trieste

I. Introduction

This report contains an account of some attempts to unfurl the intricacies of the mass spectrum of elementary particles. Although the discussion is limited to ideas that work, in the sense that they are not immediately refuted by experiments, we are not yet convinced that any of them are correct. We have gone off in several different directions; in some cases we have carried out enormous calculations. We hope that others may use the results to better advantage.

We have posed, and answered, the following questions:

A. Is there a sense in which the mass operator for baryons and for mesons can be said to be equal?

This was first considered by Harari and Lipkin [1], who compared the mass operator for the 56 baryons with the mass-squared operator for the 35 mesons, and failed to find a similarity. Now it seems quite obvious that the two operators should be confronted by each other only if they represent the same quantity; that is, either the mass in both cases, or the mass-squared. We find (Section II) that to the extend that they

---

†Lecture given by C. Fronsdal at the V. Internationalen Universitätswochen für Kernphysik, Schladming 24 February - 9 March 1966.

can be compared, the operators for mass agree to within a few percent. (Concerning the extent to which they can be compared, see below). This prompted us to develop a point of view according to which linear mass operators are relevant objects. In fact, mass plays the dual role of inertial mass and gravitational charge. In its role as inertia, the mass is associated with the Hamiltonian, or a term in the free particle Lagrangian, or the Lehmann spectral function. However, as a gravitational charge the mass is associated with the energy-momentum density tensor $T_{\mu\nu}$. The energy density T_{oo} represents the mass for any particle regardless of its spin. From this point of view the mass operator is a close analogue of the electric charge operator. Energy-momentum conservation might imply some kind of non-renormalization of mass, and thus justify the expectation that it have simple transformation properties.

B. Does the mass operator transform like a component of an irreducible SL(6,C) tensor?

This question makes sense if the elementary particles are assigned to unitary representations of SL(6,C). The only irreducible SL(6,C) tensor that suggests itself as a mass operator is M_{AB}^{CD} [2]. The SU(6) reduction of this tensor includes the 405, 189, 35 and 1-dimensional representations. This then has the obvious merit of having just enough neutral components to allow a fit to the experimental mass spectrum. Perhaps one should not play down the fact that this hypothesis allows to fit eight baryon masses with only five parameters. (For the mesons, eight masses are fitted with eight parameters).

Within the strict framework of SU(6) theory, the only sense in which the baryon and meson mass operators can be compared is with respect to the relative contributions of M^1_{405}, M^8_{405} and M^{27}_{405} [1]. However, if SL(6,C) is invoked, then the relative contributions of M^1_1, M^8_{35} and M_{405} may be compared as well. Our conclusion is that equality of the two mass operators is incompatible with their irreducibility. (Section III). Thus "A" and "B" can be answered in the affirmative if considered separately, but not if taken together.

C. How does the mass operator transform under Lorentz transformations?

We suppose that the relevant operator - be it mass or $(\text{mass})^2$ or $(\text{mass})^{-2}$ - is a sum of the neutral components of an SL(6,C) tensor $M^{C\dot{D}}_{A\dot{B}}$. In terms of the reduction of SL(6,C) to SU(3) × SL(2,C) this is a set $(T^{cd}_{ab})_{\mu\nu}$ of SL(2,C) tensors. The question is whether the mass operators are related to the components $\mu = \nu = 0$ or perhaps to the trace $g^{\mu\nu}T_{\mu\nu}$. From the point of view outlined above, according to which the mass is regarded as a coupling constant, we expect that the $\mu=\nu=0$ component represents the mass linearly. A different criterion is found as follows. Let us suppose that it is possible to write down a Lorentz-invariant Klein-Gordon equation

$$\Gamma_{\mu\nu}p^{\mu}p^{\nu} - m^2 = 0 \quad ,$$

where $\Gamma_{\mu\nu}$ is a constant SL(2,C) tensor, and that the physical masses are determined thereby. In the frame $\vec{p} = 0$, $p_0^2 = \Gamma_{00}^{-1}m^2$. There are two simple possibilities: (a) that $\Gamma_{\mu\nu} = g_{\mu\nu}$ and the symmetry breaking is determined by the SL(2,C) invariant operator m^2, or (b) that m^2 is a universal constant and the operator for $(\text{mass})^{-2}$ transforms like the component Γ_{00} of a tensor $\Gamma_{\mu\nu}$. Our main conclusion is: Any simple SL(2,C) behaviour implies that for mesons the relative sizes of the M^1_{189}, M^8_{189}, M^{27}_{189} contributions be the same as the relative sizes of the M^1_{405}, M^8_{405}, M^{27}_{405} contributions. We find this criterion to be satisfied extremely well by the $(\text{mass})^{-2}$ operator.

D. In the framework of SL(6,C), can mixing between different SU(6) representations be avoided?

We find that if we insist on avoiding mixing, then we are led to consider the mass as the solution of a second order equation, as above. Lorentz invariance of this equation leads to three conditions on the parameters of the $(\text{mass})^{-2}$ operator, and these are all very well satisfied by the experimental masses. (See section V, where the less successful aspects of this idea are discussed also.)

E. What are the qualitative features of the mass spectrum for the higher SU(6) multiplets?

In the SL(6,C) models the mass increases rapidly and without bound as one passes to the higher SU(6) subspaces, suggesting an infinite number of particles. Another fascinating possibility arises out of the use of a $(\text{mass})^{-2}$ operator. This is discussed in section V.

II. Phenomenology

For the 35 negative parity mesons the most general mass (or mass squared) operator is of the form [3]

$$M = \rho_1 M_1^1 + \rho_{35} M_{35}^1 + \rho_{405}^1 M_{405}^1 + \rho_{405}^8 M_{405}^8 + \rho_{405}^{27} M_{405}^{27} + \rho_{189}^1 M_{189}^1 + \rho_{189}^8 M_{189}^8 + \rho_{189}^{27} M_{189}^{27} \quad \text{(II.1)}$$

where the ρ's are real parameters and M_i^j is an operator that belongs to an i-dimensional representation of SU(6), a j-dimensional representation of SU(3), and is invariant with respect to spin, isospin and hypercharge. Each M_i^j may be written as a bilinear expression in the SU(6) generators (see Appendix).

We shall ignore mass splittings within the isotopic spin multiplets. To take these into account we should have included in M terms that transform like vectors or tensors under isospin. Such operators being traceless in each isospin multiplet, it is clear that the appropriate mass values to use here are the means within each isospin multiplet. The operator (II.1) then reduces to an 8 × 8 matrix

$$m_i = \sum_{j=1}^{8} A_i^j \rho_j \quad , \quad i = 1,\ldots 8 \tag{II.2}$$

where the ρ_j are the parameters in (II.1). Among the eight m_i seven are the masses (or masses squared) of $\phi,\omega,\pi,K,\eta,K^*,\rho$ and one is the ϕ-ω mixing parameter (see Appendix).

For the 56 positive parity baryons the most general mass operator has the form [3,4]

$$M = \rho_1 M^1_1 + \rho_{35} M^8_{35} + \rho^1_{405} M^1_{405} + \rho^8_{405} M^8_{405} + \rho^{27}_{405} M^{27}_{405} + \rho^8_{2695} M^8_{2695} +$$
$$+ \rho^{27}_{2695} M^{27}_{2695} + \rho^{64}_{2695} M^{64}_{2695} \tag{II.3}$$

Again we write this in the form (II.2), where now the eight M_i are the masses of $N,\Sigma,\Lambda,\Xi,N^*,Y^*_1,\Xi^*$ and Ω.

Harari and Lipkin [2] first suggested that a direct comparison between the mass operators of mesons and of baryons might be possible. This is, of course, model dependent, for the normalization of M_{35} say, is arbitrary in both formulae. Nevertheless, we shall attempt to compare the relative contributions of M^1_{405}, M^8_{405} and M^{27}_{405}

$$M^i_{405} = P^i M_{405} \tag{II.4}$$

where M_{405} is an SU(6) tensor and P^1, P^8, P^{27} are projection operators. There is only one M_{405} in the meson mass operator, and only one in the baryon mass operator. The normalization of P^i, determined by $P^iP^i = P^i$ therefore yields an unambiguous definition of the ratios $\rho^1_{405}:\rho^8_{405}:\rho^{27}_{405}$. Another model leading to a meaningful comparison between the two mass operators consists of writing each M^j_i as a bilinear expression in terms of SU(6) generators. The comparable elements are, in this case as well, the ratios $\rho^1_{405}:\rho^8_{405}:\rho^{27}_{405}$. Harari and Lipkin [1] calculated the first of these ratios, using a mass formula for baryons and a mass squared formula for mesons, and found widely different values. It seems obvious, however, that if the operators are comparable at all, then they have to represent the same quantity; either mass or mass squared in both cases. Using mass squared in both cases leads to essentially the same result as that of Harari and Lipkin. For this reason, and for other reasons mentioned in the introduction, we shall consider linear mass formulae.

Using the latest values of the masses [5], and an $\omega - \phi$ mixing angle of arcsin $\sqrt{1/3}$, [6] we find

$$\rho^1_{405} : \rho^8_{405} : \rho^{27}_{405} = \begin{cases} 1: -0.823: 0.074 & \text{Mesons} \\ 1: -0.879: 0.025 & \text{Baryons} \end{cases} \qquad \text{(II.5)}$$

A mixing angle of 39° gives the ratios 1:-0.766:0.048 for mesons. In addition to the near equality in the first ratio, one should not dismiss as insignificant the smallness of the second ratio in both cases. The Gell-Mann-Okubo hypothesis, according to which ρ^{27}_{405} should vanish, has scant theoretical justification, and we do not feel that it should be invoked here. The last ratio is small but not zero; the fact that it is small in both cases constitutes supplementary evidence of the comparability of the two mass operators.

One may try to develop this idea further, by writing

$$M^j_i = P^j_i \mathcal{M}$$

for all the terms in (II.1) and (II.3). This still requires, however, a prescription for writing down $\mathcal{M}$ for both the meson 35 and the baryon 56. For an example, $\mathcal{M}$ might be required to transform irreducibly with respect to a larger algebra. One such model is considered below.

III. An SL(6,C) Irreducible Operator

Turning now to a special model, we assume that the baryons and the mesons belong to irreducible, unitary representations of SL(6,C) [2]. The simplest assumption that can be made for the universal mass operator $\mathcal{M}$ is that it is a component of a tensor T that transforms irreducibly under that group. The representation $T_A^{\dot{B}}$ may be discarded immediately because (i) its SU(6) reduction contains a singlet and a 35 only, (ii) under Lorentz transformations $T_A^{\dot{B}} = T_{a\alpha}^{b\dot{\beta}} = (\sigma_\mu)_\alpha^{\dot{\beta}} (T_\mu)_a^b$ transforms like a fourvector, whereas we expect to deal with a tensor $T_{\mu\nu}$ (see introduction), (iii) since $T_A^{\dot{B}}$ is not selfadjoint it cannot be constructed symmetrically from Ψ and $\overline{\Psi} \sim \Psi^+$. The representation $T_{A\dot{B}}^{C\dot{D}}$ (this is irreducible if $T_{A\dot{B}}^{A\dot{D}}=T_{A\dot{B}}^{C\dot{B}}= 0$), on the other hand, satisfies all these tests: (i) its reduction to SU(6) gives the representations 405, 280, 280^*, 189, 35 and 1, (ii) under Lorentz transformations it behaves like the energy-momentum tensor, (iii) it can be constructed from ψ and $\overline{\psi}$ in both the baryon and meson representations.

In the case of baryons [7], the most general irreducible tensor that transforms in this way is

$$\mathcal{M}_{A\dot{B}}^{C\dot{D}} = M_A^C N_{\dot{B}}^{\dot{D}} \tag{III.1}$$

We have calculated all the matrix elements of this operator (see Appendix). There are nonzero matrix elements with $\Delta t=0$, i.e. "diagonal" matrix elements between the states of the same SU(6) representation, and with $\Delta t = \pm 2$, i.e. mixing between different SU(6) representations with the same parity such as 56^+ and 4536^+. It is possible that this mixing is important (in fact it is possible that the spectrum of our mass operator is continuous), but here we shall have to ignore it.

The $(\overline{56}, 56)$ matrix elements of $\mathcal{M}_{A\dot{B}}^{C\dot{D}}$ may be expressed in terms of products of two compact SL(6,C) generators: that is SU(6) generators. The result is (ignoring the 280 which has no neutral component):

$$<56|\mathcal{M}^{C\dot{D}}_{A\dot{B}}|56> = [\tfrac{3}{4}P_1 + \tfrac{3}{8}P_{35} + \tfrac{9}{40}P_{405}](M+N)^C_A(M+N)^{\dot{D}}_{\dot{B}} \qquad \text{(III.2)}$$

Here P_i are projection operators, and

$$P_{405} = P^1_{405} + P^8_{405} + P^{27}_{405}$$

In the case of mesons, the most general operator that transforms like $\mathcal{M}^{C\dot{D}}_{A\dot{B}}$ involves one more parameter:

$$\mathcal{M}^{C\dot{D}}_{A\dot{B}} = \mathcal{M}^{C\dot{D}}_{1A\dot{B}} + \gamma\mathcal{M}^{C\dot{D}}_{2A\dot{B}} = \frac{1}{2} M^C_A N^{\dot{D}}_{\dot{B}} + (\gamma-1)\mathcal{M}^{C\dot{D}}_{2A\dot{B}} \qquad \text{(III.3)}$$

where

$$<\mathcal{M}^{C\dot{D}}_{1A\dot{B}}> = P\phi^{C\ldots,\dot{\cdot}\ldots}_{\ldots,\dot{B}} \; \phi^{\ldots,\dot{D}\ldots}_{A\ldots,\ldots}$$

$$<\mathcal{M}^{C\dot{D}}_{2A\dot{B}}> = -P\phi^{C\ldots,\dot{D}\ldots}_{\ldots,\ldots} \; \phi^{\ldots,\dot{\cdot}\ldots}_{A\ldots,\dot{B}\ldots} \qquad \text{(III.4)}$$

and P means "traceless part of". We have calculated the $(35^-, 35^-)$ matrix elements of this operator (see Appendix), with the result

$$<35^-|\mathcal{M}^{C\dot{D}}_{A\dot{B}}|35^-> = \left[C_1P_1 + C_{35}P_{35} + C_{405}P_{405} + C_{189}P_{189}\right] \times$$

$$\times \; (M+N)^C_A(M+N)^{\dot{D}}_{\dot{B}} \qquad \text{(III.5)}$$

where

$$C_1 = -\frac{1}{12} Q + \frac{1}{2} + (\gamma-1) \left[-\frac{1}{72} Q + \frac{5}{36} M - \frac{1}{6} Q'\right]$$

$$C_{35} = -\frac{1}{48}Q + \frac{1}{4} + (\gamma-1)\left[-\frac{1}{144}Q + \frac{1}{36}M - \frac{1}{24}Q'\right]$$

$$C_{189} = -\frac{1}{72}Q + \frac{1}{2} + (\gamma-1)\left[-\frac{1}{72}Q - \frac{1}{36}Q'\right]$$

$$C_{405} = \frac{13}{432}Q + \frac{11}{36} + (\gamma-1)\left[\frac{17}{1296}Q + \frac{1}{108}M - \frac{17}{648}Q'\right]$$

$$Q = 2M(M+5) \qquad Q' = \frac{M(M-1)}{2M+4}$$

and $-3<M<-2$.

Comparing (III.2) and (III.5) with the phenomenological values for the ρ_i^j we see that, to represent both mass spectra by the same neutral components of $\mathcal{M}_{A\dot{B}}^{C\dot{D}}$ amounts to:

$$\left(\frac{\rho_{35}}{C_{35}} : \frac{\rho_{405}}{C_{405}}\right)_{\text{mesons}} = \left(\frac{\rho_{35}}{3/8} : \frac{\rho_{405}}{9/40}\right)_{\text{baryons}}$$

or

$$C_{35} : C_{405} = \frac{5}{3}\left(\frac{\rho^1_{405}}{\rho_{35}}\right)_{\text{baryons}} \Big/ \left(\frac{\rho^1_{405}}{\rho_{35}}\right)_{\text{mesons}}$$

or $C_{35}:C_{405} \approx \frac{3}{4}$. If $M = -\frac{5}{2}$, then this means that $\gamma \approx -\frac{3}{5}$ which in turn leads to $C_1:C_{35} \approx 5$. The experimental value of this last ratio is about 1.5 . We thus reluctantly conclude that, if the mass operator is the same for baryons as for mesons, then it is not irreducible under SL(6,C). The situation may be saved by supposing that our mass operator is the symmetry breaking part of the total mass operator, thus allowing the addition of an invariant part. In that case the comparison between baryon and meson mass operators fixes all the parameters.

Finally we may apply our results to the higher baryon resonances. The center of mass of the t'th SU(6) multiplet ($t = 0,1,2,\ldots$ corresponds to 56^+, 700^-, $4536^+,\ldots$) is given by

$$m_t = \lambda + \rho\left[t^2 + 8t + \frac{135}{8}\right]$$

If the baryon mass operator is irreducible, $\lambda=0$ then m_1 (the mean mass of the 700^- multiplet) is 2020 MeV, and m_2 = 2880 MeV.

It should be remembered that all these conclusions may be materially altered when inter-SU(6) multiplet mixing is taken into account.

IV. Lorentz Transformation Properties

It is possible to discuss Lorentz transformations of the mass operator in the framework of a relativistic generalization of SU(6) only. For definiteness we shall assume that the mass operator transforms irreducibly under SL(6,C) as indicated by (III.1) for baryons and by (III.3) for mesons.

The most general "neutral" component of the mass operator may be written

$$\begin{aligned} \mathcal{M}_{\text{neutral}} = {} & \alpha_1 \mathcal{M}^{ab\alpha\dot\beta}_{ba\beta\dot\alpha} + \alpha'_1 \mathcal{M}^{ab\alpha\dot\beta}_{ab\beta\dot\alpha} + \alpha_8 \mathcal{M}^{3b\alpha\dot\beta}_{b3\beta\dot\alpha} \\ & + \alpha'_8 \mathcal{M}^{3b\alpha\dot\beta}_{3b\beta\dot\alpha} + \alpha_{27} \mathcal{M}^{33\alpha\dot\beta}_{33\beta\dot\alpha} \\ & + \beta_1 \mathcal{M}^{ab\alpha\dot\beta}_{ba\alpha\dot\beta} + \beta_8 \mathcal{M}^{3b\alpha\dot\beta}_{b3\alpha\dot\beta} + \beta_{27} \mathcal{M}^{33\alpha\dot\beta}_{33\alpha\dot\beta} \end{aligned} \qquad \text{(IV.1)}$$

The relationship between these eight real parameters and the eight phenomenological parameters is

$$\rho_1 = C_1(\alpha_1 + \frac{9}{35}\alpha'_1 + \frac{1}{3}\alpha_8 + \frac{3}{35}\alpha'_8 + \frac{1}{210}\alpha_{27} + \frac{16}{35}\beta_1 + \frac{16}{105}\beta_8 + \frac{4}{105}\beta_{27})$$

$$\rho_{35} = C_{35}(\alpha_8 + \frac{9}{16}\alpha'_8 + \frac{5}{16}\beta_8 + \frac{5}{8}\alpha_{27} + \frac{1}{8}\beta_{27})$$

$$\rho^1_{189} = C_{189}\left(\frac{1}{2}(\alpha'_1-\beta_1) + \frac{1}{6}(\alpha'_8-\beta_8) + \frac{1}{24}(\alpha_{27}-\beta_{27})\right)$$

$$\rho^8_{189} = C_{189}\left(\frac{1}{2}(\alpha'_8-\beta_8) + \frac{1}{5}(\alpha_{27}-\beta_{27})\right)$$

$$\rho^{27}_{189} = C_{189}\cdot\frac{1}{2}(\alpha_{27}-\beta_{27})$$

The expressions for ρ^i_{405}/C_{405} are obtained from ρ^i_{189}/C_{189} by replacing β_1, β_8 and β_{27} by their negative.

If $\mathcal{M}_{neutral}$ is Lorentz invariant, then all the five α's must vanish. For baryons this means that $\rho_1 = (320/105)\rho^1_{405}$ and $\rho_{35} = (25/24)\rho^8_{405}$. Although the first relationship fits to about 1/2 %, the second one is completely wrong. We discard this possibility.

With respect to Lorentz transformations, $\mathcal{M}$ is a symmetric second rank tensor $T_{\mu\nu}$. We have just discarded the possibility $\mathcal{M}_{neutral} \sim g^{\mu\nu}T_{\mu\nu}$ and we are left with the other simple conjecture, namely $\mathcal{M}_{neutral} \sim T_{oo}$. In terms of the parameters in (IV.1) this means that $\beta_1 = \beta_8 = \beta_{27} = 0$. This is inconsequential for the baryons, because $C_{189} = 0$, but for mesons we get

$$\frac{\rho^1_{189}}{\rho^1_{405}} = \frac{\rho^8_{189}}{\rho^8_{405}} = \frac{\rho^{27}_{189}}{\rho^{27}_{405}} = \frac{C_{189}}{C_{405}}$$

This condition is reasonably well satisfied when one uses a linear mass formula. When the mass formula is taken to be a formula for $(\text{Mass})^{-2}$ we find

$$\frac{\rho^{1}_{189}}{\rho^{1}_{405}} = 1.001 \quad , \quad \frac{\rho^{8}_{189}}{\rho^{8}_{405}} = 1.007 \quad , \quad \frac{\rho^{27}_{189}}{\rho^{27}_{405}} = 0.950 \quad . \qquad \text{(VI.2)}$$

We used the usual mixing angle of arcsin $\sqrt{(1/3)}$ but these ratios are extremely insensitive to variations from this value. There is a very great temptation to suppose that the three ratios should, ideally, be equal to 1. In that case, with the same mixing angle, one would have the sum rules (for m^{-2}):

$$2\phi + 5\omega = 3\rho + 4K^*$$

$$\phi + 3\rho = 2\omega + 2K^*$$

$$4K^* = 2\phi + \omega + \rho \quad .$$

The first is satisfied to about 1 % accuracy, the others to about 4 %. We have reasons to think that the parameter M, which determines the meson representation and may lie between, - 3 and - 2 , cannot be close to either for these limits. We therefore tentatively set M = - 5/2 . Then the ratio C_{189}/C_{405} = 1 determines the parameter γ in (III.3): $\gamma \simeq -1$. This value gives a rather attractive form to (III.3), but nothing more conclusive than the above sum rules.

V. An Operator Without Mixing

The confrontation of the models discussed so far with the experimental mass sepctrum is clouded by the fact that we were unable to take into account mixing between different SU_6 multiplets. Here we consider a mass operator that does not include any mixing terms.

The generators of SL(6,C) that span the little group

$SU(6)_p$ are

$$\lambda_A^{\dot{B}} = M_A^C \, p_C^{\dot{B}} + p_A^{\dot{C}} \, N_{\dot{C}}^{\dot{B}} \tag{V.1}$$

A mass operator constructed from $\lambda_A^{\dot{B}}$ would therefore decompose into block form under the reduction of SL(6,C) to the compact subgroup $SU(6)_p$, and no inter-$SU(6)_p$-multiplet mixing would occur. In view of the necessity of including 405, 189, 35 and 1-dimensional representations of $SU(6)_p$ in the mass operator, we may try out a second order polynomial in the λ's, in other words, the neutral components of $\lambda_A^{\dot{B}} \;\; \lambda_{\dot{C}}^{D}$ where

$$\lambda_{\dot{A}}^{B} = p_{\dot{A}}^{C} \, M_C^B + N_{\dot{A}}^{\dot{C}} \, p_{\dot{C}}^{B} = p_{\dot{A}}^{C} \, \lambda_C^{\dot{D}} \, p_{\dot{D}}^{B} \,/\, p^2$$

Now this is of the second order in p, which strongly suggests writing down a "Klein-Gordon equation":

$$\Gamma^{\mu\nu} p_\mu p_\nu - m^2 = 0 \quad , \qquad \Gamma^{\mu\nu} p_\mu p_\nu = \sum \lambda_A^{\dot{B}} \, \lambda_{\dot{C}}^{D} + c p^2 \tag{V.2}$$

where m^2 and c are constants and the summation is over neutral components, with coefficients to be determined by the experimental mass spectrum.

In the frame $\vec{p} = 0$ Eq. (V.2) defines the mass spectrum by

$$\frac{m^2}{p^2} = \Gamma_{oo} \sum (M+N)_A^{\dot{B}} (M+N)_{\dot{C}}^{D} + c \tag{V.3}$$

This operator for $(\text{mass})^{-2}$ is the same as that considered above, except for the additional constant c and the change of the position of the dot on the superscript (compare with Eq. (III.2)). It is clear that $\Gamma^{\mu\nu} p_\mu p_\nu$ must be Lorentz invariant, but because of the change in transformation properties implied by the new position of the dot, this now means

$\beta_1 = \beta_8 = \beta_{27} = 0$. Since $C_{189} = C_{405}$ for this operator we see that the requirement of Lorentz invariance is that all three ratios (IV.2) be equal to one. (This is for mesons; for baryons no new conditions are implied by Lorentz invariance because of the absence of "189".)

Thus it would seem that we have found the ideal mass operator: It contains no inter-$SU(6)_p$-multiplet mixing. - - It's SU(6) reduction contains exactly what is needed. - - It fits eight baryon masses with five parameters and the same for mesons. - - It may be applied to higher multiplets without great computational problems. Still, there are short comings: the absence of "2695" contributions in the baryon case is a good prediction for the $(\text{mass})^{-2}$ operator, but not so spectacular as for the linear mass operator. - - In our choice of meson representation there is an SU(6) multiplet of 35 positive parity mesons, for which our operator predicts the same masses as for the 35 negative parity mesons.

Operators for $(\text{mass})^{-2}$ have the interesting property that the eigenvalues of mass tend to have a point of accumulation at the origin. In that case it is not possible to interpret the eigenstates as true particles. Instead one may postulate that the operator

$$(\Gamma^{\mu\nu} p_\mu p_\nu - m^2)^{-1} \qquad \text{(V.4)}$$

is to be understood as an approximation to a two-particle Green's function in a scattering state. The poles need not all lie on the real axis; instead most may find themselves on the second sheet of the scattering amplitude. Those that lie nearest to the real axis may be associated with unstable particles. The accumulation of poles at the origin in the second sheet is known to be a real physical feature of scattering amplitudes. The only trouble with this is that, when the experimental widths of the unstable particles is interpreted as imaginary adjuncts to the masses, then we can no longer do without a "2695" contribution to the baryon mass operator. In fact, absence of "2695" is equivalent to the following sum rules. Absence of 2695-8:

$$2\Xi^* + 2N + 2\Omega + 9\Sigma = 3\Lambda + 8\Xi + 4N^*$$

Absence of 2695-27:

$$4\Xi^* + 14N + 10Y_1^* + 14\Xi = 6\Omega + 21\Lambda + 7\Sigma + 8N^*$$

Absence of 2695-64:

$$3\Xi^* + N^* = \Omega + 3Y_1^*$$

The first one is particularly badly violated by the imaginary parts of masses. As we noted, the principle objection against this mass formula arises from the fact that our choice of meson SL(6,C) representation has two 35 dimensional representations. If we ignore this difficulty, assuming that another choice of meson representation has only one 35, and if an SU(6) singulet meson κ is included, then the mean value of $(\text{mass})^{-2}$ for every SU(6) multiplet is given by

$$\frac{1}{\kappa^2} + C^{(6)}\left(\rho 1 - \frac{1}{12\kappa^2}\right) \tag{V.5}$$

According to this formula, meson multiplets with mean masses higher than the 35 will exist if the mass of the κ-meson is less than 400 MeV.

Appendix

Some of the following expressions where given by Bég and Singh [3]. The complete list is given here to clarify the choice of normalization:

$$M^{1}_{1} = C^{(6)}$$

$$M^{8}_{35} = \frac{1}{6} C^{(6)} + \frac{1}{2} \left[2S(S+1) - C^{(4)} + \frac{1}{4} Y^{2}\right]$$

$$M^{1}_{405} = - \frac{5}{7} C^{(6)} + 2J(J+1) + C^{(3)}$$

$$M^{8}_{405} = - \frac{7}{48} C^{(6)} + \frac{1}{6} \left[2J(J+1) + C^{(3)}\right] - \frac{7}{16} \left[2S(S+1) - C^{(4)} + \frac{1}{4} Y^{2}\right] - \frac{1}{2} \left[2I(I+1) - \frac{1}{2} Y^{2} + 2N(N+1) - 2S(S+1)\right]$$

$$\frac{5}{2} M^{27}_{405} = - \frac{3}{8} C^{(6)} - \frac{3}{8} \left[2J(J+1) + C^{(3)}\right] + \frac{1}{2} C^{(4)} + I(I+1) + \frac{27}{8} Y^{2} + N(N+1) + 3S(S+1)$$

$$M^{1}_{189} = \frac{1}{5} C^{(6)} + 2J(J+1) - C^{(3)}$$

$$M^{8}_{189} = - \frac{1}{24} C^{(6)} + \frac{1}{6} \left[2J(J+1) - C^{(3)}\right] - \frac{1}{8} \left[2S(S+1) - C^{(4)} + \frac{1}{4} Y^{2}\right] + \frac{1}{2} \left[2I(I+1) - \frac{1}{2} Y^{2} - 2N(N+1) + 2S(S+1)\right]$$

$$\frac{5}{2} M^{27}_{189} = - \frac{3}{8} C^{(6)} - \frac{3}{8} \left[2J(J+1) - C^{(3)}\right] + \frac{1}{2} C^{(4)} - I(I+1) - \frac{9}{8} Y + N(N+1) + 3S(S+1)$$

For the values of $S, N, C^{(3)}, \ldots$ see Bég and Singh [3]. (The value of the SU(6) Casimir operator $C^{(6)}$ is 12 for mesons and 45/2 for baryons). Inserting the numerical values we get Eq. (II.2):

$$
\begin{vmatrix} \phi_U \\ \omega_U \\ \sqrt{2}(\phi\omega)_U \\ \rho \\ K^* \\ \pi \\ \eta \\ K \end{vmatrix}
=
\begin{vmatrix}
1 & 4 & 11 & 16 & 3 & -9 & -12 & -3 \\
1 & -2 & 6 & -18 & 6 & -2 & 26 & -6 \\
0 & 0 & -10 & -8 & 6 & 14 & 16 & -6 \\
1 & -2 & 1 & 2 & 1 & 5 & -14 & -1 \\
1 & 1 & 1 & -1 & -3 & 5 & 7 & 3 \\
1 & -2 & -9 & 18 & -1 & -9 & 18 & -3 \\
1 & 2 & -9 & -18 & -9 & -9 & -18 & -27 \\
1 & 1 & -9 & -9 & 3 & -9 & -9 & 9
\end{vmatrix}
\begin{vmatrix} 12\,\rho_1 \\ \rho_{35} \\ \frac{2}{5}\rho^1_{189} \\ \frac{1}{12}\rho^8_{189} \\ \frac{1}{4}\rho^{27}_{189} \\ \frac{2}{7}\rho^1_{405} \\ \frac{1}{24}\rho^8_{405} \\ \frac{1}{4}\rho^{27}_{405} \end{vmatrix}
$$

The inverse of this matrix is

.08571	.08571	0.0	.25714	.34286	.08571	.02857	.11429
.09375	-.04688	.07031	-.14063	.09375	-.04687	.01563	.03125
.01250	.02292	-.02604	.00625	.00833	-.01875	-.00625	-.02500
.01250	-.01458	.00521	.00625	-.00417	.01875	-.00625	-.01250
.03750	.01875	.04688	.01875	-.07500	-.00625	-.01875	.02500
-.00298	-.01339	.02604	.00232	.02976	-.01339	-.00446	-.01876
-.00208	.00937	.00260	-.02187	.01458	.00937	-.00312	-.00625
-.01250	-.00625	-.01562	-.00625	.02500	-.00625	-.01875	.02500

The relationship between the physical particles ϕ, ω and the SU(3) singlet ϕ_P and octet-member ω_P is

$$
\begin{vmatrix} \phi \\ \omega \end{vmatrix} = \begin{vmatrix} \cos\theta & \sin\theta \\ -\sin\theta & \cos\theta \end{vmatrix} \begin{vmatrix} \phi_P \\ \omega_P \end{vmatrix}
$$

The states ϕ_U, ω_U are defined by the right hand side when $\theta = \theta_o = \text{arc sin} \sqrt{1/3}$.

In the case of baryons, the formula (II,2) reads

$$\begin{vmatrix} \Xi^* \\ \Omega \\ N \\ \Lambda \\ \Sigma \\ Y_1^* \\ \Xi \\ N^* \end{vmatrix} = \begin{vmatrix} 1 & 1 & 2 & 1 & -6 & -1 & -3 & 6 \\ 1 & 2 & 2 & 2 & 18 & -2 & 9 & -4 \\ 1 & -1 & -5 & 7 & 3 & -2 & 21 & 0 \\ 1 & 0 & -5 & 4 & -9 & 6 & 63 & 0 \\ 1 & 0 & -5 & -4 & -1 & -6 & 7 & 0 \\ 1 & 0 & 2 & 0 & -10 & 0 & -5 & -4 \\ 1 & 1 & -5 & -3 & 3 & 8 & -21 & 0 \\ 1 & -1 & 2 & -1 & 6 & 1 & 3 & 1 \end{vmatrix} \begin{vmatrix} \frac{45}{2} \rho_1 \\ 4 \rho_{35} \\ \frac{12}{7} \rho^1_{405} \\ \frac{1}{2} \rho^8_{405} \\ \frac{1}{4} \rho^{27}_{405} \\ \rho^8_{2695} \\ \rho^{27}_{2695} \\ \rho^{64}_{2695} \end{vmatrix}$$

The last three columns have been lifted from Harari and Rashid [4]. The inverse of this matrix is

.1430	.0714	.0714	.0357	.1070	.2140	.0714	.2860
.1670	.1670	-.0833	.0000	.0000	.0000	.0833	-.3333
.0286	.0143	-.0357	-.0179	-.0536	.0429	-.0357	.0571
.0200	.0200	.0700	.0200	-.0600	.0000	-.0300	-.0400
-.0133	.0200	.0033	-.0050	-.0017	-.0333	.0033	.0267
-.0133	-.0133	-.0133	.0200	-.0600	0.0	.0533	.0267
-.0019	.0028	-.0067	.0100	.0033	-.0047	-.0067	.0038
.0857	-.0286	0.0	0.0	0.0	-.0875	0.0	.0286

We give the numerical values of the ρ's:

	Baryons	Mesons	
	$(\mathrm{Mass})^{+1}$ MeV	$(\mathrm{Mass})^{+1}$ MeV	$(\mathrm{Mass})^{-2}$ $(\mathrm{BeV})^{-2}$
ρ_{1}	58.52	61.742	.513
ρ_{35}	38.60	52.450	-2.391
ρ^{1}_{189}		61.243	-2.592
ρ^{8}_{189}		-55.593	10.880
ρ^{27}_{189}		6.495	-1.136
ρ^{1}_{405}	19.37	59.312	-2.590
ρ^{8}_{405}	-17.03	-48.797	10.812
ρ^{27}_{405}	.478	4.370	-1.195

In this table the values for mesons have been calculated using a mixing angle of 35° which corresponds to the U-chain [3].

For baryons, we assign the SL(6,C) tensor [2]

$$\psi^{\dot{B}_1 \ldots \dot{B}_N}_{A_1 \ldots A_{N+k}} \quad , \quad N = -\frac{9}{2} \quad , \quad k = 3$$

The expansion in terms of SU(6) tensors is

$$\psi^{\dot{B}_1 \ldots \dot{B}_N}_{A_1 \ldots A_{N+k}} = S \sum_{t=o} \beta(t)\, \psi^{B_1 \ldots B_t}_{A_1 \ldots A_{t+k}}\, \delta^{B_{t+1}}_{A_{t+k+1}} \ldots \delta^{B_N}_{A_{N+k}}$$

$$\beta(t) = (-1)^t \frac{(2t+k+n-1)!}{t!(t+k)!}$$

and the normalization is given by expanding the invariant:

$$\bar{\psi}\psi = \sum_{t=o}^{\infty} \beta(t)\, \bar{\psi}^{A_1 \ldots A_{t+k}}_{B_1 \ldots B_t}\, \psi^{B_1 \ldots B_t}_{A_1 \ldots A_{t+k}}$$

$$\bar{\psi}^{A_1 \ldots A_{t+k}}_{B_1 \ldots B_t} = (-1)^t \left(\psi^{B_1 \ldots B_t}_{A_1 \ldots A_{t+k}} \right)^*$$

The generators are defined by $(n = 6)$:

$$M^C_A\, \psi^{\dot{B}_1 \ldots \dot{B}_N}_{A_1 \ldots A_{N+k}} = \sum_{t=1}^{N+k} \delta^C_{A_t}\, \psi^{\dot{B}_1 \ldots \dot{B}_N}_{A_1 \ldots A \ldots A_{N+k}} - \frac{N+k}{n} \delta^C_A\, \psi^{\dot{B}_1 \ldots \dot{B}_N}_{A_1 \ldots A_{N+k}}$$

$$N^{\dot{C}}_{\dot{A}}\, \psi^{\dot{B}_1 \ldots \dot{B}_N}_{A_1 \ldots A_{N+k}} = - \sum_{t=1}^{N} \delta^{\dot{B}_t}_{\dot{A}}\, \psi^{\dot{B}_1 \ldots \dot{C} \ldots \dot{B}_N}_{A_1 \ldots A_{N+k}} + \frac{N}{n} \delta^{\dot{C}}_{\dot{A}}\, \psi^{\dot{B}_1 \ldots \dot{B}_N}_{A_1 \ldots A_{N+k}}$$

Using these formulae we find after a long but straightforward calculation:

$$\bar{\psi} M^C_A N^{\dot{D}}_{\dot{B}} \psi = \sum_{t=o}^{\infty} \beta(t) \{ t(t+k) (\bar{\psi}^D_B\ \psi^C_A)^{189}_t - \frac{t(t+k)}{2t+k+n+1} (\bar{\psi}^D_B\ \psi^C_A)^{405}_t +$$

$$+ \frac{1}{4} \frac{2t+k+n}{2t+k+n+1} \left[t(t-1)(\bar{\psi}_{BA} \psi^{DC})^{405}_t + (t+k)(t+k-1)(\bar{\psi}^{DC} \psi_{BA})^{405}_t \right] +$$

$$+ \frac{t}{4} \left(\frac{t+k}{n-2} + \frac{t+n}{n+2}\right) \left[\delta^C_B (\bar{\psi}_A \psi^D)^{35}_t + \delta^D_A (\bar{\psi}_B \psi^C)^{35}_t - \frac{2}{n} \delta^C_A (\bar{\psi}_B \psi^D)^{35}_t - \right.$$

$$- \frac{2}{n}\,\delta_B^D(\bar\psi_A\psi^C)_t^{35}\Big] + \frac{t+k}{4}\Big(\frac{t}{n-2} + \frac{t+k+n}{n+2}\Big)\Big[\delta_B^C(\bar\psi^D\psi_A)_t^{35} + \delta_A^D(\bar\psi^C\psi_B)_t^{35} -$$

$$- \frac{2}{n}\,\delta_A^C(\bar\psi^B\psi_D)_t^{35} - \frac{2}{n}\,\delta_B^D(\bar\psi^C\psi_A)_t^{35}\Big] + \frac{1}{n^2-1}\Big[t(t+k+n+1) + (k+n)^2\,\frac{n-1}{4n}\Big]$$

$$\times\,(\delta_A^D\,\delta_B^C - \frac{1}{n}\,\delta_A^C\,\delta_B^D)(\bar\psi\psi)_t + \frac{1}{4}\,t(t+k)\,\frac{2t+k+n}{2t+k+n+1}\,\times$$

$$\times\,\Big[(\bar\psi_{AB}^{CD}\,\psi)_{t+1,t-1} + (\bar\psi\psi_{AB}^{CD})_{t-1,t+1}\Big]\Big\}$$

where by $(\bar\psi\psi)_{t,t'}$ we mean the result of contracting

$$\bar\psi^{A_1\ldots A_{t+k}}_{\dot B_1\ldots\dot B_t} \quad \text{with} \quad \psi^{\dot B_1\ldots\dot B_{t'}}_{A_1\ldots A_{t'+k}}$$ leaving uncontracted two or four indices as shown.

The result (III.2) is obtained by taking the t = 0 term, ignoring mixing, and comparing with

$$\bar\psi^{A_1A_2A_3}(M+N)_A^C(M+N)_B^D\,\psi_{A_1A_2A_3} = k(k-1)(\bar\psi^{CD}\psi_{AB})_o^{405} +$$

$$+ \frac{k(k+\frac{n}{2})}{n+2}\Big[\delta_B^C(\bar\psi^D\psi_A)^{35} + \delta_A^C(\bar\psi^D\psi_B)^{35} + \delta_B^D(\bar\psi^C\psi_A)^{35} +$$

$$+ \delta_A^D(\bar\psi^C\psi_B)^{35}\Big] + \frac{k(k+n)}{n(n+1)}\,(\delta_B^C\,\delta_A^D - \frac{1}{n}\,\delta_B^D\,\delta_A^C)(\bar\psi\psi)$$

The analogous result for mesons is obtained in substantially the same way. It is only necessary to know the relation between the complete meson tensor and the 35^- projection of it:

$$\pi_A^{-B} = \phi^{\dots,B\dots}_{A\dots,\dots} - \phi^{B\dots,\dots}_{\dots,A\dots}$$

$$\phi^{A_1\dots A_M \dot{B}_1\dots\dot{B}_M}_{C_1\dots C_M \dot{D}_1\dots\dot{D}_M} = S \sum_{s=0}^{\infty} \gamma(s)(\pi^{-B_1}_{C_1}\delta^{A_1}_{D_1} - \pi^{-A_1}_{D_1}\delta^{B_1}_{C_1}) \times$$

$$\times\, \delta^{A_2}_{D_2}\dots\delta^{A_s}_{D_s}\,\delta^{B_2}_{C_2}\dots\delta^{B_s}_{C_s}\;\delta^{A_{s+1}}_{C_{s+1}}\dots\delta^{A_M}_{C_M}\;\delta^{B_{s+1}}_{D_{s+1}}\dots\delta^{B_M}_{D_M}$$

References

1. H. Harari and H. J. Lipkin, Phys. Rev. Lett. 14, 570 (1965)
2. C. Fronsdal, "Relativistic Symmetries" Proceedings of the Seminar on Elementary Particles and High Energy Physics, Trieste, 1965, published by IAEA, Vienna, 1965.
3. M. A. B. Bég and V. Singh, Phys. Rev. Lett. 13, 418 (1964)
4. H. Harari and M. A. Rashid, SLAC PUB-153, October 1965 (Stanford University)
5. A. Rosenfeld et al., Rev. of Mod. Phys. 37, 633 (1965)
6. This corresponds to the case when the mass operator is diagonal in the U-channel; see Appendix and ref. 3.
7. We use the same irreducible, unitary representations of SL(6,C) as in ref. 2.

ON SOME APPLICATIONS OF ABSTRACT METHODS IN MATHEMATICAL PHYSICS†

(Definition of a renormalizable theory and of an unrenormalizable theory)

By

J. L. DESTOUCHES

Physique Mathématique, Institut Henri Poincaré

Paris

1. Introduction

Classical mathematical physics studies various physical problems by uniform methods deriving from classical analysis.

I consider modern mathematical physics as the study of the "schemas" [1] of the physical theories. Abstract methods issued from modern functional analysis are used, and some of them are interesting for the physicist.

2. Basic Differential Equation

The main method is called "reduction method" and consists in representing, either physical system or, according to the case, informations about a physical system, by only one point of an abstract space X, in such a way that the equations concerning the system can be written as equations governing the motion of this point in X.

When we look at all the physical theories, we see that in each theory the basic equations can be always reduced

---

† Lecture given at the V. Internationalen Universitätswochen für Kernphysik, Schladming, 24 February - 9 March 1966.

to an abstract differential equation of the form

$$\frac{dx}{dt} = F(x, t) \quad , \quad x \epsilon X$$

Where x belongs to an abstract space X.

The most important problem is the Cauchy-problem, which consists in determining the solution x(t) for a given initial element x_o at the time t_o.

Except the case of classical mechanics in which X is a finite dimensional space, in all other cases X is a functional space of the same infinity type of dimension, but the choice of spaces depends on the theory under consideration.

Sometimes there are supplementary conditions:

i) $\Phi(x) > 0$

for example in the case of the theory of heat conduction; in this case the space X is the space of continuous functions.

ii) $\Phi(x) \geqslant 0$

for example in the case of the diffusion theory; in this case the space X is the space of measurable functions.

iii) a time derivative independent equation

$$C(x) = 0$$

for example in the case of electromagnetism theory where C(x) denotes the divergence equations on fields; in this case the space X is a Hilbert space.

If the system is conservative, the time t does not appear as an argument in F. In some cases, as in quantum mechanics, F is linear in x and in this case we can write

$$F(x,t) = A(t) \, . \, x$$

or, if the system is conservative:

$$F(x) = A . x$$

where A(t), or A, is a linear operator in the space X.

3. Fundamental Theorems

During the last years many theorems have been established concerning the existence and the unicity of the solution of such abstract differential equations.

For quantum physics the most important is the theory of semigroup of Hille [2] and Yosida [3]. If the operator A generates a strongly continuous semigroup of evolution-operators U(t) for $0 \leqslant t \leqslant +\infty$ the Cauchy-problem has one and only one solution. The theorem of Hille-Yosida gives the conditions for the operator A to be the generator of a strongly continuous semigroup.

The constants of motion and all the conservative laws, with connection to group-invariance, symmetries and breakdown of symmetry, can be studied by this abstract general method.

4. The Case of Theories with Divergences

If the schema above with the basic equation has a sufficient generality for the classical theories and for quantum mechanics, it is too narrow for the case of quantum theories of field where divergences appear. This case can be described in the following way: we have an infinite sequence of functions F_k, or of operators A_k in the linear case, and thus a sequence of basic equations

$$\frac{dx}{dt} = F_k(x)$$

but this sequence has no limit element for the special value k_o corresponding to the physical reality; either k is an integer or k varies continuously for example k_o is infinite. In the most general case, k is a point of an abstract space and there is a special point k_o which corresponds to the physical reality.

But physically the F_k or A_k are approximate formulations, and only the limit for k_o corresponds to the physical reality. For example we can take k as a cutoff, or k is some parameter connected with a renormalization. This case is important in physics and must be studied in more details: in some cases one can define a differential equation and in the other cases it is impossible. This is closely connected with the notion of renormalization.

5. Sekine's Abstract Theory of Renormalization

Sekine [4] has recently given a mathematically correct method of renormalization. He will explain himself his method in details on an example [5], but I will first give a general idea of it. His method is interesting because it proves the unicity of the renormalization and it gives a general definition of a renormalizable theory and of a nonrenormalizable theory.

Given a linear theory in which we have for a system a sequence A_k or a continuous set $A(k)$ of linear in general unbounded operators with the sequence of equations

$$\frac{dx}{dt} = A_k x$$

i) If there exists a limit for the sequence

$$\lim_{k \to k_o} A_k = A$$

then we have in the limit a new differential equation with this operator A and there is no problem: we come back to the usual case.

ii) If the sequence A_k has no limit, we can consider the resolvent $R(s; A_k)$ of A_k given by [6]

$$R(s; A_k) \equiv (sI - A_k)^{-1}$$

defined for $s \epsilon S$ where S is a set in the complex s-plane. This set should be determined in each concrete case.

It is possible and it happens often that the sequence $R(s; A_k)$ has a limit R(s) when k goes to k_o (for example k_o is infinite), although the sequence A_k has no limit.

$$R(s) = \lim_{k \to k_o} R(s; A_k)$$

If R(s) fulfills 3 conditions given by Sekine [7] from the theory of pseudoresolvents, there exists one and only one operator A which is independent of s and is determined uniquely from R(s) by

$$A = s\, I - R^{-1}(s) \ .$$

In this case there exists a differential equation for the system of the basic form

$$\frac{dx}{dt} = A\, x \ .$$

That is the renormalized equation; this equation is free from divergences.

6. Renormalizable and Unrenormalizable Theories

I say that a theory is renormalizable if there exists a limit operator R(s) which fulfills the three conditions of Sekine such that an operator A exists.

In the practical cases the operator R(s) can be defined using some procedures in which the constants are considered as functions of k and, in the sequence of resolvents $R(s;A_k)$, we make some manipulations on the terms $R(s,A_k)$ to obtain by subtractions a limit operator R(s).

In some cases the three conditions are not fulfilled, but with some changes, for example taking as X a subspace of the original space, eventually with modification of topology, or considering an equivalence relation, it is possible to go back to the three conditions and in this case we have still a differential equation, but the vectorial-schema is lost: a ψ-vector does not admit in the general case a spectral decomposition like

$$\psi = \sum_i c_i \phi_i$$

but only an equivalence relation

$$\psi \equiv \sum_i c_i \ \phi_i$$

with an open schema, as this has been considered by Paulette Février [8] in her book on the structure of physical theories.

But in this case we still say that the theory is renormalizable. In this case also the ψ-vectors are elements of a vectorial space which contains a Hilbert-space as a subspace.

The simplest model of such a space is a Hilbert-space with an indefinite metric, but it may happen that we must use a vectorial space of another kind.

In the other cases the theory is called unrenormalizable. Hence the evolution equation cannot be a differentiable one.

Sometimes it is possible to separate the operators A_k into two parts

$$A_k = A_k^o + B_k$$

such that the sequence $R(s;A_k^o)$ of the resolvents admits a limit operator $R^o(s)$ fulfilling the three conditions. In this case there exists an operator A^o defined by $R^o(s)$ and a differential evolution equation playing the part of a nonperturbed system. Thus the sequence B_k represents a perturbation which cannot be described by means of a differential equation. In some cases the perturbed system obeys an equation with differences, or more generally an integral equation between operators.

If a limit operator R(s) exists, then R(s) contains in itself all the observable quantities of the system and from R an S-matrix can be defined even when R does not fulfill the three conditions.

7. Classification of Linear Theories

Finally the various possibilities for a linear theory can be summarized as follows:

i) An operator A is given directly.

ii) A sequence A_k (or a function A(k) of a real variable k) is given and has a limit A:

$$\lim_{k\to k_o} A_k = A$$

In both cases we have a differential equation without any renormalization.

iii) The sequence A_k has no limit, but the sequence $R(s;A_k)$ of the resolvents has a limit R(s) which fulfills the three conditions of Sekine:

$$\lim_{k\to\infty} R(s;\, A_k) = R(s)$$

In this case we have an operator A, a differential equation and a vectorial schema with a Hilbert space. The theory is simply renormalizable.

iv) The sequence A_k has no limit, the sequence of the resolvents has a limit R(s) and R(s) does not fulfill the three conditions, but with some changes the three conditions are fulfilled.

In this case we have still an operator A and a differential equation, but an open-schema with a vectorial space X which contains a Hilbert space (this space X is not a Hilbert space, but sometimes it reduces to a Hilbert space with an indefinite metric).

The theory is still renormalizable.

v) The sequence A_k has no limit, the sequence of the resolvents has a limit R(s), but by all means it is impossible to connect an operator A with R(s).

In this case there is no operator A, no differential equation, the theory is unrenormalizable, but it contains an S-matrix,and the theory obeys an integral equation.

vi) The sequence A_k has no limit neither the sequence of the resolvents $R(s;A_k)$.

In this case again there is no operator A, no differential equation, the theory is essentially unrenormalizable, it can obey an integral equation. A theory of this kind is a very weak theory.

Sometimes the operators A_k can be divided into two parts:

$$A_k = A_k^o + B_k$$

such that the sequence $R(s;A_k^o)$ has a limit $R^o(s)$ for which we have one of the cases among the (iii), (iv), (v) cases above. In the cases $(iii)^o$ and $(iv)^o$ we have a differential equation for an operator A^o and a perturbation connected to the sequence B_k which obeys an integral equation.

The case of a nonlinear theory is much more complicated, but a classification can be done using the classification above for linear theories; it will be given somewhere else.

8. Case of a Theory with Several Sequences A_k

All the considerations above concern the case where there is only one sequence A_k which is issued from the approximate formulation of the theory before renormalization.

But it is often possible to construct several distinct sequences

$$\{A_k^{(1)}\} , \{A_k^{(2)}\} , \ldots\ldots \{A_k^{(p)}\} ,$$

and even sometimes infinite in number from the approximate formulation of the theory, and the renormalization in the naive sense seems to be not unique. This case appears also when k is a point in a space K of several or an infinite number of dimensions: in this case the sequences depends on the sequence of points k which is selected.

In these cases we can consider the resolvents and thus we have several sequences of resolvents $R(s; A_k^{(1)})$, $R(s;A_k^{(2)})$,.. .., $R(s,A_k^{(p)})$. Several cases can happen:

a) No sequence of resolvents has a limit; we are in the case (vi) above. The theory is essentially unrenormalizable.

b) At least one sequence of resolvents has a limit.In this case the sequences which have no limit are rejected as inadequate, and only the sequences which have a limit are maintained.

This decision comes from the fact that we want a physical theory as strong and complete as possible. More generally, each time when an indeterminacy gives us an opportunity to make a choice, we apply the principle of the strongest theories.

b_1) If all the sequences of resolvents have the same limit element R(s), we have a well defined theory and it is one among the three cases (iii), (iv), (v) above. In the cases (iii) and (iv) the theory is renormalizable, and in the case (v) the theory is unrenormalizable.

It can happen that several distinct sequences $A_k^{(i)}$ give several distinct sequences of resolvents $R(s; A_k^{(i)})$ which have the same limit element R(s).

b_2) If there are several sequences of resolvents which have distinct limit-elements $R^{(1)}(s)$, $R^{(2)}(s),\ldots,R^{(m)}(s)$ and if at least one of them fulfills the three conditions of Sekine, the limit operators $R^{(i)}(s)$ which do not fulfill the three conditions are rejected as inadequate and only the operators which fulfill the three conditions are maintained, in accordance with the principle of the strongest theories.

b_{21}) If there is only one such operator $R(s)^{(\alpha)}$, the theory is well defined, and we are in the case (iii), the theory is simply renormalizable.

b_{22}) If there are several such distinct operators $R(s)^{(\alpha)}$, $R^{(\beta)}(s),\ldots$, all of them are suitable, the theory is not well-defined, i.e. noncomplete, i.e. non saturated, i.e. noncategorical: one can add a postulate (or principle) to achieve the definition (or univocity) of the theory. Something remains undetermined in the theory. But after achievement we shall be in the case (iii). The theory is renormalizable.

b_3) If there are several sequences of resolvents which have distinct limit elements $R^{(1)}(s)$, $R^{(2)}(s),\ldots$, and none of them fulfilling the three conditions, but if, with some change, at least one of them fulfills the three conditions, then using the principle of the strongest theory, we reject the limit operators $R^{(i)}(s)$ which cannot in any way fulfill the three conditions.

b_{31}) If there is only one such operator $R^{(\alpha)}(s)$, the theory is well defined, and we are in the case (iv), the theory is renormalizable.

b_{32}).If there are several such distinct operators $R^{(\alpha)}(s)$ $R^{(\beta)}(s),\ldots$, all of them are suitable, the theory is not well defined, i.e. non categorical as in the case b_{22}). After achievement of the theory with a supplementary principle, we shall be in the case (iv). The theory is renormalizable.

b_4) If there are several sequences of resolvents which have distinct limit elements $R^{(1)}(s)$, $R^{(2)}(s),\ldots$, but no one fulfilling the three conditions even with change and it is impossible to define any A operator, then they are all suitable, the theory is not well-defined, i.e. not categorical as in the case b_{22}). After achievement of the theory by admitting a supplementary principle, we shall be in the case (v) the theory is unrenormalizable.

In this case there are several S-matrices definable from the operators: $S^{(1)}$ is defined by $R^{(1)}(s)$, and $S^{(2)}$ by $R^{(2)}(s)$,..., $S^{(m)}$ by $R^{(m)}(s)$,...; the theory is not categorical. After the choice of only one $R^{(\alpha)}(s)$ by a supplementary principle, this choice determines univocally the S-matrix: it shall be $S^{(\alpha)}$.

9. Final Remarks

Thus one can see how involved is the renormalization of a physical theory. But finally, roughly speaking there are three main cases:

1) the theory is unrenormalizable
2) the theory is not well-defined, not categorical.
3) the theory is well-defined, sometimes using the principle of the strongest theory, and is renormalizable. In this case the renormalization can be made in a unique manner by Sekine's method.

When one operator R(s) exists, it contains all the observable physical quantities concerning the studied system; in particular an S-matrix is defined from R(s).

To conclude, the abstract methods of modern mathematical physics seem to be useful, in particular to formulate the renormalization method in a rigourous mathematical form, to discuss the properties of unrenormalizability and of renormalizability of a physical theory, to prove the uniqueness of the renormalization when the theory is well-defined (i.e. categorical) and renormalizable.

I stop now and Sekine will explain his method with more details, taking a model as an example of application.

References

1. J. L. Destouches:
   a) Cours de Physique mathématique appliquée, Faculté des Sciences de l'Université de Paris (Cours polycopié, Association corporative des étudiants en Sciences, Paris 1962).
   b) Les notions fondamentales de la Physique mathématique, Cours à la Faculté des Sciences de Paris, 1e partie: 1er semestre 1964-1965, 2e partie: 1er semestre 1965-1966 (Cours polycopié, Association Corporative des étudiants en Sciences, Paris 1965 et 1966).
   c) La Physique mathématique (Coll. Que sais-je, Presses Unitersitaires de France, Paris 1965).
   d) Qu'est-ce que la Physique mathématique? (Gauthier-Villars, Paris 1966).
2. E. Hille, Functional Analysis and semi-groups, (Am. Math. Soc. Coll. Publ. Vol. 31, 1948; 2nd ed. with R. S. Phillips, 1957, New-York).
3. K. Yosida, Journ. Math. Soc. Japan, 1, 1948, 15, Functional Analysis, Springer 1965.
4. K. Sekine, Comptes Rendues, Acad. Sc. Paris, t, 261, 4996 (1965); Comptes Rendus, Acad. Sc. Paris,t. 262A, 158 (1966).
5. K. Sekine, in this issue, p. 440.
6. M. Stone, Linear transformations in Hilbert space (Am. Math. Soc. Coll. Publ. Vol. 15, New-York 1932).
7. K. Sekine, in this issue, p. 456.
8. P. Février, La structure des théories physiques (Presses Universitaires de France, Paris 1951).

FINITE FORMULATION OF RENORMALIZATION METHOD[†]

By

K. SEKINE

Institut Henri Poincaré
Paris

1. Introduction

We all know that quantum field theory in its present form is not quite free from a number of difficulties in principle, among which the so-called divergence difficulty is certainly the most serious one.

When we calculate a physical quantity to be observed, that is to say, whose value must in any way be finite, we find infinity.

The method of renormalization has been proposed in order to eliminate this difficulty. In some general terms, we can summarize the method as follows.

There appears some divergent integral over momentum variable $\underline{k}$. One introduces a cut-off K in order to make this integral finite. On the other hand, in the equations from which we start, we have some given constant called unrenormalized. Denote it by μ_o just symbolically. One then considers this μ_o as a function of K, $\mu_o(K)$, which tends to infinity with K going to ∞ in such a way that infinity of this function just cancels the contribution of the divergent integral. In this way we arrive at a finite result for observable quantities, such as physical mass.

[†] Lecture given at the V. Internationalen Universitätswochen für Kernphysik, Schladming, 24 February - 9 March 1966.

2. Simple Model

To illustrate the situation. I take here a simple model [1]. I don't write its Hamiltonian. It will be better to explain in words some qualitative features of the model. The model consists of two kinds of particles, which we call A and C, both bosons, without spin and any other internal degrees of freedom, interacting through the following processes: $A + A \rightleftarrows C$

Just like the Lee model, owing to this superselection rule, the whole Hilbert space is decomposed into separate sectors, namely mutually orthogonal invariant subspaces of the Hamiltonian. In what follows we are concerned only with the lowest nontrivial one, containing one C or two A particles. We further take the center-of-mass system. Thus, we have a C particle at rest at the origin, which transforms into two A particles of momentum $\underline{k}$ and $-\underline{k}$, and reciprocally. The state vector is of the form

$$|\Phi\rangle = (\alpha c^*(0) + \int d^3\underline{k}\beta(\underline{k})\ a^*(\underline{k})\ \underline{a}^*(-\underline{k}))|0\rangle \tag{1}$$

The true vacuum coincides with the free one in the present model. The system can be represented by two wave functions α and $\beta(\underline{k})$.

Now I make explicit the equations written for these c-number functions. The time-independent Schrödinger equation reads

$$E\alpha = \mu_o\alpha + 2g_o \int d^3\underline{k}\ v^*(\underline{k})\beta(\underline{k})$$

$$E\beta(\underline{k}) = g_o v(\underline{k})\alpha + \frac{k^2}{2m}\ \beta(\underline{k}) \tag{2}$$

where $k^2/2m$ is the non-relativistic kinetic energy of one A particle, while I give a C particle a rest mass μ_o. $v(\underline{k})$ is a cut-off function. In the case of straight cut-off, $v(\underline{k}) = \theta(K-k)$, $k = |\underline{k}|$. It is a routine to solve this eigenvalue problem. We get an equation whose roots are eigenvalues:

$$\frac{E - \mu_o}{g_o^2} - 8\pi \int_o^K dk \frac{k^2}{E - k^2/m} = 0 \quad . \tag{3}$$

The integral is linearly divergent as K tends to infinity. To eliminate this, I assume μ_o to be a function of K such that $\mu_o(K)$ behaves for large K as

$$- \frac{\mu_o(K)}{g_o^2} + 8\pi m K \to C \tag{4}$$

where C is a finite real constant. Put $h_K(E)$ the expression of the left hand side of (3), i.e.

$$h_K(E) = \frac{E - \mu_o(K)}{g_o^2} - 8\pi \int_o^K \frac{dk\ k^2}{E-k^2/m} \quad . \tag{5}$$

Then, under the assumption made, the limit

$$\lim_{K\to o} h_{(K)}(E) = h(E) \tag{6}$$

exists. The infinity of $\mu_o(K)$ when $K \to \infty$ and the divergent part of the integral cancel out with one another to give a finite constant C, and we are left with a finite expression. Thus, the limit function h(E) is well-defined function free from divergence. In fact, we get

$$h(E) = \frac{E}{g_o^2} + C - 8m\pi E \int_o^\infty \frac{dk}{E-k^2/m} \quad . \tag{7}$$

As a root of the equation h(E) = 0, we obtain a finite eigenvalue. Then, from the second of the equations (2) (now with v = 1) we get the corresponding eigenfunction which also is well defined.

Now, a question of fundamental importance arises. With $K \to \infty$ we get a finite result for observable quantities, here the physical mass of the C particle and the corresponding wave function, while in the original equations appears an infinite constant $\mu_o(\infty) = \infty$. Furthermore, it is not assured that the integral in the first of the equations (2) converge when we substitute in it the eigenfunction obtained above. In fact, the eigenfunction $\beta(\underline{k})$ is of the form

$$\beta(\underline{k}) = \frac{g_o\alpha}{E-k^2/m} \tag{8}$$

where α is a finite constant. The integral $\int d^3\underline{k}\beta(\underline{k})$ diverges, although $\beta(\underline{k})$ is a square-integrable function. The last condition, square-integrability, is sufficient for the interpretation of $|\beta|^2$ as a probability density. In passing, I want to remark that, if the region of integration is finite, every square-integrable function is also integrable itself. If the region is infinite, this is not always the case. We have here an example of functions square-integrable, but not integrable over the whole momentum space.

The situation described above is typical of renormalization procedures. This is the situation which Dyson [2] called paradoxical; in order to deduce finite answers one should start from infinite quantities. Rigorously, however, we should rather say that the fundamental equation is lost, or ceases to exist in the limit, because the equation then contains infinite constants and integrals which may diverge. Such an equation is deprived of mathematical existence.

A way out this difficulty can, however, be found though in a heuristic manner, for our special model because of its simplicity. Solve $\mu_o(K)$ from the equation (5), then we get

$$\mu_o(K) = E - g_o^2\, h_K(E) - 8\pi g_o^2 \int_0^K dk \frac{k^2}{E-k^2/m}$$

Put back this expression into the first of the equations (2). Then

$$E\alpha = [E-g_o^2 h_K(E)-8\pi g_o \int_0^K dk \frac{k^2}{E-k^2/m}]\alpha + 2g_o \int d^3\underline{k}\,\theta(K-k)\beta(\underline{k})$$

which can be written as

$$0 = -g_o^2 h_K(E)\alpha + 2g_o \int d^3\underline{k}\,\theta(K-k)\left[\beta(\underline{k}) - \frac{g_o\alpha}{E-k^2/m}\right]$$

Here, we can pass to the limit $K \to \infty$. The first term gives a finite function $- g_o h(E)\alpha$. As to the integral term, the expression makes sense if we assume

$$\int d^3\underline{k}\left[\beta(\underline{k}) - \frac{g_o\alpha}{E-k^2/m}\right] < \infty \tag{9}$$

that is to say, if we limit ourselves to $\alpha,\beta(k)$ such that this condition is satisfied. This is a restriction imposed on the domain of definition of the operator defined by the right hand side of the new system of equations in the case of infinite cut-off. Restoring the term $E\alpha$, let us define the operator which transform

$$\begin{pmatrix} \alpha \\ \beta(\underline{k}) \end{pmatrix} \to \begin{pmatrix} \alpha_E \\ \beta_1(\underline{k}) \end{pmatrix}$$

where

$$\alpha_E = [E - g_o^2 h(E)]\alpha + 2g_o \int d^3\underline{k}\left[\beta(\underline{k}) - \frac{g_o\alpha}{E-k^2/m}\right]$$

$$\beta_1(\underline{k}) = g_o\alpha + \frac{k^2}{m}\beta(\underline{k}) \tag{10}$$

The domain of this operator is restricted by the above inequality (9). (More exactly, by an inequality a little more stronger, as we shall see later).

It should here be remarked that the eigenfunction (8) is certainly contained in this domain. In fact for $\alpha,\beta(\underline{k})$ related by (8), the inequality (9) holds in a trivial way, when we understand by E everywhere the eigenvalue obtained.

Now we have a new system of equations free from divergences.

$$E\alpha = \alpha_E$$

$$E\beta(\underline{k}) = \beta_1(\underline{k}) \qquad (11)$$

or explicitly

$$0 = - g_o^2 h(E)\alpha + 2g_o \int d^3\underline{k}\left[\beta(\underline{k}) - \frac{g_o\alpha}{E-k^2/m}\right]$$

$$E\beta(\underline{k}) = g_o\alpha + \frac{k^2}{m}\beta(\underline{k}) \qquad (11a)$$

Here one cannot divide the integral into two parts, each of which diverges. Starting from these well-defined equations, with the above domain condition (9), we get directly the finite results which have been obtained earlier by cancellation of infinities. In fact by solving $\beta(\underline{k})$ from the second equation of (11a) and putting the result into the first one, we see that the integral term vanishes, and we find $h(E) = 0$. The eigenvalue is obtained as a root of this equation, and the corresponding eigenfunction is determined by using again the second equation.

Although this may seem trivial, this is a mathematically well-defined scheme. Starting from a system of equations free from divergences we arrive at a finite result. In any step of the solution, divergence does not appear.

Here, I wish to add a remark, which is important. The eigenoperator that is the operator transforming $(\alpha,\beta(\underline{k}))$ into $(\alpha_E,\beta_1(\underline{k}))$ with the expressions for $\alpha_E,\beta(\underline{k})$ given in the above, is actually independent of E. Presence of E in α_E is merely apparent. To see this consider the operator transforming $(\alpha,\beta(\underline{k})$ into $\alpha_{E'},\ \beta_1(\underline{k}))$. Then

$$\alpha_E - \alpha_{E'} = \alpha\Big[E - E' - g_o^2\,(h(E) - h(E')) +$$

$$+ 2g_o^2\,(E-E')\int d^3\underline{k}\,\frac{1}{(E-k^2/m)(E'-k^2/m)}\Big] = 0 \quad ,$$

in view of the expression for h(E). In fact, from (7) we get the relation

$$h(E)-h(E') = (E-E') \left[\frac{1}{g_o^2} + 8\pi \int_o^\infty \frac{k^2 dk}{(E-k^2/m)(E'-k^2/m)}\right]. \quad (12)$$

Under these circumstances, we can define the eigenoperator at one point, say E_o. The eigenvalue problem then takes the form

$$E\alpha = \alpha_{E_o}$$

$$E\beta(\underline{k}) = \beta_1(\underline{k}) \quad (13)$$

where E_o is a fixed value, and only E in the left hand side is unknown. It is easy to confirm that the solution of this system of equations is the same as that obtained before. It is here to be remarked that, to define the equations, it suffices to fix E_o and to give $h(E_o)$. The two constants E_o and $h(E_o)$ define completely the operator and the equations. Here, reality of the two constants is required for hermiticity of the operator. In any way, it is not necessary to know the whole function h(E) at the out set. After solving the equations, we find the function h(E). This is a quite reasonable situation. If on the contrary we had not an operator independent of E, we should have given the whole function h(E) at the beginning in order to get h(E). That would be nonsense. The fact that the eigenoperator actually does not depend on E, is therefore very important.

Let us summarize. Although in a heuristic manner, we obtained a reasonable scheme. The system of equations from which we start is (13) or explicitly

$$E\alpha = [E_o - g_o^2 h(E_o)[\alpha + 2g_o \int d^3\underline{k}]\beta(\underline{k}) - \frac{g_o\alpha}{E-k^2/m}]$$

$$E\beta(\underline{k}) = g_o\alpha + \frac{k^2}{m}\beta(\underline{k}) \quad (13a)$$

This is a system free from divergences. There appears no infinite constant, and the integral makes sense if we add a condition (9):

$$\int d^3\underline{k}\left[\beta(\underline{k}) - \frac{g_o\alpha}{E-k^2/m}\right] < \infty$$

that is to say, we consider only the class of pairs $(\alpha, \beta(\underline{k}))$ satisfying the above condition.

We seek the solution in this class, and we do find the solution. The physical mass of the C particle is calculated as a root of the equation $h(E) = 0$. The corresponding eigenfunction

$$\left(\alpha,\ \beta(\underline{k}) = \frac{g_o\alpha}{E-k^2/m}\right)$$

satisfies the above condition.

If analytically continued, $h(z)$ on the upper boundary of the cut enables us to calculate all quantities related to the A-A scattering. One can show that also the wave functions for the scattering states satisfy the condition given above.

In this way, we arrive at finite results for all observable quantitites, and in any step of the solution, we do not encounter divergence any longer.

Finally, the out-put information is really richer than the in-put information. To define the equations, we need only two constants E_o, $h(E_o)$ without knowing the whole function $h(E)$ which we obtain by solving the problem.

3. General method

However, the derivation of the new system of equations (13a) which has been made in the above, is not quite rigorous.

I now propose a general abstract method, which enables us to justify the above result with mathematical rigour. The general method not only reproduces the result already known, but also answers some delicate questions about existence and unicity, by giving precisely the conditions for them. Moreover, because of its generality, the method is applicable to any other theories if the conditions required are fulfilled.

3.1 Abstract Formulation of the Problem

I propose to formulate the problem in an abstract language [3]. Quite generally, consider a Hilbert space X, whose element we denote by x. The space which we have constructed in the above as the collection of pairs $(\alpha, \beta(\underline{k})) = x$ is an example. More precisely, α is a complex number, $|\alpha| < \infty$, and $\beta(\underline{k})$ is a function square-integrable in the 3-dimensional momentum space, i.e. $\int d^3\underline{k} |\beta(\underline{k})|^2 < \infty$. Let us define the norm of x by

$$||x|| = \left(|\alpha|^2 + 2 \int d^3\underline{k} |\beta(\underline{k})|^2\right)^{1/2} \qquad (14)$$

The factor 2 is a matter of convention, but then we have $||x||^2 = \langle\Phi|\Phi\rangle$ with the expression (1) for $|\Phi\rangle$. Let A_K be a linear operator defined in X, to be more precise, A_K transforms each x of some set $D \subseteq X$ into $A_K x \epsilon X$. The set D is called domain of definition of the operator A_K. In general, A_K is an unbounded operator whose domain of definition does not cover the whole space X.

An operator, say T in general, is said to be bounded and defined everywhere in X if there exists a constant M such that

$$||Tx|| \leq M \, ||x||$$

for all $x \epsilon X$. For A_K we do not assume this property.

We suppose that A_K is defined for sufficiently large K and we are interested in the limit $K \to \infty$. Consider the initial value problem of the equation

$$\frac{dx}{dt} = A_K x \ , \quad x = x(t) \tag{15}$$

The time-dependent Schrödinger equation for our model can be put into this general form. Let us define A_K by

$$A_K x = x_1 = \binom{\alpha_1}{\beta_1(\underline{k})} \tag{16}$$

with

$$\alpha_1 = (-i)[\mu_0(K)\alpha + 2g_0\int d^3\underline{k}\ \theta(K-k)\beta(\underline{k})]$$

$$\beta_1(\underline{k}) = (-i)[g_0\theta(K-k)\alpha + (k^2/m)\beta(\underline{k})]$$

Then the equations read

$$\frac{d\alpha}{dt} = \alpha_1$$

$$\frac{d\beta(\underline{k})}{dt} = \beta_1(\underline{k})$$

or $dx/dt = A_K x$. Let us study first these concrete equations. To solve the initial value problem for these equations, we can use the Laplace transform method. The transformed equations are

$$s\tilde{\alpha}(s) - \alpha_0 = (-i)[\mu_0(K)\tilde{\alpha} + 2g_0\int d^3\underline{k}\theta(K-k)\tilde{\beta}(\underline{k},s)]$$

$$s\tilde{\beta}(\underline{k},s) - \beta_0(\underline{k}) = (-i)[g_0\theta(K-k)\tilde{\alpha} + \frac{k^2}{m}\ \tilde{\beta}(\underline{k},s)] \tag{17}$$

where s is a complex variable, α_0 and $\beta_0(\underline{k})$ are the initial values of $\alpha(t)$ and $\beta(\underline{k}, t)$. We solve $\tilde{\alpha}(s)$, $\tilde{\beta}(\underline{k},s)$ as function of α_0, $\beta_0(\underline{k})$. Then by using the well-known inversion formula, we find the original function $\alpha(t)$, $\beta(\underline{k},t)$. We see here that, as the first step to the solution, one solves the Laplace-transformed equations. In an abstract language, this step is nothing but the construction of the resolvent of the operator A_K. In fact, the above system of transformed equations can be written as $s\tilde{x} - x_0 = A_K\tilde{x}$ or $(sI - A_K)\tilde{x} = x_0$

Solving $\tilde{x}$ we have

$$\tilde{x} = (sI - A_K)^{-1} x_o \tag{18}$$

where the operator $(sI-A_K)^{-1}$ is called resolvent of A_K. I give the precise definition.

3.2 Notion of Resolvent

Consider the complex s plane. The set of s for which the inverse $(sI-A_K)^{-1}$ exists as a bounded operator defined everywhere in X, is called resolvent set of A_K, and then $(sI-A_K)^{-1} \equiv \equiv R(s;A_K)$ is called resolvent of A_K. The complement of the resolvent set is the spectrum of A_K which comprises all discrete eigenvalues and the continuous spectrum. It should be remarked that the resolvent is by definition a bounded operator defined for all $x \epsilon X$ and for s belonging to the resolvent set.

Suppose now that the sequence of bounded operators $R(s;A_K)$ has a limit in some sense, and defines a new operator $R(s)$. On the other hand, the question as to whether the limit of A_K exists or not, remains quite open. This is just the situation encountered in renormalization theory.

I illustrate this on our example. By solving the system of the Laplace-transformed equations (17), we get

$$\tilde{\alpha}(s) = \frac{1}{g_o^2 H_K(s)}\left[\alpha_o - 2ig_o \int d^3\underline{k}\,\theta(K-k)\,\frac{\beta_o(\underline{k})}{s+i\;k^2/m}\right]$$

$$\tilde{\beta}(\underline{k},s) = \frac{1}{s+i\;k^2/m}\left[\beta_o(\underline{k}) - ig_o\theta(K-k)\tilde{\alpha}(s)\right] \tag{19}$$

where

$$H_K(s) = \frac{s+i\mu_o(K)}{g_o^2} + 2\int d^3\underline{k}\;\frac{\theta\;(K-k)}{s+i\;k^2/m} \tag{20}$$

Note that $\tilde{\alpha}, \tilde{\beta}(\underline{k})$ in function of arbitrary α_o, $\beta_o(\underline{k})$, define an operator, which transforms x_o into $\tilde{x}$, that is the resolvent

$R(s;A_K)$ (see Eq. (18)).

Take the limit $K \to \infty$ in the above expressions. Then, all $\theta(K-k)$ are replaced by 1, $H_K(s)$ by its limit function $H(s)$ which exists. In fact, just as in $h(E)$ the infinity of $\mu_o(K)$ cancels the divergent part of the integral, and we are left with a finite expression. Thus, we have a new operator defined by $x_2 = R(s)x$ with

$$\alpha_2 = \frac{1}{g_o^2 H(s)}\left[\alpha - 2ig_o \int d^3\underline{k} \frac{\beta(\underline{k})}{s+i\ k^2/m}\right]$$

$$\beta_2(\underline{k}) = \frac{1}{s+i\ k^2/m}[\beta(\underline{k}) - ig_o\alpha_2] \tag{21}$$

where

$$H(s) = \lim_{K\to\infty} H_K(s) = \frac{s}{g_o^2} - i\left[C - 8\pi ms \int\frac{dk}{s+i\ k^2/m}\right] \tag{22}$$

Put $z = is$, $h(z) = iH(s)$. Then $h(z)$ is just the function obtained by analytic continuation from $h(E)$ which we encountered earlier in solving the eigenvalue problem by the conventional method.

The operator $R(s)$ defined by (21) is a good operator. It is a bounded operator defined everywhere in X and for s belonging to a subset S of the complexe s plane. The subset S consists of all s for which $H(s)$ is defined and different from zero. It is not difficult to confirm that $H(s)$ is analytic in the s plane cut along the lower imaginary axis, with eventual exception of one point on the upper imaginary axis, i.e. zero of $H(s)$. (If $C > 0$ there is one and only one zero of $H(s)$ on the upper imaginary axis. If $C < 0$, no zero. For $C = 0$, $H(0) = 0$).

Now under the condition that $||x|| < \infty$, which is equivalent to $|\alpha| < \infty$ and $\beta(\underline{k}) \in L^2$, we can prove that

$$|\alpha_2| < \infty\ ,\quad \beta_2(\underline{k}) \in L^2\ ,\quad \text{i.e. } ||x_2|| < \infty\ .$$

For this purpose, we note first the fact that

$$\frac{1}{s+i\ k^2/m} \epsilon L^2$$

Hence, according to the Schwarz inequality

$$\left|\int d^3\underline{k}\ \frac{\beta(\underline{k})}{s+i\ k^2/m}\right| \leqslant \left|\left|\frac{1}{s+i\ k^2/m}\right|\right|_{L^2} ||\beta(\underline{k})||_{L^2} < \infty$$

On the other hand, the first term in the bracket for α_2, that is α, is finite. Therefore we see that $|\alpha_2| < \infty$ if H(s) is defined and different from zero, that is for $s\epsilon S$. To prove that β_2 is square-integrable we note that

$$\left|\frac{1}{s+i\ k^2/m}\right| \leqslant \frac{1}{d_s} < \infty$$

where d_s is the distance of the point s to the cut, which is > 0 as far as $s\epsilon S$. Thus

$$\left|\frac{\beta(\underline{k})}{s+i\ k^2/m}\right| \leqslant \frac{1}{d_s} |\beta(\underline{k})|$$

Hence, the first term of β_2 is square-integrable. The second term is also square-integrable. Therefore, by the triangle inequality,the sum is square-integrable. Thus we have proved that $\beta_2 \epsilon L^2$. Summarizing, $||x_2|| < \infty$ if $||x|| < \infty$. That is to say, the operator R(s) for $s\epsilon S$ is defined for all $x\epsilon X$ and the result R(s)x belongs also to X.

To prove that R(s) is a bounded operator, we need some more refined estimations of the norms. I don't give the proof here.

On the other hand, we know that the A_K for K = ∞ is not a well-defined expression, firstly because of the infinity of $\mu_0(\infty)$ and secondly because of the integral which may diverge. In the beginning of the lecture, I suggested some manipulations to get a well-defined "limit" operator, whose justification is now the matter of subject. Therefore, the question whether $\lim_{K\to\infty} A_K$ exists is still open.

Now I give a precise formulation of what we wish. We wish to find a well-defined operator A such that by solving the equation for A, we just arrive at R(s). Is it possible or not? If possible, under what conditions?

Generally, one believes that this is impossible. Although one succeeds in obtaining finite answers for observable quantities, the fundamental equations now suffer from infinity of renormalization constants. One is then inclined to think that a theory totally free from divergence does not exist.

But in the beginning of this lecture, I suggested, although in a heuristic manner, that the situation is not always so hopeless.

It may be that if one takes some caution to specify the domain of definition of operators, one arrives at a well-defined theory which gives directly finite results obtained usually by the renormalization procedure cancellation of divergences in $R(s,A_K)$).

I now prove that such a finite theory is possible under certain conditions. For this purpose, I make use of some results recently established in functional analysis.

3.3 Theory of pseudo-resolvent [4]

I define the notion of pseudo-resolvent. This is a bounded linear operator defined for all $x \in X$, R(s) depending on s. As a function of s, it is defined on a certain subset S of the complex s-plane. One assumes as its fundamental property the equation

$$R(s) - R(s') = (s'-s)\, R(s)\, R(s') \qquad (23)$$

for all s, $s' \in S$

I remark that every resolvent satisfies all the conditions enumerated above, thus is a pseudo-resolvent. In particular, the last equation called frequently Hilbert's identity, can be proved by a simple operator calculus, using the expression $R(s) = (sI-A)^{-1}$. For a pseudo-resolvent we do not require that there exists an operator A such that R(s) is put into this form.

Now I establish some properties of a pseudo-resolvent.

1) From the equation (23) defining a pseudoresolvent, it follows immediately that R(s) and R(s') commute for all s, s' $\in S$.

2) One calls null space of an operator, say T in general, the collection of x such that $Tx = 0$.

We now prove that all R(s), $s \in S$, have a common null space, independent of S. Denote by N, N', the null spaces of R(s), R(s'). Then from (23) we have

$$R(s)x = [R(s') + (s'-s) R(s) R(s')]x = 0$$

if $x \in N'$ because of $R(s')x = 0$. The result shows in turn that $x \in N$. Hence, $N' \subseteq N$. By a similar reasoning we have also $N \subseteq N'$. Therefore, $N = N'$.

3) Further, all R(s), $s \in S$, has a common range, independent of s. Denote by Y, Y', the ranges of R(s), R(s'). That is Y is the collection of y such that $y = R(s)x$ with a certain $x \in X$. Then

$$y = R(s)x$$

$$= R(s')x + (s'-s) R(s) R(s') x$$

by virtue of (23). Since R(s) and R(s') commute, we have

$$y = R(s') [x + (s'-s) R(s) x]$$

This equation shows that there exists

$$x' = x + (s'-s) R(s) x$$

with which y can be written as $y = R(s')x'$, that is to say,

$y \in Y'$. Thus $y \in Y$ implies $y \in Y'$, that is $Y \subseteq Y'$. Similarly we have $Y' \subseteq Y$, and finally $Y = Y'$.

Now denote by N and Y the common null space and the common range of R(s) for all $s \in S$. We have the following theorem.

Theorem: Given a pseudo-resolvent R(s). If $N = \{0\}$, then R(s) is a resolvent. In other words, there exists A such that

$$R(s) = (sI-A)^{-1}$$

The proof is the following:

The condition means that if $R(s)(x_1 - x_2) = 0$ then $x_1 - x_2 = 0$, or if $R(s)x_1 = R(s)x_2$ then $x_1 = x_2$. Then, R(s) establishes a one-to-one correspondence between X, domain of R(s), and Y, range of R(s), both common to all s. Under these circumstances, one can define the inverse $[R(s)]^{-1}$. This is an operator depending on s. But we can show that the combination $sI-[R(s)]^{-1}$ does not depend on s. That is to say

$$sI - [R(s)]^{-1} = s'I - [R(s')]^{-1}$$

for all $s, s' \in S$. In fact

$$R(s)\ R(s')\{(sI - [R(s)]^{-1}) - (s'I - [R(s')]^{-1})\}$$

$$= (s-s')\ R(s)\ R(s') - R(s)\ R(s')\{[R(s)]^{-1} - [R(s')]^{-1}\}$$

$$= (s-s')\ R(s)\ R(s') - R(s') + R(s)$$

where we used the commutativity between R(s) and R(s'). Now by virtue of (23), the last expression is zero. Multiplying from the left by $[R(s')]^{-1}[R(s)]^{-1}$ which exists, we obtain the identity desired.

The above observation permits us to put

$$sI - [R(s)]^{-1} = A \tag{24}$$

where A is an operator independent of s. Solving R(s) we find

$$R(s) = (sI-A)^{-1} \equiv R(s;A)$$

The pseudo-resolvent R(s) is now proved to be the resolvent of A.

From the equation (24) defining A, we know that the domain of A is the domain of $[R(s)]^{-1}$, that is the range of R(s), i. e. Y. Therefore $D[A] = Y$.

The equation (24) determines A from R(s) uniquely; there is one and only one A which gives R(s) as its resolvent.

This theorem just answers our question. I summarize here the conditions required under which we can get a new operator A .

1°) R(s) is a bounded operator defined for all $x \in X$ and for $s \in S$ a subset of the complex s plane.

2°) R(s) satisfies the equation of pseudo-resolvent (23).

3°) $R(s)x = 0$ implies $x = 0$.

Then we can put the new equation

$$\frac{dx}{dt} = Ax \quad , \qquad A = sI-[R(s)]^{-1} \tag{25}$$

The operator A is independent of S, and is uniquely determined by R(s). By solving the new equation (25),we find R(s) as the resolvent of A.

3.4 Application to our model

Let us examine these conditions on our model.

We have already shown that the condition 1°) is realized. By some lengthy but straightforward calculations using the explicit expression (21) for R(s), we can verify the condition 2°).I don't give here the calculations, but I wish to point out that the equation satisfied by H(s)

$$H(s) - H(s') = (s-s')\left[\frac{1}{g_o^2} - 8\pi m^2 \int \frac{k^2 dk}{(k^2- ims)(k^2-ims')}\right] \tag{26}$$

which corresponds to eq. (12) for h(E), plays an important role in the proof of the condition 2°.

Finally, the condition 3° is easily checked.

Therefore, we can construct A. In order to find its explicit form, we first solve the system of equations (21), which is $x_2 = R(s)x$, to get $\alpha, \beta(\underline{k})$ in function of $\alpha_2, \beta_2(\underline{k})$. The result gives the expression for $[R(s)]^{-1}$. Then, by (24) we obtain the expression for A. This operator is given by $x_3 = Ax$ with

$$\alpha_3 = [s - g_o^2 H(s)]\alpha - 2ig_o \int d^3\underline{k}[\beta(\underline{k}) + \frac{ig_o\alpha}{s+i\ k^2/m}]$$

$$\beta_3(\underline{k}) = (-i)[g_o\alpha + \frac{k^2}{m}\ \beta(\underline{k})] \qquad (27)$$

In the above expression for α_3 appears s, but the operator A actually does not depend on S, as was shown in the proof of the general theorem. Therefore, to define the operator A, we can use the expressions (27) for some fixed s, say s_o.

By putting z = is, h(z) = iH(s) and taking its real values z = E, we see that -iA is identical with the eigenoperator (10) which we obtained earlier by a less rigorous method. In this way, not only we have justified the earlier result with mathematical rigour, but also we can say now something more. The new system of equations is the only one which reproduces R(s). This operator, obtained from $R(s;A_K)$ by the usual method of renormalization, contains the function H(s), or h(z), whose analytic property in the complex s plane determines the subset S on which R(s) is defined. On the other hand, all observable quantities can be calculated from the functions h(z)

I now make precise the domain of the new operator A in the Hilbert space X. The domain D[A] is characterized by the two following conditions:

$(i)_s$

$$\int d^3\underline{k}[\beta(\underline{k}) + \frac{ig_o\alpha}{s+i\ k^2/m}] < \infty \qquad \text{for all } s\varepsilon S$$

(ii)

$$\int d^3\underline{k}\,|g_o\alpha + \frac{k^2}{m}\beta(\underline{k})|^2 < \infty$$

It is easy to see that $(i)_{s_o}$ verified for some s_o (which does not necessarily belong to S) implies $(i)_s$ for all s. Take in particular $s_o = 0$. Then we have

$(i)_o$

$$\int d^3\underline{k}\,\frac{1}{k^2}\,|\frac{k^2}{m}\beta(\underline{k}) + g_o\alpha| < \infty$$

Taking into account the fact that $1/k^2 \epsilon L^2$ the second condition (ii), which states that $g_o\alpha + (k^2/m)\beta(\underline{k})$ is square-integrable, implies $(i)_o$ by the Schwarz inequality. Therefore, (ii) implies $(i)_s$ for all $s \epsilon S$. In conclusion, (ii) is the only condition which defines the domain of A.

On the other hand, for A_K defined by (16), there is no condition corresponding to the above $(i)_s$, because under the integral sign we have always the cut-off function $\theta(K-k)$. In the place of (ii), we have

$$g_o\alpha\;\theta(K-k) + (k^2/m)\beta(\underline{k})\;\epsilon L^2$$

But the first term of this is square-integrable because of the factor θ . Therefore, it is sufficient to assume $k^2\beta(\underline{k})\epsilon$ L^2 . The condition is also necessary. Thus the domain of A_K is defined by this last condition, independent of s, and also independent of K.

It should be noted that $D[A]$ and $D[A_K]$ are different domains. But we confirm that the intersection $D(A) \cap D(A_K)$ is not trivial. For example, our eigenfunction (8) belongs to this common domain. We can now prove that for all $x \epsilon D(A) \cap D(A_K)$, $\lim_{K\to\infty} A_K x$ exists and coincides with Ax. (This is, however, not to say that $\lim_{K\to\infty} A_K = A$ because the domains of the two operators are not the same).

It is interesting to see how the new equations transformed into the ordinary space look like. By Fourier transformation, we find

$$\frac{d}{dt}\,\alpha(t) = [s-g_o^2H(s)]\alpha\,(t) - 2ig_o(2\pi)^3\int d^3\underline{x}\delta(\underline{x})\hat{\phi}(\underline{x}t,s)$$

$$\frac{d}{dt}\,\hat{\beta}(\underline{x}t) = (-i)\left[g_o\delta(\underline{x})\alpha(t) + \frac{-\Delta}{m}\,\hat{\beta}(\underline{x}t)\right] \qquad (28)$$

where $\hat{\beta}(\underline{x}t)$ and $\hat{\phi}(\underline{x}t,s)$ are Fourier transforms of $\beta(\underline{k}t)$ and

$$\phi(\underline{k}t,s) = \beta(\underline{k}t) + \frac{ig_o\alpha\,(t)}{s+i\;k^2/m}$$

In this system of equations, the presence of the δ-function accompagnied by g_o shows that the interaction is local, which corresponds to the fact that we passed to the infinite cut-off in the momentum space.

The δ-function does no harm in the equations (28). Firstly, we remark that

$$\int d^3\underline{x}\delta(\underline{x})\hat{\phi}(\underline{x}t,s) = \hat{\phi}(0t,s)$$

is a finite quantity because

$$\hat{\phi}(0t,s) = \int d^3\underline{k}\phi(\underline{k}t,s)$$

and $\hat{\phi}(\underline{k}t,s)$ is an integrable function by virtue of the condition $(i)_s$.

Secondly, the δ in the second of the equations (28) can be eliminated. In fact,

$$\hat{\beta}(\underline{x}t) = \hat{\phi}(\underline{x}t,s) - \frac{m}{4\pi r}\,e^{-\sqrt{m}(-is)^{1/2}\,r}\,g_o\alpha(t)$$

and

$$[\Delta + ims]\ \frac{e^{-\sqrt{m}(-is)^{1/2}r}}{4\pi r} = -\ \delta(\underline{x}) \qquad r = |\underline{x}|$$

Therefore

$$(-\ \frac{\Delta}{m})\ \hat{\beta}(\underline{x}t) = (-\ \frac{\Delta}{m})\ \hat{\phi}(\underline{x}t,s) - ims\ \frac{e^{-\sqrt{m}(-is)^{1/2}r}}{4\pi r}\ g_o\alpha(t) -$$
$$-\ g_o\delta(\underline{x})\alpha(t)$$

The last term containing δ just cancels the term $g_o\delta(\underline{x})\alpha(t)$ in the second of the equations (28).

4. Conclusion

Now as concluding remarks I wish to say the following.

In the above I examined in details a special simple model. But I took this example only to illustrate the general method, and the method itself can be applied to many other theories, as far as the conditions required are satisfied. I formulated the conditions precisely.

If there exists, or better, if we succeed in constructing an operator R(s) satisfying the conditions $1^o) - 3^o)$, then we can put the new equation with A,operator uniquely determined from R(s), independent of s, and which reproduces R(s) as its resolvent. This is the equation free from divergence and giving directly all renormalized finite results for observable quantities.

What we have to do in each case is therefore, first of all, to arrange the cancellation of infinities in the usual sense of renormalization theory in order to get a good expression of R(s), and then to verify whether the above three conditions are realized on this R(s).

The special model which we have studied is simple, among other things, in the fact that it contains only one linearly divergent integral. Therefore, we need only the renormalization of mass μ_0.

A slightly more complicate model in this respect is the Lee model [5] which contains one linearly and one logarithmically divergent integral. To eliminate these divergences, we need the renormalization of both mass and coupling constant.

Heisenberg [6] showed that in order to eliminate both of the divergences by the cancellation method, we are forced to introduce an indefinite metric. Under these circumstances, some arrangement should be made before applying the general method. Our theory is formulated in a Hilbert space in the classical sense, i.e. of positive definite metric, and cannot be applied directly to the space of the Lee model. But we can construct an auxiliarly Hilbert space of positive definite metric in order to carry out our general program.(This is the case (iv) of Destouches' lecture [7]). In such a way, I applied the above method also to the Lee model with indefinite metric and I obtained a new equation free from divergence and reproducing the renormalized finite results. The equation coincides with the one obtained earlier, by some what heuristic manner, by Haag and Luzzatto [8].

This is a differential equation, not an equation containing a finite difference Δt as Heisenberg wishes [9]. According to our general theory, the equation reproducing all the renormalized results is unique, in other words, the differential equation of Haag-Luzzatto is the only one which meets the above requirement. But the question is open whether there exists or not a formalism containing Δt which is equivalent to the differential equation of Haag-Luzzatto, of course under certain conditions which are, however, practically not very important; also the question as to which one is more reasonably interpreted from the physical point of view.

In more realistic theories, we have to solve a series of questions concerning approximation methods.

Already in non-relativistic cases, the whole Hilbert space is not always decomposed into separate sectors, and we have in general an infinite system of coupled equations. The Tamm-Dancoff method reduces the system to a finite one, which is certainly more manageable. It will be possible to apply our method in the subspace corresponding to this Tamm-Dancoff system of equations. Even then, in general, some further approximations will be required to obtain the explicit form of the resolvent $R(s;A_K)$. The resolvent may contain many different types of divergence. If a finite number of subtractions are not sufficient for elimination of these divergences, we can call the theory non-renormalizable. Then some new idea will be necessary to get R(s) if it exists.

In all such situations where the problem is not exactly solved and we have to make approximations, there arises a question of fundamental importance as to whether the approximations made are compatible with the general principle.

All these problems are out of the scope of my today's lecture. But I hope that just in order to attack these yet unsolved difficult problems, the knowledge of the general structure of renormalization theories which we have revealed, will be of some use.

References

1. K. Sekine, Cahiers de Phys. 18 (1964) 177; Nuclear Phys. 76, 513 (1966).
2. F. J. Dyson, Phys. Rev. 75, 1736, Sec. IX (1949).
3. K. Sekine, C. R. Acad. Sci. Paris 261, 4995 (1965); 262A, 158 (1966).
4. K. Yosida, Functional Analysis, Springer, 1965, p. 215.
5. T. D. Lee, Phys. Rev. 95, 1329 (1954); G. Källén and W. Pauli, Mat. Fys. Medd. Dan. Vid. Selsk. 30, no. 7 (1955).
6. W. Heisenberg, Nuclear Phys. 4, 532 (1957).
7. J. L. Destouches, in this issue, p. 428.
8. R. Haag and G. Luzzatto, Nuovo Cim. 13, 415 (1959).
9. W. Heisenberg, Nuclear Phys. 5, (1957) 195; K. Sekine, Nucl.

Phys. 23, 245 (1961); Cahier de Phys. 16, 261 (1962); Communication au Colloque à la mémoire d'E. W. Beth, Inst. Henri Poincaré, mai 1964, texte sous presse.